Emerging Technologies and the Application of WSN and IoT

The Internet of Things (IoT) has numerous applications, including smart cities, industries, cloud-based apps, smart homes, and surveillance.

The Internet of Things (IoT) enables smarter living by connecting devices, people, and objects. As networking became a crucial aspect of the Internet, rigorous design analysis led to the development of new research areas.

The Internet of Things has revolutionized daily living in countless ways. It enables communication between buildings, people, portable gadgets, and vehicles, facilitating mobility. Smart cities and cloud-based data have transformed corporate practices. With billions of connected gadgets, everything will soon be able to communicate remotely. IoT networks, whether public or private, rely significantly on machine learning and software-defined networking. Indian and other governments have approved various research projects on IoT-based networking technologies. This field of study will significantly impact society in the future.

Researchers are concerned about the many application areas and driving forces behind smart cities. The authors aim to provide insights into software-defined networking, artificial intelligence, and machine learning technologies used in IoT and networking. The framework focuses on practical applications and infrastructures. The books includes practical challenges, case studies, innovative concepts, and other factors that impact the development of realistic scenarios for smart surveillance. It also highlights innovative technology, designs, and algorithms that can accelerate the creation of smart city concepts.

This resource includes real-world applications and case studies for smart city technology, enormous data management, and machine learning prediction, all with confidentiality and safety problems.

Prospects in Smart Technologies

Series Editors:
Mohammad M. Banat, Jordan University of Science and Technology, Irbid, Jordan
banat@just.edu.jo

Sara Paiva, Instituto Politécnico de Viana do Castelo, Viana do Castelo, Portugal
sara.paiva@estg.ipvc.pt

Published Titles
Emerging Technologies for Sustainable and Smart Energy
Edited by: Anirbid Sircar, Gautami Tripathi, Namrata Bist, Kashish Ara Shakil, and
Mithileysh Sathiyanarayanan

CYBORG: Human and Machine Communication Paradigm
Kuldeep Singh Kaswan, Jagjit Singh Dhatterwal, Anupam Baliyan, and Shalli Rani

Emerging Technologies and the Application of WSN and IoT: Smart Surveillance,
Public Security, and Safety Challenges
Edited by: Shalli Rani

For more information on this series, please visit: https://www.routledge.com/
Prospects-in-Smart-Technologies/book-series/CRCPST

Emerging Technologies and the Application of WSN and IoT

Smart Surveillance, Public Security, and Safety Challenges

Edited by Shalli Rani

CRC Press
Taylor & Francis Group
Boca Raton London New York

CRC Press is an imprint of the
Taylor & Francis Group, an **informa** business

First edition published 2024
by CRC Press
6000 Broken Sound Parkway NW, Suite 300, Boca Raton, FL 33487-2742

and by CRC Press
4 Park Square, Milton Park, Abingdon, Oxon, OX14 4RN

CRC Press is an imprint of Taylor & Francis Group, LLC

ISBN: 978-1-032-56685-6 (hbk)
ISBN: 978-1-032-57180-5 (pbk)
ISBN: 978-1-003-43820-5 (ebk)

DOI: 10.1201/9781003438205

Typeset in Times LT Std
by KnowledgeWorks Global Ltd.

Contents

Preface

During the past decades, Internet of Things (IoT) offer a wide range of uses in many different industries and fields, including smart cities, smart industries, cloud-based applications, smart home scenarios, smart surveillance, and many more areas. Through connectivity between devices, people, and objects, the IoT has made it possible to live a smarter life. Furthermore, when networking became an essential component of the Internet, rigorous examination of the design issues in this field prompted the emergence of new research pathways.

The IoT has altered daily life in numerous ways. It has offered communication between numerous items, including buildings, people, portable devices, and vehicles for mobility. By transforming cities into smart ones where data is accessible on the cloud, it has altered how business is done. All things and objects will eventually be able to converse remotely thanks to billions of connected devices. Both public and private networks that support IoT heavily rely on machine and deep learning, software-defined networking. Governments in India and other nations have approved numerous research projects on IoT-based different technologies of networking. This area of study will have a significant influence on society in the future.

The concerns of researchers have risen as a result of the various application categories and forces that are driving the smart cities phenomena. It is anticipated that the collaborating authors would offer insights into the software-defined networking, artificial intelligence, or machine learning technologies employed in IoT and networking. Practical applications and infrastructures are at the center of the framework for smart cities, which is centered on the needs of the people. For observing, comprehending, and managing the better learning environment, it needs the deep association of cyber and physical aspects.

The editors of the book have conducted a top-down investigation into the aspects that are influencing the development of smart cities, taking into account key elements including "smart surveillance", "public security and strategic issues", and "science, engineering, and development". We examine key areas and problems with practical hurdles, use case illustrations, unique ideas and possibilities, and other elements influencing the creation of realistic situations for smart surveillance. This book also emphasized new technologies, designs, and algorithms that might hasten the development of smart city ideas. Additionally, it contains extensive information on the range of real-world applications and case studies pertaining to certain smart city technologies, massive data handling, and machine learning prediction approaches with confidentiality and safety challenges.

Shalli Rani
*Chitkara University Institute of Engineering and Technology,
Chitkara University, Rajpura, Punjab, India*

Acknowledgements

First and foremost, I would like to thank my husband, **Shvet Jain**, for standing beside me throughout my research career so far and editing this book. He has been my motivation for continuing to improve my knowledge and move forward in my career. He is my lifetime achievement, and I dedicate this book of mine to him.

Moreover, without the prayers and best wishes of my parents, my sister, and both of my sons, none of my achievements would have been possible. I am also thankful to the CRC Press/Taylor & Francis Group editorial team, who really helped me in every aspect in the preparation of this book. Their prompt response and care to the literature and contribution is remarkable. She is great and responsible researcher indeed.

I would also like to thank each and every one behind the completion of this book who actually helped a lot, and without their quick and efficient efforts, it wouldn't have been possible to get our book published. They motivated and gave me this platform to go ahead.

Last but not least, my gratitude is due to my inspiration **Dr. Archana Mantri**, Vice Chancellor, Chitkara University, Punjab, India for her trust in me and for making me confident that I can put efforts to be successful in each and every field.

Editor Biography

Dr. Shalli Rani is pursuing postdoctoral from Manchester Metropolitan University, UK, from July 2022. She is professor in CSE at Chitkara University Institute of Engineering and Technology, Chitkara University, Punjab, India. She has 18+ years of teaching experience. She received an MCA degree from Maharishi Dyanand University, Rohtak in 2004, and the MTech degree in Computer Science from Janardan Rai Nagar Vidyapeeth University, Udaipur in 2007 and PhD degree in Computer Applications from Punjab Technical University, Jalandhar in 2017. Her main areas of interest and research are WSNs, underwater sensor networks, machine learning, and Internet of Things. She has published/accepted/presented more than 100+ papers in international journals/conferences (SCI + Scopus) and edited/authored five books with international publishers. She is serving as the associate editor of IEEE Future Directions Letters. She served as a guest editor in *IEEE Transaction on Industrial Informatics*, *Hindawi WCMC*, and *Elsevier IoT Journals*. She has also served as a reviewer in many repudiated journals of IEEE, Springer, Elsevier, IET, Hindawi, and Wiley. She has worked on big data, underwater acoustic sensors, and IoT to show the importance of WSN in IoT applications. She received a young scientist award in Feb. 2014 from Punjab Science Congress, Lifetime Achievement Award, and Supervisor of the Year award from Global Innovation and Excellence, 2021.

Contributor Biographies

Dr. Satyam Kumar Agrawal is a biotechnologist with more than 17 years' experience in Mammalian Cell Culture, designing and performing experiments with a focus on cell-based assays to identify the mechanism of action of various plant isolates and synthetic molecules, anticancer drug delivery, apoptosis, transfection, and cell line development. He is well versed with the latest molecular biology, nanotechnology, genomics, flow cytometry, and microscopic techniques.

Dr Satyam Kumar Agrawal completed his master's in Biotechnology from the University of Allahabad and earned his PhD from Guru Nanak Dev University, Amritsar, while working as a CSIR fellow from the Indian Institute of Integrative Medicine, Jammu. Later, he worked as a research associate at CDRI, Lucknow as post-doc and as Scientist II at NIPER, Mohali. He also has industrial experience as Principal Scientific Officer—Oncology, at a Private Biotech company. He further joined Panjab University as a Scientific Assistant and later as a project scientist at IGIB, Delhi, and then as Scientist C at Division of RBMCH, at ICMR, New Delhi, in project-based positions. Until most recently, he worked as professor at the School of Pharmacy, Baddi University, HP. Currently, he is working as professor (research) at the Centre for in vitro Studies and Translational Research, CURIN, Chitkara University Institute of Engineering and Technology, Chitkara University, Punjab, India. His current focus areas are in vitro cell biology, mechanistic profiling of lead compounds with emphasis on nutraceuticals, and development of target-specific nanotherapeutics from natural products alone and in combinatorial mode against cancer. His research interests are also in Artificial Intelligence, machine learning, and data sciences.

Dr. Rakesh Ahuja secured a PhD degree in the field of Computer Science and Engineering. He is having experience of 26 years of experience in Academics, Research, Administration, and Industries. He is currently working as a professor in the Department of Computer Science and Engineering, at Chitkara University Institute of Engineering Technology, Chitkara University, Punjab, India. He has published more than 50 papers in international journals and reputed conferences. He also filed 27 national and international patents. His research area includes machine learning, digital rights management, multimedia security, pattern recognition, and information hiding. His research interest is in the development schemes of information and multimedia security for and database management systems. He has filed more than 25 patents. Currently, he is working on solving problems of the healthcare sector through machine learning and deep learning technologies.

Mohd. Akram received a Master's degree in Computer Engineering from the University of Jammu. He is currently pursuing PhD in the Computer Science Department, Rayat Bahra University, Mohali, Punjab, India. His research interests includes machine learning.

Dr. Himanshi Babbar is assistant professor of research in CSE with Chitkara University, Rajpura, Punjab, India. She has two years of teaching experience, CGC, Landran, Mohali, Punjab. She received an MCA (Master's in Computer Applications) degree from Chitkara University, Punjab Campus, in 2015 and completed her PhD and post-doctoral fellowship in Computer Applications from Chitkara University, Punjab Campus and UAE in 2021 and 2022, respectively. Her area of research is software-defined networking, load balancing, deep learning, intrusion detection, and Internet of Things. She has served as a reviewer in many conferences and journals of the *International Conference on Intelligent and Innovative Technologies in Computing, Electrical and Electronics (ICIITCEE 2023), IEEE International Conference on Current Regards Development in Engineering and Technology (CCET-2022), Peer-to-Peer Networking and Applications, Ad Hoc Networks*, Elsevier, Springer's *Journal of Network and Systems Management*, and so on. She has published/accepted/presented many papers in national and international conferences, published more than 25 papers in SCI-indexed journals, and filed/published/granted more than 10 patents.

Dr. Ali Kashif Bashir (Senior Member, IEEE) is a Reader at the Department of Computing and Mathematics, Manchester Metropolitan University, UK. He is also affiliated with the University of Electronic Science and Technology of China (UESTC), China, National University of Science and Technology, Islamabad (NUST), Pakistan, and the University of Guelph, Canada in honorary roles. He received his PhD from Korea University, South Korea in 2012. His previous assignments include the National Fusion Research Institute, South Korea, Osaka University, Japan and the University of the Faroe Islands, Denmark. He has obtained funding from several international bodies, accumulatively over 3 million USD. He has obtained over 4 million USD funding from Korean, Japanese, European, Asian and Middle Eastern bodies to solve interdisciplinary research problems in the field of wireless sensor networks, internet of things/ cyber-physical systems, cyber security and smart infrastructures. Along with his students and colleague, he has published over 200 high impact articles in the top venues. He has chaired several international conferences and has delivered over 40 invited and keynote talks. He is editor of several journals and Editor in Chief of the IEEE Technology, Policy and Ethics newsletter.

Dr. Gurbinder Singh Brar is an academician and administrator with around 15 years of experience in teaching, administration, and research and development. Presently he is associate professor in School of Engineering Technology at CT University, Punjab. Dr. Brar has more than 50 research papers in journals (SCI/Scopus/UGC Care) and conference proceedings to his credit. Dr Brar specializes in wireless networks, soft computing techniques, operating systems, network security, GPU, etc. He supervised 25 MTech scholars. Currently he is guiding eight PhD scholars.

Harshvardhan Singh Chauhan is currently pursuing BE degree with the Institute of Technology, Chitkara University Institute of Engineering Technology, Chitkara University, Punjab, India. His current research interests include software-defined networking, digital twin, edge computing, and metaverse.

Muskan Dixit is a meritorious undergraduate student and presently perusing BE (Computer Science and Engineering) from Chitkara University Institute of Engineering and Technology, Chitkara University, Punjab, India. She has strong conceptual and analytical skills. Besides, she has the ability to comprehend and critically evaluate scientific literature. She actively looks for opportunities to expand her research skills. Her area of interest are artificial intelligence, the Internet of Things, and data analytics.

Kanwal Garg is an assistant professor at the Department of Computer Science and Applications, Chitkara University Institute of Engineering and Technology, Chitkara University, Punjab, India with expertise in various fields like databases, data warehousing, data mining, web mining, text mining, pattern identification, data streams, OLAP technology, and multi-dimensional technology. With a passion for innovative research, he has made significant contributions to the academic community and guided numerous students in their quest for knowledge. His dedication to advancing cutting-edge technologies and empowering future generations is commendable.

Dr. Deepali Gupta is currently working as a research professor in Chitkara University Research and Innovation Network (CURIN) at Chitkara University Institute of Engineering Technology, Chitkara University, Punjab, India. Dr. Deepali Gupta specialises in software engineering, cloud computing, and genetic algorithms. She has published more than 50 research papers in national and international journals and conferences. Dr Deepali has worked at various administrative positions including principal, head (CSE), dean academics, IBM (SPOC), remote centre coordinator (IITB), coordinator for IITB spoken tutorial, executive committee member in Computer Science Division of Haryana State Centre (IEI), president of Sangam Kala Group (Kurukshetra, Mohali and Chandigarh Chapter), member of Anti-Ragging Committee, Academic Council, Faculty of Engineering and Technology, Board of Management, chairman SC/ST Cell of MMU, Sadopur and principal (MMGI, Sadopur). She is an active member of various professional bodies like IEI (India), IETE, ISTE, and so on. Apart from being editor-in-chief of MMU journal, she is an editorial board member and reviewer of various journals.

Her fields of interest include software engineering, cloud computing, machine learning, and blockchain.

Dr. Divya Gupta received her PhD degree in computer science and engineering from Chitkara University Institute of Engineering Technology, Chitkara University, Punjab, India. She is currently working with the Department of Computer Science and Engineering, Chandigarh University, Mohali, Punjab, India. She has more than 10 years of teaching experience. She has published more than 20 research papers in reputed journals, books, and conferences. Her research interests include information centric networking, edge computing, and the Internet of Things.

Dr. Kamali Gupta did her BTech from YMCA Institute, Faridabad in 2007, and received her MTech and PhD degree from Maharishi Markandeshwar University, Mullana. She initiated her career with Infosys Technologies Limited as System Engineer. Subsequently, she extended her services to Geeta Institute of Management and Technology, Kurukshetra and Maharishi Markandeshwar University, Sadopur. Currently, she is working as an associate professor, DCSE, Chitkara University. She has guided many M Tech thesis and has undergone numerous MOOC course certifications. Presently, five

PhD research scholars are pursuing their PhD research work under her guidance. She has many professional affiliations to her credit. She has a rich experience in domain of university accreditations, event management, and other institutional workloads. She has hosted two TEDx conferences. She has published around 40 research publications, majority of which have been published in Scopus and SCI databases. Dr. Kamali has filed more than 25 patents, out of which five are published and one is granted. She has applied and presented one project in DST. Her research interests include cloud computing, security, IOT, computer architecture, and data structures.

Sushmita Jain is a highly motivated research scholar at CVSTR, CURIN, Chitkara University Institute of Engineering Technology, Chitkara University, Punjab, India. With a strong background in Forensic Science and Biotechnology, she earned first-class distinctions during her bachelor's and distinction in her master's studies. Her expertise includes DNA extraction, HPLC, PCR, and molecular techniques. She also possesses knowledge in cyber forensics, including Kali Linux and forensic tools like FTK and autopsy with basic knowledge in blockchain, AI, IoT, IoMT, and WSN. Sushmita's inquisitive nature led her to explore psychological aspects, exemplified in her bachelor's project on child abuse. With her dedication and diverse skill set, she is poised to make significant contributions to the fields of forensic biotechnology and cyber forensics.

Anirudh K.C. is a PhD Scholar at the ICAR—National Dairy Research Institute, India, specializing in Agricultural Economics. He has been the gold medal winner of postgraduation and completed the International Agriculture and Rural Development course from Cornell University, Ithaca. Anirudh has also worked as an assistant professor at the Division of Social Sciences at the Kerala Agricultural University's Regional Agricultural Research Station. Significant areas of interest are agriculture and policy, sustainable agriculture, and bio-circular economy.

Dr. Isha Kansal is currently working as assistant professor in Chitkara University Institute of Engineering and Technology, Chitkara University, Punjab, India. Dr. Isha has attained her doctorate degree from Thapar Institute of Engineering and Technology, Punjab. Her both MTech (CSE) (from Thapar Institute of Engineering and Technology) and BTech (CSE) (from BBSBEC Fatehgarh Sahib) degrees are with distinction. She has about seven SCI publications and a number of publications in renowned international journals and fully refereed international conferences with an experience of more than nine years. Recently, she has been certified under the Wipro Talent Next Program in JAVA. Her main areas of research are image/video processing, machine and deep learning.

Shilpa Karat is PhD Scholar at Kerala Agricultural University, specializing in Agricultural Extension. She has worked on farmer psychology and adoption of technology through her postgraduate research. She has also participated and gained recognition in various endeavours related to climate change, artificial intelligence through Agriculture Science Congress, national workshops, and other scientific networks. She possesses numerous publications in the same field. She has worked as a coordinator in the flagship Young Innovators' Programme.

Saurabh Manoj Kothari is a computer science student, studying at Brunel University, London. He is currently pursuing master's in Data Science and Analytics and completed bachelor's in Computer Engineering from Savitribai Phule Pune University, Pune. His research interests are in the field of software engineering and information systems, including WSN, blockchain, data science, and more.

Dr. Kanwal Preet Kour pursued her BTech in 2013 from Jammu University and her MTech in 2016 from Guru Nanak Dev University, Amritsar. She completed her PhD from Chitkara University Research and Innovation Network (CURIN) at Chitkara University Institute of Engineering Technology, Chitkara University, Punjab, India. Dr. Kanwal specialises in IoT, fog computing, and data security. She has various publications to her account in various national and journals and conferences in Scopus and SCI databases. She has filed eight patents based on IoT applications. She has contributed many articles related to IoT and its vast applications like smart agriculture, hydroponics, smart saffron cultivation, and use international of fog computing in IoT. Her areas of interest include IoT, precision farming, machine learning, deep learning, and blockchain.

Dr. Rajeev Kumar received his BTech and MTech degrees in Electronics and Communication Engineering from Kurukshetra University, Kurukshetra, India in 2008 and 2010, respectively. He completed his PhD degree in Electronics Engineering from Banasthali University, Rajasthan, India in 2017. He is currently working as an assistant professor in the Department of Electronics and Communication Engineering, Chitkara University Institute of Engineering and Technology, Chitkara University, Punjab, India. His research interests include reconfigurable antenna, ultra-wideband antennas, dual-band/triple-band microstrip antennas for wireless communication, smart and MIMO antenna systems and also include Internet of Things (IOT).

Shailja Kumari is pursuing her PhD degree in Computer Science and Engineering from Chandigarh University, Mohali, Punjab, India. She is currently working as an assistant professor in the Department of Computer Science, Government PG College, Sec. 1, Panchkula. Her research interests include wireless sensor networks, Internet of Things, and artificial intelligence.

Kamini Lamba is pursuing a PhD in Computer Science and Engineering from Chitkara University Institute of Engineering and Technology, Chitkara University, Punjab, India from September 2022. She attended various sessions such as National Seminar on "Publications and Research Ethics," "FDP Design Thinking Webinar," "International Level Workshop on Cyber Security and FPGA," "Medical Emergency: Understanding Signs and Management" organized by Chitkara University, Punjab, and "Panel Discussion & Ideation Session on Strengthening Logistics & Warehousing for Sustainable Growth" organized by ASSOCHAM, India. She successfully completed module of "Creating a Spark for Artificial Intelligence"—powered by mindSpark Learning. Apart from this, she also participated in the workshop on "Antariksh Prodyogiki Aur Iske Anuprayog" organized by the Indian Institute of Remote Sensing, ISRO, Department of Space, Government of India. Her research interests include machine learning, deep learning, image processing, and artificial intelligence.

Sumit Kumar Mahana is pursuing his PhD from the National Institute of Technology, Kurukshetra. He has completed his BTech (Computer Engineering) and MTech (Software Engineering) from Kurukshetra University, Kurukshetra. He has also qualified National Eligibility Test (NET) conducted by the Central Board of School Education (CBSE) in 2017. He is in the teaching profession for more than 12 years and has several research publications to his credit. He has one book to his credit. His research interests include image processing, cryptography, and multimedia security.

Imran Memon is editor-in-chief of *Journal of Network Computing and Applications*. He is also an editor of *JDCTA*. He received Academic Achievement Award 2011–2012 and Excellent Performance Award 2011–2012 from UESTC China. He is serving as an organizing committee chair and TPC member of 350 international conferences. He is editorial board member of 20 international journals.

Dr. Nagma received her PhD degree in computer science and engineering from Chitkara University Institute of Engineering Technology, Chitkara University, Punjab, India in 2019. She is presently working with Algoma University, Brampton Campus, Brampton, Ontario, Canada. Her research interests include cloud computing and the Internet of Things.

Dr. Renu Popli is currently working as an assistant professor at the Department of Computer Science and Engineering, Chitkara University Institute of Engineering and Technology, Chitkara University, Punjab, India. He has nine years of experience in research and academia. She received her MTech and PhD degrees in Computer Science from Kurukshetra University, Kurukshetra, India in 2012 and 2018, respectively. She has published more than 30 publications in renowned international journals and fully refereed international conferences. She has filed more than 10 patents in multiple domains. Her research interest includes mobile ad-hoc, wireless sensor networks, ML, DL, and IoT.

Dr. Vyasa Sai is currently with the Visual and Machine Learning IP Team, Intel, Santa Clara, CA, USA. Sai received a PhD degree in Computer Engineering from the Department of Electrical and Computer Engineering, University of Pittsburgh, Pittsburgh, PA, USA, in 2013. He is the lead series editor for the Design and Implementation of Devices, Circuits, and Systems Series for IEEE Communication Magazine. He is also a Technical Committee member for the IEEE Circuits and Systems for Communications, editorial board member for the *International Journal of Radio Frequency Identification Technology and Applications*, associate editor for *IEEE Access, Elsevier International Journal of Computers and Electrical Engineering, IEEE Communications Magazine*, guest editor for *Elsevier Computer Communication*, among others.

Dr. Preeti Saini is an esteemed academician with a 19-year-long career, currently holding the position of assistant professor at Chitkara University Institute of Engineering Technology, Chitkara University, Punjab, India in the field of Computer Science and Engineering (CSE). With a profound passion for technology, she pursued a PhD in CSE from the same institution, focusing on digital rights management, multimedia security, pattern recognition, machine learning, and information hiding. In her role as an assistant professor, she excel in both teaching and mentoring students while maintaining an active research presence, evident through numerous publications in distinguished conferences and journals.

Dr. Syed Hassan Shah is a Wi-Fi connectivity subject matter expert with the Qualcomm Inc. product management team, where he is involved in Consumer and Compute Wireless products with a focus on Mi-Fi, CPE, and UWB technologies. In addition to that, Dr. Shah is also an adjunct faculty member at California State University, Fullerton Campus, where he teaches computer science courses to graduate classes. Over the past decade, Dr. Shah held multiple industrial and academic roles such as a product specialist for Distributed Antenna Systems (DAS), CBRS, Private LTE, Digital Electricity, and open RAN product lines. Dr. Shah was an assistant professor in the Department of Computer Science at Georgia Southern University, Statesboro, GA, followed by a post-doctoral fellowship at the University of Central Florida, Orlando, FL. Before moving to the United States, he completed his BS with honors in CS from Kohat University of Science & Technology (KUST), Pakistan, and his PhD degree (combined with Masters) from the School of Computer Science and Engineering (SCSE), Kyungpook National University (KNU), Republic of Korea (South Korea). During the summer of 2015, he was invited as a distinguished visiting researcher at Georgia Tech, Atlanta, GA to investigate MAC layer findings in IEEE 1609.4 protocol stacks. Overall, Dr. Shah has authored/co-authored over 250 peer-reviewed international publications including journal articles, conference proceedings, book chapters, and five books. In 2016, his work on robust content retrieval in future vehicular networks won the Qualcomm Innovation Award at KNU, South Korea. Dr. Shah's research interests include wireless and ad hoc networks, cyber-physical systems, smart cities, connected vehicles, and future internet architectures. Furthermore, Dr Shah is a Senior IEEE and ACM member, served as a TPC member or reviewer in over 100 international conferences and workshops including IEEE Globecom, IEEE ICC, IEEE CCNC, IEEE ICNC, IEEE VTC, IEEE INFOCOM, ACM CoNEXT, ACM MobiHoc, ACM SAC, and many more. Furthermore, he has been reviewing papers for over 50 different international journals including IEEE magazines on wireless communications, networks, communications, IEEE communications letters, IEEE sensors letters, and IEEE transactions on industrial informatics.

Ankita Sharma received BTech and MTech degree from Ganpati Group of Institutions affiliated from KUK University in 2016 and 2019, respectively, pursuing her PhD degree from the Chitkara University Institute of Engineering Technology, Chitkara University, Punjab, India., since 2020, in Computer Science and Engineering. Her research interests include security domain in Internet of Things using machine learning and deep learning.

Dr. Ashutosh Sharma is an assistant professor specializing in Reliability and Risk Engineering, Computer Networks, AI/ML Applications, Internet of Things, Cryptography and Network Security, and Digital Logic and Computer Organization. He obtained his postdoctoral degree from Southern Federal University, Russia in 2022, with his research focused on "Risk Aware Communication Network Architecture and Planning." Prior to that, he completed his PhD from Jaypee University of Information Technology, India in 2020. Dr. Sharma holds an MTech degree from Jaypee University of Information Technology and a BTech from IK Gujral Punjab Technical University. He has an impressive publication record of 110 papers in *SCI/SCOPUS* journals, with 3665 citations, an H-index of 34, and an i-10 index of 61. Currently, he serves as an assistant professor at the School of Computer Science, Chitkara University Institute of Engineering Technology, Chitkara University, Punjab, India and has previously held positions at Southern Federal University, Russia and Lovely Professional University, Punjab, India. Dr. Sharma's research interests include data communication and network, Internet of Things and AI/ML applications, intelligent transport system, smart cities, and image processing.

Dr. Pooja Sharma is an associate professor in Computer Science Department, Rayat Bahra University since 2007. Her area of specialization is computer networks for high-performance computing. As an expert in computer networks, she is guiding the next generation of engineers and network administrators as well as contributes to moving the research front in computer networks. She has many good publications and patents titled AI-Powered Virtual Interior Design System and method using generative models.

Dr. Shruti received her BTech degree in Computer Science from UIET, Panjab University, Chandigarh, and has done MTech in Computer Science from PEC University of Engineering and Technology, Chandigarh. Currently, she is serving as assistant professor in Goswami Ganesh Dutta Sanatan Dharma College, Chandigarh, from the past eight years and pursuing her PhD in Computer Science from Chitkara University Institute of Engineering and Technology, Chitkara University, Punjab, India. Her main area of interest is distributed networks. She has published/accepted/presented many papers in international journals/conferences both SCI and Scopus indexed and also authored and edited books with international and national publishers.

S. Kanwal Deep Singh pursued his BTech in 2018 from Jammu University and is currently pursuing his MS from the Department of Computer Science, Federation University, Australia. His area of specialisation is network security and IoT. He is currently working on different projects related to IoT. His areas of interest include IoT, networking, security, smart city, and blockchain.

Shivani Wadhwa is pursuing her PhD degree in computer science and engineering from Punjabi University, Patiala. She is currently working with the Department of Computer Science and Engineering, Chitkara University Institute of Engineering Technology, Chitkara University, Punjab, India. She has more than 10 years of teaching experience. She has published more than 20 research papers in reputed journals, books, and conferences. Her research interests include blockchain, security in WSN, and the Internet of Things.

1 Transforming Urban Spaces and Industries

The Power of Machine Learning and Deep Learning in Smart Cities, Smart Industries, and Smart Homes

Ankita Sharma and Shalli Rani
Chitkara University Institute of Engineering and
Technology, Chitkara University, Punjab, India

INTRODUCTION

When authors in Ref. [1] handed Procter & Gamble a study on radio-frequency identification (RFID) in 1999, the phrase "Internet of Things" (IoT) was first used. The idea of autonomous data collecting via RFID and sensing technology, along with developments in machine-to-machine (M2M) topologies, wireless sensor networks (WSNs), artificial intelligence (AI), and semantic technologies, has encouraged the growth of the IoT. By 2030, according to Cisco's predictions, 80 billion connected devices are expected to be available, which is 6.58 times the estimated world population [2]. The researchers in Ref. [3] noted that the need for accessible enabling technologies was the main factor fuelling the IoT's explosive expansion. An intelligent computer server, commonly referred to as a smart machine, is able to create personalised content for a dynamic web page and send it to a particular user depending on their browsing history. Networked computers can now be included into a variety of items because of Moore's law and the ongoing miniaturisation of electrical components. As a result, things are getting more intelligent, computerised, and internet-connected. Wesier predicted the concept of ubiquitous computing more than two decades ago [4]: The idea that computers will be present everywhere, linked, and effortlessly incorporated into everyday life. Thing-to-thing or M2M interaction is the core technology behind the IoT, a concept that aims to interconnect physical things to the internet. These IoT devices have transformed the equation of human-computer interaction (HCI) and developed new strategies for incorporating technology into our daily lives. Today's HCI technology, in contrast to the past, lays more emphasis on being human-centric than computer-centric [2]. HCI has significantly advanced in the last ten years, leading to expanded applications and more efficient use of HCI [3].

DOI: 10.1201/9781003438205-1

1

Using sophisticated data analytics to grasp, monitor, regulate, and manage the city is the underlying concept of smart cities, industries, and residences from a data-centric viewpoint [5]. It is well acknowledged that there are four layers to the data analysis process, as depicted in Figure 1.1, despite slight discrepancies [6–9]. These layers include data collection, data preparation, analysis of the data, and service supply. The data preliminary processing layer performs initial calculations (such as cleaning the information, making decisions, and interpolation) to acquire higher quality data

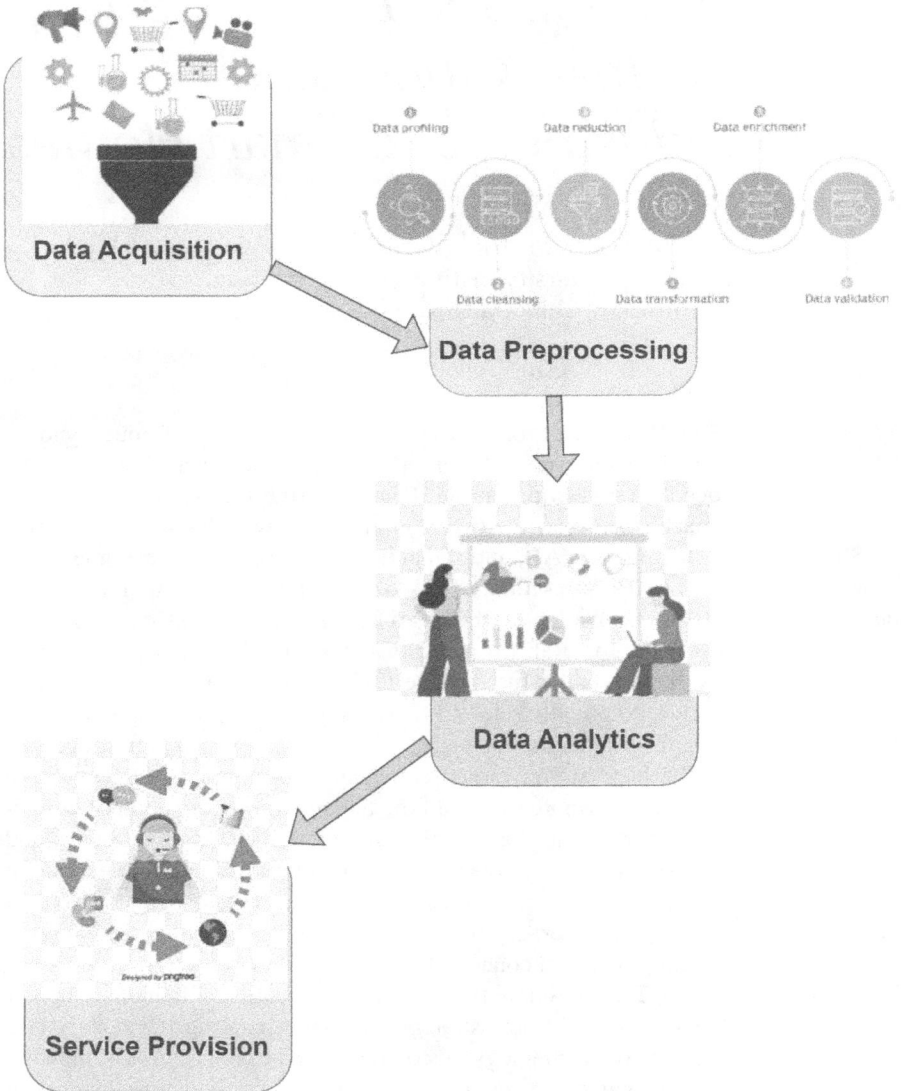

FIGURE 1.1   Steps for data analysis.

before statistics, and the data statistical analysis layer performs intelligent evaluation using various types of machine learning (ML). The data acquisition layer gathers and stores reliable city information from various fields and sources. Deep learning has lately received a lot of interest from the scientific community due to its amazing capabilities. It excels at managing massive quantities of data, reducing the need for a lot of manual data engineering work. Deep learning models beat shallow neural networks (NNs) in learning complex structures from large-scale datasets. Due to the wide nature of deep learning, its distinct characteristics enable it to handle enormous datasets without the need for additional dimensionality reduction procedures. Moreover, it surpasses conventional deep learning methods when applied to real data and eliminates the laborious data engineering processes they need. In contrast to traditional ML methods, deep learning provides the capacity to model extremely complex functions using layers of non-linear transformations that can be learnt from beginning to end. Deep learning-based methods have made significant strides in a number of fields, including neural machine translation and natural language processing. In addition, deep learning has significantly improved speech recognition and computer vision, outpacing state-of-the-art techniques in all of the areas [10].

OUR CONTRIBUTIONS

1. This chapter comprises motivation along with the development of smart cities, automatic homes, and intelligent industrial systems.
2. The fundamental prerequisites for smart cities, automatic homes, and intelligent industrial systems.
3. The corresponding attacks on smart cities, automatic homes, and intelligent industrial systems.
4. ML and DL in smart cities, automatic homes, and intelligent industrial systems.
5. Security solutions: Effective measures have been taken.

MOTIVATION

Automation has become more prevalent in many applications because of its capacity to process immense datasets with greater precision than conventional methods. These applications produce information gathered from a range of sources such as images, videos, text, and audio. As a result, this data is complex and extremely large. There are numerous formats for the data, including arranged, semi-arranged, and unstructured. Therefore, depending on the data analysis problem the SC is trying to solve, it may be acceptable to analyse the data using deep learning. Classification, clustering, and regression are three data analytics methods that may be coupled with deep learning. The SC may employ the deep learning architecture to execute duties such as object and voice recognition. Authors in Ref. [11] used ConvNet, for example, to classify images acquired by intelligent CCTV cameras in parking lot spaces. It has been discovered that ConvNet outperforms the conventional procedure. Comparable research conducted by researchers in Ref. [12] demonstrates that ConvNet outperforms ANN. In order to predict electrical energy usage in smart cities contexts, authors in Ref. [13] compare deep learning algorithms to established techniques like hidden Markov, support vector machines (SVM), and factored hidden Markov. The examination of the days and weeks and the energy predictions reveal the accuracy of the various prediction approaches that vary considerably, with the DBN providing the most dependable performance when compared to the older methods. In NLP and language modelling applications, recurrent NNs have demonstrated encouraging performance compared to conventional ML techniques [8].

DEVELOPMENT OF SMART CITIES, AUTOMATED HOMES, AND INDUSTRIAL INTELLIGENT SYSTEMS

Government agencies, urban planners, technology suppliers, industry specialists, and community involvement must work together to create smart cities, smart industries, and smart households. The transformation of conventional urban settings and industries into intelligent and sustainable entities depends on the deployment and use of cutting-edge technology, networking, and data-driven decision-making. The fusion of numerous technology and infrastructural elements results in smart cities, smart industries, and smart households. The creation of a smart city begins with a strategic plan that lays out the project's goals, objectives, and expected results. Collaboration between local government representatives, urban planners, IT companies, and community stakeholders is required for this. Physical and digital infrastructure, such as sensors, cameras, IoT devices, communication networks, and data centres, must be deployed in order for smart cities to function. The foundation for gathering and distributing data around the city is provided by this infrastructure. To acquire insights into municipal operations, infrastructure performance, and citizen behaviour, data is gathered from a variety of sources, including sensors, social media, and public records. To analyse and extract useful insights from the gathered data, advanced analytics methods are used, including ML and AI [14]. Different systems and elements of the city, such as public safety, waste management, transportation, and energy, are connected to allow for coordinated and effective operations. Through data exchange and interoperability standards, many systems are connected and integrated. Smart cities place a strong emphasis on including citizens in the decision-making process. Platforms and apps are created to let residents and the municipal government communicate, enabling comments, reporting problems, and accessing services. Smart industries improve production processes, boost productivity, and save costs by using automation technology, IoT gadgets, and sensor networks. To allow real-time monitoring and control, this entails integrating sensors, actuators, and control systems into industrial contexts [15].

Industrial IoT (IIoT) technologies are used to link network systems, equipment, and devices in the industrial sector. In addition to enabling centralized monitoring, control, and analysis of industrial activities, this facilitates seamless data interchange. Advanced analytics methods are used to examine data gathered from industrial machinery and processes. ML techniques are used to find trends, forecast faults, increase output, and allow predictive maintenance [16].

To safeguard industrial systems and data from online attacks, strong cybersecurity measures are put in place. This includes network segmentation, encryption, intrusion detection systems, access controls, and regular security audits. Smart homes come with a variety of networked appliances and systems, including smart thermostats, security cameras, lighting systems, and entertainment systems. These gadgets are designed to be wired into a home network and automated or remotely operated according to set rules. To enable device connection and data exchange inside the house, a dependable and secure communication infrastructure has been established. To centralise control and administration of smart home devices, home automation systems and platforms are employed. Homeowners may now monitor and control a range of services, such as temperature, lighting, security, and entertainment, using voice commands or mobile apps. To safeguard the private information that smart home gadgets gather, privacy and security safeguards have been put in place. This covers data security, user authentication, firmware upgrades, and encryption. Utilizing information and services from the infrastructure of the city, smart houses may be connected to larger smart city programs. For instance, a smart house may obtain data on current energy usage or take part in demand-response initiatives provided by the city's smart grid [5].

Government agencies, urban planners, technology suppliers, industry specialists, and community must work together to create smart cities, smart industries, and smart households. The transformation of conventional urban settings and industries into intelligent and sustainable entities depends on the deployment and use of cutting-edge technology, networking, and data-driven decision-making.

IoT-SMART CITIES, SMART HOMES, AND SMART INDUSTRIES

The IoT plays a significant role in transforming smart cities, smart homes, and smart industries by connecting various devices and systems to the internet and enabling them to communicate, collect data, and make informed decisions. Here's how IoT is utilized in each context: IoT is used to monitor and manage critical infrastructure like traffic lights, streetlights, waste management systems, and utilities. This enables better resource management and optimization, reducing energy consumption and enhancing overall efficiency. IoT-based surveillance systems, smart cameras, and sensors are deployed to enhance public safety and emergency response by detecting and responding to incidents in real time. IoT enables smart parking systems, traffic management, and public transportation solutions, enhancing mobility and reducing congestion in cities. IoT sensors are deployed to monitor air quality, noise levels, and other environmental factors, helping cities to address pollution and climate change challenges. IoT-enabled devices like smart thermostats, smart lighting, and smart appliances allow homeowners to control and manage their home environments remotely, optimizing energy usage and enhancing comfort. IoT-based security systems, such as smart door locks, surveillance cameras, and motion sensors, provide real-time monitoring and alerts to enhance home security. IoT-powered virtual assistants, like Amazon Echo or Google Home, enable voice commands to control various smart devices, set reminders, and answer queries. IoT is integrated into manufacturing processes to enable automation, remote monitoring, predictive maintenance, and real-time data analysis to optimize production and reduce downtime.

IoT devices and sensors are used to track and monitor goods throughout the supply chain, ensuring transparency, efficiency, and minimizing inventory errors. IoT is utilized for tracking and managing industrial assets, such as machinery and equipment, ensuring they are efficiently utilized and maintained.

The deployment of IoT in these contexts leads to enhanced efficiency, cost savings, and improved decision-making. However, it also presents security and privacy challenges, which need to be addressed through robust security measures and privacy protocols to ensure the safe and reliable operation of IoT-enabled smart environments.

COLLABORATION BETWEEN IoT AND AI

AI and IoT have reached advanced stages of development, and their collaboration holds tremendous potential for numerous advantages. IoT, often regarded as the catalyst for the Fourth Industrial Revolution, has spurred technological advancements and transformations across diverse domains. It is widely acknowledged that AI is crucial for the progress of IoT, with many experts asserting that AI is futuristic. In fact, the integration of IoT and AI has been a prevalent practice in various industries and sectors for quite some time. IoT serves as a data collection mechanism, generating vast volumes of data, while AI serves as the ideal tool for effectively interpreting and extracting insights from such extensive data. AI acts as the decisional engine used to carry out analysis and different data pre-processing techniques. By leveraging AI, patterns can be understood, leading to more informed decision-making. The utilization of ML, combined with the power of big data, has created new prospects and possibilities within the realm of IoT. Amazon's voice assistant, Alexa [9], exemplifies the process of data collection. However, the subsequent steps of organizing, analysing, and making informed decisions based on that data are distinct challenges. It is evident that for AI to enhance its utility in IoT, there is a need for the development of more precise and efficient algorithms and tools [17]. By combining IoT with AI, enterprises can leverage the most effective approach to optimize their store operations and ensure long-term sustainability. The integration of IoT and AI enables retailers to achieve various benefits, including minimizing theft and maximizing sales through techniques such as cross-selling. This synergy empowers enterprises to enhance their overall performance and achieve greater success in the retail industry. The study [18] explores the interaction between AI and IoT, emphasizing that AI and ML in the realm of data science go beyond merely applying statistical predictive algorithms to IoT. The study stated that intelligent system for IoT differs from existing databases, as it involves specialized techniques tailored to handle time series data, including average methods and similar approaches. To achieve significant outcomes, AI relies on big data. In fact, AI has the ability to address the challenges associated with big data analytics.

Due to the vast quantity of data generated by numerous smart devices/objects, humans face limitations in comprehending and effectively managing such data using traditional methods. Therefore, it becomes imperative to explore innovative approaches for analysing performance data and information. To fully harness the potential of IoT data, there is a pressing need to significantly enhance the performance. Furthermore, the ongoing progress in AI is leading to a convergence between AI and IoT. The fundamental description of IoT will ultimately necessitate the intelligence of nearly all devices. In simpler terms, IoT relies on smart devices and machines. The convergence of both is driving the continuous expansion of IoT, influenced by six key factors, with the most influential factor being the emergence of big data and cloud/fog computing.

FUNDAMENTAL PREREQUISITES FOR SMART CITIES, AUTOMATED HOMES, AND INDUSTRIAL INTELLIGENT SYSTEMS

Building and running smart cities, smart businesses, and smart homes all depend on deep learning models. In order to enhance various aspects of these settings, these models may manage complex data, discover pertinent patterns, and provide insightful forecasts. Some required fundamentals required for smart cities, smart homes and intelligent industries are shown in Figure 1.2. A few deep learning models that have been used in each domain are listed below:

1. Traffic management: Deep learning models can analyse real-time traffic data from cameras and sensors to enhance traffic flow, predict congestion, and offer efficient routes [19].
2. Energy management and efficiency: Models can anticipate energy use trends, enhance energy distribution, and assist demand-response systems in reducing energy waste. Algorithms using deep learning can learn and anticipate household energy usage trends, enhance electronic devices for smart homes, and recommend energy-saving activities [9].
3. Public safety: Deep learning algorithms can analyse video feeds and spot irregularities, such as unusual crowd behaviour or accidents, to enhance security and emergency response.
4. Trash management: Models may study past data and sensor inputs to enhance trash collection routes, save costs, and minimize environmental impact.
5. Predictive maintenance: Deep learning models can examine sensor data from machinery to forecast equipment faults, enabling proactive maintenance and reducing downtime.

FIGURE 1.2 Key requirements for smart cities, automatic homes, and intelligent industrial systems.

6. Quality control: Using visual or sensor data, models may evaluate and categorize items in real time, assuring consistency and finding faults.
7. Supply chain optimization: Deep learning algorithms may increase overall operational efficiency by optimizing inventory management, demand forecasting, and logistics.
8. Process optimization: Sensor data and historical records may be analysed by models to detect bottlenecks, adjust production parameters, and boost overall productivity.
9. Security: Models can scan video feeds, identify and categorize questionable activity, and provide warnings to homes or security agencies [20].
10. Personalized assistance: Deep learning algorithms can interpret and react to voice instructions, learn personal preferences, and deliver tailored suggestions for different home automation chores.
11. Environmental monitoring: Models may evaluate sensor data to monitor air quality, humidity, and temperature, enabling homeowners to keep their homes healthier and more pleasant.

ATTACKS ON SMART CITIES, SMART INDUSTRIES, AND SMART HOMES

Smart homes, smart businesses, and smart cities all confront different security risks and possible intrusions. Following are a few typical attack methods for each domain that are vulnerable to cyber-attacks due to their reliance on interconnected networks and technologies as shown in Figure 1.3. Examples comprise:

1. DDoS: Distributed denial of service (DDoS) attacks that target essential public services or infrastructure; intrusion into municipal communication networks or control systems with the potential to impact emergency services, power infrastructures, and transportation [8].

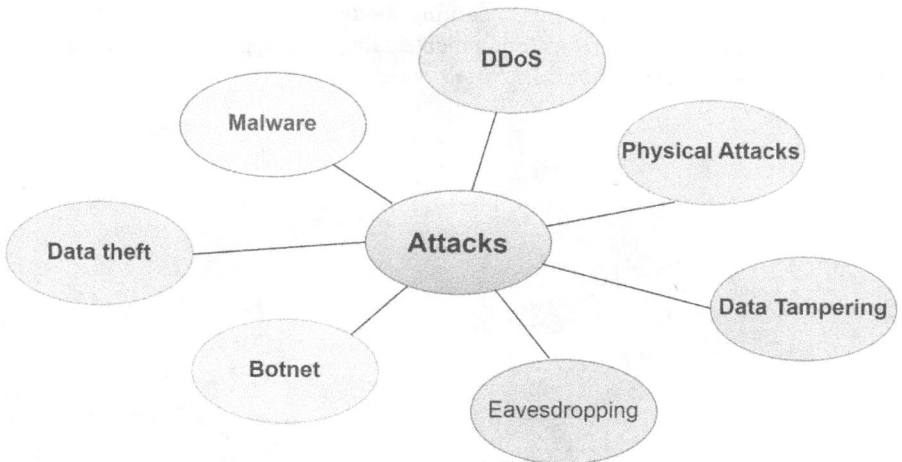

FIGURE 1.3 Attacks on smart cities, automatic homes, and intelligent industrial systems.

2. Malware: Malware-based attacks that compromise private information or obtain unauthorized access to municipal systems. The collection of huge amounts of data by cities raises concerns regarding privacy violations.
3. Data theft: Theft or illicit access to personal information extracted from citizen databases or surveillance systems. The improper handling or storage of sensitive data, resulting in privacy breaches or identity theft [20]. Interception of data communications within the city's communication infrastructure, potentially jeopardizing private data, attempts to compromise the physical infrastructure. Attacks on the infrastructure of smart cities may impede operations significantly [21].
4. Physical attacks: Vandalism or destruction of surveillance cameras, sensors, or IoT equipment interference with traffic control systems, resulting in collisions or traffic congestion. Manipulation or interruption of water supply or electricity infrastructures, resulting in disruption of essential services [17].
5. Eavesdropping: The act of manipulating or hijacking connected equipment, such as smart locks or security cameras, to gain access or eavesdrop on individuals. Eavesdropping on communication lines in smart residences in order to obtain sensitive information or learn more about the residents. Smart home–connected devices may be exploited for malicious purposes [18].
6. Botnet: Utilizing IoT device vulnerabilities to seize control of them and potentially grant unauthorised users access to the home network. Utilize compromised equipment to launch botnet attacks that disrupt internet connectivity or service availability. Smart home technologies collect a great deal of data, which raises concerns about intrusions of privacy and surveillance.
7. Data tampering: Abuse of the personal information collected by smart devices, such as voice assistants and health monitoring systems. Using vulnerabilities in cloud storage or data-sharing systems to expose confidential information to the public. Tampering with surveillance cameras or infant monitors, compromises privacy and makes intrusive monitoring possible [22].

ROLE OF MACHINE LEARNING AND DEEP LEARNING IN SMART CITIES, SMART HOMES, AND SMART INDUSTRIES

The authors in Ref. [8] have compiled a brief analysis of numerous significant and vital elements of a smart city. A few of the important issues that have been highlighted include protecting privacy and data security, network protection from potential cyber-attacks, promotion of ethical information-sharing knowledge, and practical use of automation and ML.

The researchers of Ref. [14] conducted a detailed analysis to look at several research questions and related solutions to the concept of designing smart cities from the viewpoints of interaction, anonymity, and security. They mainly concentrated on examining the large variety of difficult problems that appear throughout the configuration of contemporary communication protocols for technologies like sensors and actuators. In Ref. [15], the authors studied the function of ML and automation algorithms from a sophisticated security perspective of IoT and recently identified safety issues. The authors made recommendations for future research directions and assessed the advantages, disadvantages, and potential IoT security of ML and deep reinforcement learning (DRL) protocols. The role and prospective advantages of using ML and DRL approaches in the bioinformatics and healthcare sectors were scheduled to be examined in Ref. [16], according to the authors. They investigated numerous issues and put up ideas for effectively using automation-based technologies in several industries mainly in the healthcare sector. The authors in Ref. [5] examined the underlying structure from their initial dataset and distinguished among regular as well as unexpected traffic using the self-taught potential and capabilities of ML techniques. They recommended an open-source NN-based method to detect and recognize cyber-attacks for IoT applications in smart cities. The performance of the proposed approach beats that of shallow models. The intrusion detection system for IoT was proposed by the authors of Ref. [9] as a random forest ML-based architecture to address the security of IoT devices in the smart city. The suggested method employs ML-based dataset assessment to effectively identify any kind of suspicious behaviour occurring at the dispersed fog nodes. To protect a smart city's network of computers against any kind of cyber incursion, the authors of Ref. [20] suggested a revolutionary DRL-based design. The overall credits to the recommended prototype's easy identification of attacks whose data behaviour was quite abnormal, the network may be safeguarded beforehand. The aforementioned paradigm could help in the development of many private and secure applications to make the smart city concept a reality. In order to reduce delays and consumption of energy in the Fog-Cloud-IoT context, the authors of Ref. [21] introduced a technique based on ML for safe cognitive unloading system. The suggested architecture, with neuro-fuzzy model, guarantees privacy of information at the entry point and allows IoT devices to choose the best nodes in the fog for outsourcing process using techniques known as particle swarm optimization techniques. The difference between normal and abnormal service requests depends upon the three phases of the intrusion mechanism— traffic analysis of data, enlargement, and classification techniques—are employed. The researchers in Ref. [17] looked at a number of security concerns with aerial vehicles used for transportation systems and provided an artificial dependent solution to solve these concerns. The developed architecture enables aerial vehicles to make frequent use of system resources while maintaining their ongoing safety while they carry out different assigned activities, including ITs, immediate information streaming, and unmanned aerial vehicle (UAV) assisted package delivery. The effectiveness of the prior strategy has been proven by simulation results.

The GPS spoofing attack, which enables the use of false signals to avoid (UAVs) and their ground controllers, was the focus of Ref. [18]'s writers. For the purpose of exposing GPS spoofing efforts, they recommended employing ML-based ANN. Under difficult and unfavourable conditions, the ML model effectively maintains system performance and efficiently allocates electricity to important loads. The authors in Ref. [8] proposed the concept of a fully autonomous ML platform that supports in the construction of criteria for making decisions during the creation of ML-based applications. The prior platform may be used to promote high-level learning by increasing the number of sophisticated designs and informed interruptions. The recommended platform's performance is particularly beneficial for ML storage applications used in smart cities.

As a method for intrusion detection and position determination, the authors of Ref. [7] advocate using ML and power-line communication (PLC) modems in SGs. The PLC modems continuously monitor the CSI and alert the system administrator to any CSI deviation that may have been caused by the claimed intrusion. Tracking energy usage may be made easier using the provided method. The authors of Ref. [23] proposed a proposal for the ISAAC security testbed for SG systems. It is a distributed, cross-domain platform that replicates the data of a functioning power plant. Researchers may test and evaluate their cyber-security solutions using the ISAAC platform. The authors of Ref. [10] have produced a study that outlines the many challenges that ML- and cybersecurity-based methods in SGs face. The authors of Ref. [11] published deep-Q-network detection, a DRL-based technique, as a safeguard against data integrity attacks in AC power systems. The recommended protocol is used throughout the training phase across the central and targeted networks to figure out the optimum defence method. The results of the experiments show that the aforementioned technique works much better than benchmark methods in the following parameters such as speed and detection accuracy.

For the purpose of detecting security and privacy vulnerabilities in IoT networks, the developers of Ref. [12] developed the long short-term memory and k-nearest neighbour algorithms. A bot-IoT dataset is used to assess the effectiveness of both the deep and ML algorithms used in the attack detection module. The results show that the recurrent NN performs better in terms of accuracy than the kappa statistic values. The long short-term memory classifier is excellent in identifying attacks when utilized in the IoT environment.

The authors [13] recommended a DL-based intrusion detection system (IDS) as a defence against DoS assaults on IoT networks. The evaluation of the novel model was done by the NSLKDD dataset. The authors compared the proposed IDS with the more conventional shallow model method. The recommended IDS also employs a controlled and autonomous detection strategy. The comparison results show that a geographically scattered assault detection system is more accurate than a centralized intrusion detection system. Similar to how performance measurements favour the deep model over the shallow one, researchers have described a technique for systematically building a large-scale CNN on a heterogeneous IoT computing environment [4]. The author developed a framework that examined the device's operating frequencies, pruned the system architecture, and selected the best computer configuration in order to meet the system's objectives of low power/high performance. Researchers have suggested an original, quick method for detecting DDoS malware in IoT devices [6]. The author extracted malware photos (i.e., a one-channel grey-scale image modified from a malicious binary) and categorized their families using the original dataset of 500 malware samples. For the secondary classification of unidentified system types, the author in Ref. [3] developed a hybrid mix of a supervised and unsupervised learning technique. Our method uses a deep NN and clustering to classify both visible and invisible gadgets. Using an automated encoding approach minimizes the size of the dataset while maintaining a high level of classification precision. The comparison of different ML and DL classifiers are shown in Figure 1.4.

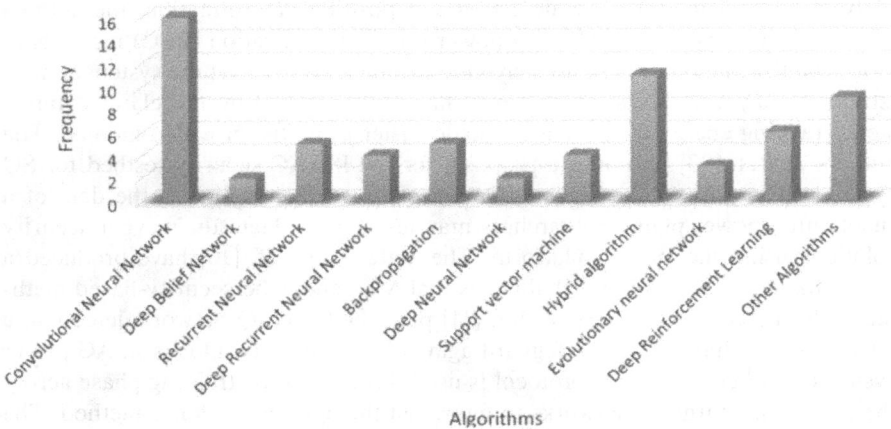

FIGURE 1.4 Comparison of different ML and DL classifiers.

SECURITY SOLUTIONS

Securing smart cities, smart homes, and smart industries is crucial to ensure the safety, privacy, and reliability of the interconnected systems and devices. Here are some key points that need to be covered in the security of these domains:

1. **Network security**
   Securing communication channels and data flows between devices, sensors, and systems within the smart infrastructure; implementing encryption and authentication mechanisms to protect sensitive data and prevent unauthorized access; ensuring the integrity and availability of the network infrastructure to prevent disruptions [7].
2. **Data privacy and protection**
   Developing robust data protection policies to safeguard the privacy of citizens, residents, or employees whose data is collected by smart devices and systems; implementing data anonymization techniques to minimize the risk of personal information exposure; complying with data protection regulations and standards to avoid legal and ethical issues [11].
3. **Device security**
   Ensuring the security of individual smart devices and endpoints by regularly updating firmware and software patches; implementing strong authentication and access control mechanisms to prevent unauthorized access to devices [24]; mitigating the risk of physical attacks on devices in public areas.
4. **Cloud security**
   Securing cloud-based services used in smart city infrastructure, smart home applications, and IIoT systems; ensuring data stored in the cloud is encrypted and properly managed to prevent data breaches; monitoring and controlling access to cloud resources [15].

5. **IoT security**

   Implementing security measures for the IoT devices to prevent potential exploitation by attackers; employing IoT device management systems to monitor device health, perform updates, and detect anomalies; addressing IoT-specific security challenges like resource constraints and lack of standardization [16].

6. **Incident response and disaster recovery**

   Developing comprehensive incident response plans to detect, respond to, and recover from security breaches or cyber-attacks promptly; creating disaster recovery strategies to ensure the continuity of critical services in case of system failures or attacks [25].

7. **Physical security**

   Protecting physical access points to critical smart infrastructure, such as power stations and water supply facilities; implementing surveillance systems to monitor public spaces and deter potential threats [12].

8. **Security awareness and training**

   Conducting security awareness campaigns to educate citizens, employees, and users about potential risks and best practices; providing specialized training for personnel involved in the maintenance and operation of smart systems [6].

9. **Interoperability and standards**

   Encouraging the use of standardized security protocols and frameworks to ensure seamless and secure communication between different smart devices and systems; addressing potential security gaps that arise from the integration of diverse technologies [26].

10. **Collaborative approach**

    Fostering collaboration between government authorities, industry stakeholders, and security experts to collectively address security challenges; establishing public-private partnerships to share threat intelligence and best practices [27].

By addressing these topics, stakeholders can work towards establishing robust and resilient security measures in smart cities, smart homes, and smart industries, thereby enabling the full potential of these technological advancements while minimizing risks and threats. When it pertains to intelligent towns, smart residences, and smart industries, security is crucial [28]. Here are some common security measures utilized in these situations:

- Implementation of different authentication methods like fingerprints, duality factor authorization, and also encrypted certificates for authorized access to the devices [11].
- Protocols that use encryption methods (e.g., SSL/TLS) to secure the transmission of data between networks, devices, and servers. This prevents unauthorized data interception and guarantees data privacy [29].
- Intrusion monitoring and detection systems: The installation of the following system to track networks, systems, and devices for intrusions, attacks,

and suspicious activity. These systems are capable of identifying and react-ing in real time to security incidents [9].

- Utilize firewalls to monitor and control traffic on the network, screening out illicit activities. The segmentation of network assists in isolating essential net-works, systems, and devices in order to contain any possible breaches [30].
- Physical security measures: To safeguard the infrastructure of smart cities, smart residences, and smart industries, installation of security equipment such as cameras for surveillance, control systems for entry, and alarms [20].
- Utilize advanced security analytics tools for recognizing abnormalities, trends, and potential hazards from large volumes of data. Monitoring in real-time enables preventive actions regarding safety-related incidents.
- Programming and applications updates: Update and restore the hardware and application of smart devices on a regular basis to resolve vulnerabilities and guarantee that they are protected from the most recent security threats [9].
- Privacy of data and enforcement: To protect sensitive and identifiable infor-mation gathered through smart systems, implementation of privacy con-trols, and compliment with applicable data protection regulations.
- Incident response planning: Establishing extensive reaction plans for suc-cessfully dealing with security incidents, including damage mitigation, incident containment, and normal operations restoration [31].
- Security education and formation: Educating users, employees, and influ-encers on best security practices, such as password strength, safe browsing behaviours, and identifying social engineering attacks [32].

It is essential to keep in mind that safety standards may vary depending on the context and implementation of intelligent solutions. Consultation with security spe-cialists and professionals to tailor security measures to the specific requirements of each smart city, smart household, or smart industry implementation.

NEED OF SECURE MODELS FOR SMART CITIES, SMART HOMES, AND SMART INDUSTRIES

Securing models for smart homes, smart cities, and smart industries is crucial to protect sensitive data, ensure privacy, and maintain the integrity of these systems. Implement strong encryption protocols to protect data both in transit and at rest. This ensures that data remains confidential even if intercepted. Enforce robust authen-tication mechanisms to verify the identity of users, devices, and applications [7]. Additionally, define appropriate access controls to restrict unauthorized access to critical systems and data. Use secure communication protocols, such as HTTPS and VPNs, to safeguard data transmitted between devices and servers. Keep all devices and systems up-to-date with the latest security patches and firmware to mitigate vul-nerabilities. Utilize secure boot processes and device integrity verification to prevent unauthorized modification or tampering of devices. Implementing multifactor autho-rization to add an extra layer of protection by requiring users to provide additional authentication factors beyond just passwords. Ensuring that application programming interfaces are securely designed and authenticated to prevent unauthorized access to back-end services [33]. Deploy anomaly detection systems to identify unusual behaviour or patterns that may indicate potential security threats. Comply with data privacy regulations and obtain explicit user consent when collecting and using personal data. Regularly conduct security testing, vulnerability assessments, and third-party audits to identify and address potential weaknesses. Implement resilient data backup and disaster recovery strategies to minimize data loss caused by cyber threats to vital infrastructure, and use physical security measures to safeguard [19]. By adopting these security measures, smart homes, smart cities, and smart industries can be better equipped to withstand potential cyber threats and protect the privacy and safety of their users and assets.

INTRUSION DETECTION SYSTEM FOR AUTOMATIC SYSTEM

Initially, there exist many research methods for industrial control systems (ICSs) to deal with security and privacy methods. The intrusion detection system is very much popular for detecting intrusions and malicious activities present in industrial control systems. As per the knowledge, there are many types of intrusion detection systems available for the industries such as Bro and Snort. The Bro Intrusion detection system is initialized by Berkeley University and a network-based ID having a set of rules parser for parsing the packets in the network. Cisco came up with the other industrial IDS which is known as Snort, an open-source ruled-based Network IDS (NIDS). Also first, the traffic is analysed; if violated by the set of rules, it alerts the network administrator of suspicious activities [31].

Specification-based: The knowledge-based approach or specification-based approach of the construction of a model is expert-defined specification-based so that the legitimate behaviour of a system can be characterized which results in the reduction of the number of false positives. Various types that come under the specification-based intrusion detection system are finite automata, formal methods, state diagrams, and so on [12].

Based on the user data being used, an anomaly-based Intrusion Detection System can be divided. Just as in IT systems, there exists some intrusion detection system that works in the system of the cyberpart which focuses on the detection of change in the format of any protocol or the transmitted traffic in the network. As a result, the automatic networks are the cyber-physical system that is used to interact with the physical process.

Machine learning-based: Algorithms work on building a mathematical model where the events will be analysed, and after analysis, the respective events will be categorized. There are two types of ML models such as supervised and unsupervised. The supervised learning method works on the labelled training data. The training set has the input data which can be categorized as normal or abnormal. In the unsupervised method of learning, no labelling of input data but based on the analysis of data characterization, the machine will learn and the respective classifier will be constructed. There exist various types of ML algorithms such as deep learning, artificial neural networks, Bayesian networks, decision trees, support vector machines (SVM), clustering and classification, fuzzy logic, random forest, and so on.

SECURE DATASETS FOR SMART NETWORKS

For the development and evaluation of security solutions in smart cities, smart households, and smart industries, it is essential to get high-quality security datasets. Here are some examples of security datasets connected to these categories, while individual datasets may differ depending on the application and context:

- Datasets including video feeds from security cameras recording different traffic circumstances, such as incidents, traffic jams, and unusual happenings, are known as traffic surveillance [34].
- Network traffic datasets containing various assaults, such as DoS (Denial of Service), DDoS, and intrusion attempts, acquired from smart city networks are used for intrusion detection.
- Social media monitoring refers to datasets that include social media postings and comments on security issues, emergency situations, and public safety in smart cities.
- Datasets including sensor readings, network activity, and user behaviour information are used to create intrusion detection systems for smart homes, which may be used to identify illegal entry or odd activity; datasets including information from smart home ambient sensors, such as temperature, humidity, and gas concentrations, are used for anomaly identification and safety monitoring; datasets recording communication data between different smart devices connected to a home network enable investigation for possible security flaws or unusual behaviour.
- Industrial control systems (ICS): Datasets used to create and test security solutions for industrial contexts by simulating ICS networks with data packets and instructions sent between devices [33].
- Manufacturing process data: Sets of data from smart factories that comprise sensor readings, machine status, and production records, enabling the creation of algorithms for anomaly identification and models for proactive maintenance.
- Cyber-physical systems: Datasets that simulate interactions between hardware and software in smart industries, allowing the assessment of security precautions against possible intrusions.

Data privacy and legal compliance must be taken into consideration while searching for security datasets [19]. To safeguard privacy, certain publicly accessible datasets could be anonymized or stripped of personally identifying information. When managing sensitive data, keep in mind all applicable laws and ethical standards [35, 36]. Security datasets are made available for academic and research use by a number of organizations and research institutes. Several noteworthy sources are as follows:

1. NIST stands for National Institute of Standards and Technology. NSRL, the National Software Reference Library
2. Industrial Control System Security Datasets from MIT Lincoln Laboratory

3. Cybersecurity from the University of New Brunswick's AI Data Repository (CADR)
4. The University of California, Irvine's Internet of Things Data Repository (IoTDS)

CONCLUSION

The combination of ML and deep learning approaches has had a significant impact on the development of smart cities, smart industries, and smart households. These cutting-edge technologies eventually enhance decision-making, operational efficiency, and user experiences by facilitating data analysis, pattern recognition, and predictive capabilities. In smart cities, automation has proven indispensable for managing the waste, energy optimization, and traffic management. By analysing vast quantities of data, these models may uncover vital information, which aids in the more efficient allocation of resources, improvement of sustainability, and enhancement of the urban environment as a whole. Deep learning and ML are essential components of predictive maintenance, quality assurance, supply chain optimization, and process enhancement in smart industries. These models may identify trends, anticipate problems, optimize production processes, and increase output, resulting in eventual cost savings and enhanced operational efficacy. They accomplish this by utilizing sensor data and historical records.

Smart homes use ML and deep learning algorithms to control energy consumption, provide individualized assistance, enhance security, and recognize speech and gestures. These models provide homeowners with energy savings, personalized recommendations, enhanced security, and simple administration of smart home devices, resulting in comfort, convenience, and energy efficiency. This is accomplished by analysing user preferences, behaviour, and contextual information. However, privacy, security, and ethical concerns must be addressed when ML, and deep learning solutions are implemented in these fields. Building trust and maintaining the integrity of these intelligent ecosystems requires the protection of data privacy, the assurance of secure data transmission, and the development of transparent and ethical processes. The use of automation in smart cities, smart businesses, and smart homes has tremendous potential to enhance urban environments, optimize business operations, and design intelligent dwellings. As technology continues to advance, developments in these areas will pave the way for wiser, more sustainable, and more interconnected communities.

FUTURE SCOPE

Developing novel and innovative IoT network layouts with secure needs like authorization, controlling access, confidentiality, and intrusions which is a potential strategy for providing end-to-end security. New IoT designs may use cutting-edge concepts like networking defined by software- and cloud-enabled IoT to provide quality of service without compromising security requirements. Optimization algorithms may be used in feature extraction, selection, and parameter tweaking techniques. Accuracy may be increased using hybrid deep learning techniques without lengthening calculation times. In terms of IoT security, deep learning–based block chain solutions are a promising area. A relatively new method that integrates the security provision with that of the data integrity is blockchain technology.

REFERENCES

1. A. Zanella, N. Bui, A. Castellani, L. Vangelista, and M. Zorzi, "Internet of things for smart cities," IEEE Internet Things J., vol. 1, no. 1, pp. 22–32, 2014.
2. (a) G. E. Hinton, and R. R. Salakhutdinov, "Reducing the dimensionality of data with neural networks," Science, vol. 313, no. 5786, pp. 504–507, 2006; (b) Q. Zhang, L. T. Yang, Z. Chen, and P. Li, "A survey on deep learning for big data," Inf. Fusion, vol. 42, pp. 146–157, 2018; (c) H. M. Fayek, M. Lech, L. Cavedon, "Evaluating deep learning architectures for speech emotion recognition," Neural Netw., vol. 92, pp. 60–68, 2017.
3. Y. Zhang, X. Li, Z. Zhang, F. Wu, and L. Zhao, "Deep learning driven blockwise moving object detection with binary scene modeling," Neurocomputing, vol. 168, pp. 454–463, 2015.
4. W. Zhang, L. Chen, W. Gong, Z. Li, Q. Lu, and S. Yang, "An integrated approach for vehicle detection and type recognition," in: Paper Presented at the 2015 IEEE 12th International Conference on Ubiquitous Intelligence and Computing and 2015 IEEE 12th International Conference on Autonomic and Trusted Computing and 2015 IEEE 15th International Conference on Scalable Computing and Communications and Its Associated Workshops (UIC-ATC-ScalCom), 2015.
5. P. Li, Y. Zang, C. Wang, J. Li, M. Cheng, L. Luo, and Y. Yu, "Road network extraction via deep learning and line integral convolution," in: Paper Presented at the 2016 IEEE International Geoscience and Remote Sensing Symposium (IGARSS), 2016.
6. Y. Lv, Y. Duan, W. Kang, Z. Li, and F.-Y. Wang, "Traffic flow prediction with big data: a deep learning approach," IEEE Trans. Intell. Transp. Syst., vol. 16, no. 2, pp. 865–873, 2015.
7. L. Atzori, A. Iera, and G. Morabito, "The internet of things: a survey," Comput. Netw., vol. 54, no. 15, pp. 2787–2805, 2010.
8. B. T. Ong, K. Sugiura, and K. Zettsu, "Dynamically pre-trained deep recurrent neural networks using environmental monitoring data for predicting PM2.5," Neural Comput. Appl., vol. 27, no. 6, pp. 1553–1566, 2016.
9. J. H. Park, M. M. Salim, J. H. Jo, J. C. S. Sicato, S. Rathore, and J. H. Park," CIoT-Net: a scalable cognitive IoT based smart city network architecture," Hum. Centric Comput. Inf. Sci., vol. 9, no. 1, p. 29, 2019. https://doi.org/10.1186/s13673-019-0190-9
10. J. Zhang, Y. Zheng, and D. Qi, "Deep spatiotemporal residual networks for citywide crowd flows prediction," in: Proc. AAAI, pp. 1655–1661, 2017.
11. U. Varshney, "Pervasive healthcare and wireless health monitoring," Mobile Netw. Appl., vol. 12, no. 2–3, pp. 113–127, 2007.
12. R. Miotto, L. Li, B. A. J. Kidd, and J. T. Dudley, "Deep patient: an unsupervised representation to predict the future of patients from the electronic health records," Sci. Rep., vol. 6, p. 26094, 2016.
13. Z. C. Lipton, D. C. Kale, C. Elkan, and R. Wetzell, "Learning to diagnose with LSTM recurrent neural networks," in: Presented at Int. Conf. Learn. Representations (2016), 2015.
14. X. Yi, J. Zhang, Z. Wang, T. Li, and Y. Zheng, "Deep distributed fusion network for air quality prediction," in: Proc. 24th SIGKDD Int. Conf. Knowledge Discovery Data Mining, 2018, pp. 965–973.
15. Z. Qi, T. Wang, G. Song, W. Hu, X. Li, and Z. M. Zhang, "Deep air learning: Interpolation, prediction, and feature analysis of fine-grained air quality," IEEE Trans. Knowledge Data Eng., vol. 30, no. 12, pp. 2285–2297, 2018.

16. X. Liu, W. Liu, T. Mei, and H. Ma, "A deep learning-based approach to progressive vehicle re-identification for urban surveillance," in: Computer Vision—ECCV 2016, pp. 869–884, 2016.

17. J. Guo, J. Lu, Y. Qu, and C. Li, "Traffic-sign spotting in the wild via deep features," in: Paper Presented at the 2018 IEEE Intelligent Vehicles Symposium (IV), 2018.

18. S. Badotra, and S. N. Panda, "SNORT based early DDoS detection system using open-daylight and open networking operating system in software defined networking," Cluster Comput., vol. 24, pp. 501–513, 2021.

19. U. Aguilera, O. Peña, O. Belmonte, and D. López-de-Ipiña, "Citizen-centric data services for smarter cities," Fut. Gener. Comput. Syst., vol. 76, pp. 234–247, 2017

20. H. Gong, R. Li, Y. Bai, J. An, and K. Li, "Message response time analysis for automotive cyber-physical systems with uncertain delay: an M/PH/1 queue approach," Perform. Eval., vol. 125, pp. 21–47, 2018. https://doi.org/10.1016/j.peva.2018.07.001

21. J. Lwowski, P. Kolar, P. Benavidez, P. Rad, J. J. Prevost, and M. Jamshidi, "Pedestrian detection system for smart communities using deep convolutional neural networks," in: Paper Presented at the 2017 12th System of Systems Engineering Conference (SoSE), 2017.

22. S. Sood, H. Singh, M. Malarvel, and R. Ahuja, "Significance and limitations of deep neural networks for image classification and object detection," in: 2021 2nd International Conference on Smart Electronics and Communication (ICOSEC). IEEE, 2021, October, pp. 1453–1460.

23. G. Pan, G. Qi, W. Zhang, S. Li, Z. Wu, and L.T. Yang, "Trace analysis and mining for smart cities: issues, methods, and applications," IEEE Commun. Mag., vol. 51, no. 6, pp. 120–126, 2013.

24. D. Haifeng, and H. Siqi, "Natural scene text detection based on yolo v2 network model," J. Phys. Conf. Ser., vol. 1634, p. 012013, 2020.

25. W. Zhihuan, C. Xiangning, G. Yongming, and L. Yuntao, "Rapid target detection in high resolution remote sensing images using yolo model," Int. Arch. Photogrammetry, Remote Sens. Spatial Inf. Sci., vol. 42, p. 3, 2018.

26. R. Cheng, "A survey: comparison between convolutional neural network and yolo in image identification," J. Phys. Conf. Ser., vol. 1453, p. 012139, 2020.

27. N. Zhang, Y. Liu, L. Zou, H. Zhao, W. Dong, H. Zhou, H. Zhou, and M. Huang, "Automatic recognition of oil industry facilities based on deep learning," in: IGARSS 2018—2018 IEEE International Geoscience and Remote Sensing Symposium, IEEE, 2018, pp. 2519–2522.

28. T. Li, B. Wang, Y. Jiang, Y. Zhang, and Y. Yan, "Restricted Boltzmann machine-based approaches for link prediction in dynamic networks," IEEE Access, vol. 6, pp. 29940–29951, 2018.

29. J. Redmon, S. Divvala, R. Girshick, and A. Farhadi, "You only look once: unified, real-time object detection," in: Proceedings of the IEEE Conference on Computer Vision and Pattern Recognition, 2016, pp. 779–788.

30. M. H. Rafiei, W. H. Khushefati, R. Demirboga, and H. Adeli, "Supervised deep restricted Boltzmann machine for estimation of concrete," ACI Mater. J., vol. 114, no. 2, 2017.

31. E. O'Dwyer, I. Pan, S. Acha, and N. Shah, "Smart energy systems for sustainable smart cities: Current developments, trends and future directions," Appl. Energy, vol. 237, pp. 581–597, 2019.

32. R. M. A. Mohammad, and M. M. Abdulqader, "Exploring cyber security measures in smart cities," in: 2020 21st International Arab Conference on Information Technology (ACIT), Giza, Egypt, 2020, pp. 1–7. https://doi.org/10.1109/ACIT50332.2020.9300050

33. R. Petrolo, V. Loscri, and N. Mitton, "Towards a smart city based on cloud of things, a survey on the smart city vision and paradigms," Trans. Emerg. Telecommun. Technol., vol. 28, no. 1, p. e2931, 2017.

34. Y. Liu, C. Yang, L. Jiang, S. Xie, and Y. Zhang, "Intelligent edge computing for IoT-based energy management in smart cities," IEEE Netw., vol. 33, no. 2, pp. 111–117, 2019.
35. P. Neirotti, A. De Marco, A. C. Cagliano, G. Mangano, and F. Scorrano, "Current trends in smart city initiatives: some stylised facts," Cities, vol. 38, pp. 25–36, 2014.
36. F. Al-Turjman, and I. Baali, "Machine learning for wearable IoT-based applications: a survey," Trans. Emerg. Telecommun. Technol., vol. 33, no. 8, p. e3635, 2019.

2 Integration of Fog Computing and Wireless Sensor Network in Smart Cities

Shruti[1,2] and Shalli Rani[2]
[1]Goswami Ganesh Dutta Sanatan Dharma College, Chandigarh, India
[2]Chitkara University Institute of Engineering and Technology, Chitkara University, Punjab, India

ENLIGHTENMENT OF FOG COMPUTING

Cloud computing is a concept that enables online access to computer resources such as servers, storage, databases, and applications. Rather than building and maintaining their own computing infrastructure, organizations can rent access to these resources from a cloud service provider, who maintains the infrastructure and provides support. The cloud computing model offers several advantages over traditional computing models like organizations can scale their computer capabilities according to demand without spending money on infrastructure. This is particularly useful for organizations with changing demands, such as seasonal businesses or startups experiencing rapid growth.

Organizations can access a wider variety of computing resources (specialized software and tools) because of cloud computing than they might be able to on their own as they would be expensive or difficult to buy and maintain. There are various types of cloud computing services (Figure 2.1), such as

1. Infrastructure-as-a-Service (IaaS): It enables pay-on-demand access to computer infrastructure, like storage, software, server, or machines.
2. Platform-as-a-Service (PaaS): It offers the creation, evaluation, and installation of applications over the Internet. PaaS is managed by a third party and consists of tools and frameworks required for creating an application.
3. Software-as-a-Service (SaaS): It offers paid subscription access to software applications over the Internet.

Since cloud computing offers greater flexibility, scalability, and cost savings than traditional computing models, it has emerged as a crucial tool for many enterprises and organisations, though it is very important that organisations must carefully assess their demands for cloud computing and select a service provider that can satisfy those objectives.

DOI: 10.1201/9781003438205-2

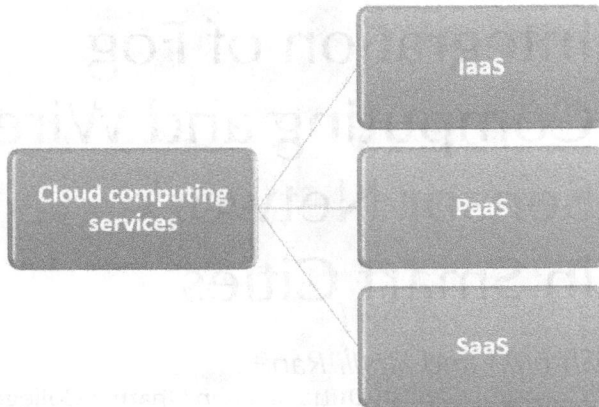

FIGURE 2.1   Various cloud computing services.

LIMITATIONS OF CLOUD COMPUTING

While there are many benefits of cloud computing, there may also be some disadvantages to consider. IoT applications require capabilities like mobility, geographical distribution, awareness of location, and low latency which are generally not provided by cloud computing [1, 2]. Few limitations of cloud computing are discussed below:

1. Security and privacy risks: Storing data and applications in the cloud can increase the chances of security threats like data breaches, as sensitive data are transmitted and stored over the internet. Therefore, organizations must come up with some security measures like encryption and access control methods to reduce these risks.
2. Service provider dependence: Organisations are becoming more and more dependent on third-party service providers who offer cloud computing services. This can make it challenging to change providers or bring services back, if required.
3. Interruption and accessibility: Due to interruptions in cloud computing services, applications and services may not be available. Therefore, organizations need to ensure that their service agreements with cloud providers include appropriate uptime guarantees and compensation for interruptions.
4. Compliance and regulatory issues: When storing and processing data in the cloud, organizations may need to follow industry-specific rules or data protection laws. This can add complexity and cost to cloud operations.
5. Cost: While cloud computing can offer cost savings over traditional infrastructure, it can also be expensive for organizations that require large amounts of computing resources or high-performance applications. Therefore, organizations need to make careful assessments while comparing the price of cloud services to the price of internal infrastructure.
6. Dependency on Internet: Cloud computing heavily relies on a stable and high-speed Internet connection, without which accessing cloud services

and data can become challenging or even impossible. This can be a limitation in areas with limited or unreliable internet connectivity, during network outages, or in situations where low-latency access is crucial.

7. Limited control and customization: Cloud computing involves outsourcing infrastructure and services to third-party providers, which means users have limited control over the hardware, software, and configurations. This can restrict customization or delay the ability to meet requirements that may require specific software configurations, hardware optimizations, or specialized environments.

8. Vendor lock-in: Moving applications and data to a specific cloud provider's infrastructure can create a dependency on that provider's ecosystem. Switching to a different provider or moving data, back-on premises can be complex, time-consuming, and costly. This can limit flexibility, negotiate power, and hinder the ability to take advantage of changing business needs.

9. Data transfer and bandwidth: Moving large volumes of data to and from the cloud can be time-consuming and may face bandwidth limitations. Uploading or downloading large datasets can take significant time, effecting data-intensive operations that require frequent data transfers.

10. Scalability: Cloud computing is capable of offering scalability but organizations need to plan and manage resources effectively to ensure proper scalability. Poorly designed applications may not take full advantage of the scalability benefits, leading to under-utilized resources or performance bottlenecks.

WHAT IS FOG COMPUTING?

Fog computing is a distributed computing architecture that brings the capabilities of cloud computing closer to the edge of the network resulting in low latency [3], high bandwidth, improved security, and the ability to process in real-time. Instead of performing data processing in a centralized cloud, fog computing performs it in sensors, IoT devices, and other edge devices that are closer to the data source. In short, we can say fog computing brings the source devices and users closer to the IoT applications [4]. This results in improved response time, a reduction in network congestion, and an increase in the system's performance. There are fog nodes in fog environment that comes in various shapes and sizes distributed over the network providing heterogeneity to the environment [5]. It is frequently used in applications like real-time processing-required smart cities, driverless vehicles, and industrial automation.

Overall, fog computing is a complementary technology to cloud computing that can improve the efficiency in terms of data processing by bringing computation near to generation and consumption of data. It is a paradigm that has the capability to extend the services of cloud computing towards network's edge [6], its layout has been discussed in Figure 2.2.

HISTORY OF FOG COMPUTING

The history of fog computing can be traced back to the early 2000s when the concept of edge computing started to emerge. In 2009, a paper titled "The Case for VM-Based

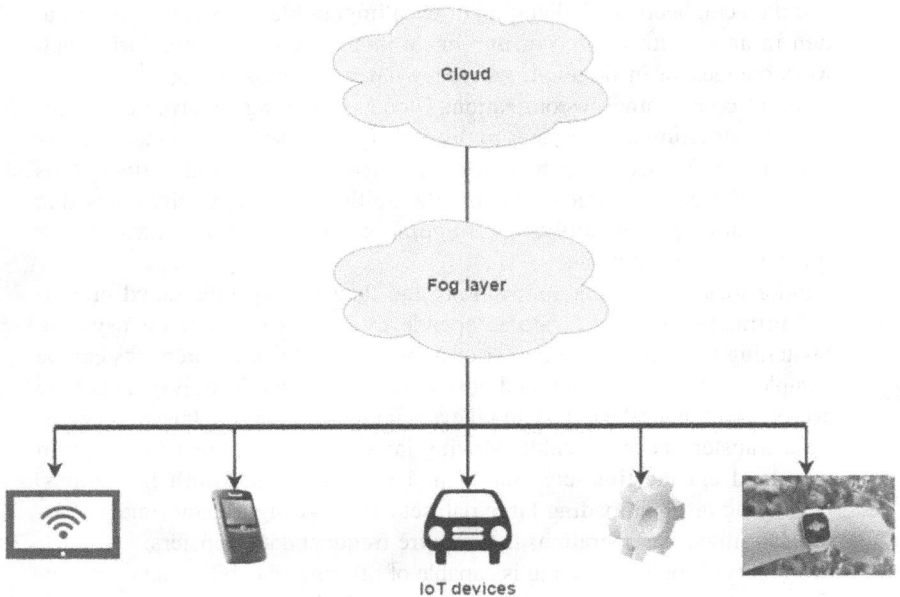

FIGURE 2.2   Layout of fog computing model.

Cloudlets in Mobile Computing" by Satyanarayanan et al. [7] laid the foundation for the concept of cloudlets, which are small-scale cloud data centres located at the network edge. These cloudlets aimed to provide computational resources closer to mobile devices, enabling offloading of resource-intensive tasks and reducing network latency.

The term "fog computing" was first used by Cisco in 2012; it is relatively a new idea in the area of distributed computing. This new architecture shows how cloud computing capabilities were extended to the network's edge. With the increase in demand for real-time processing and low-latency communication in Internet of Things (IoT), the concept of fog computing was developed. As the requirements of IoT applications for large connected devices and the volume of data created by these devices could not be met by traditional cloud computing systems. The architecture was initially called "IOx" that enables communication between devices and cloud resources resulting in better performance and faster decision-making process. It encourages developers to bring applications and connectivity to network edges by combining the network operating systems, Cisco IOS and open-source Linux in a single networked device [8].

In 2014, the European Union's Horizon 2020 research and innovation program launched the Mobile Cloud Networking (MCN) project. The MCN project aimed to advance the concept of fog computing and develop practical solutions for seamless integration between mobile networks and cloud computing resources. Furthermore, the rise of 5G networks accelerated the adoption of fog computing. 5G networks offered significantly higher speeds and lower latency, making it even more crucial to process and analyse data at the network edge. Fog computing became an essential component in enabling the full potential of 5G networks, supporting real-time applications, autonomous vehicles, smart cities, and other bandwidth-intensive use cases.

Over time, other companies and organizations began to adopt fog computing and develop their own solutions. Various researches were also done in this field [9–14]. In 2015, the OpenFog Consortium was formed to advance the development and adoption of fog computing technologies. The consortium included companies like Cisco, Intel, Microsoft, and ARM, along with academic institutions and organizations. In recent years, fog computing has evolved to embrace a wider range of technologies and concepts. Edge AI (Artificial Intelligence) has become an integral part of fog computing, enabling intelligent decision-making and real-time analytics at the network edge. The convergence of fog computing, IoT, and AI has paved the way for innovative applications such as smart homes, industrial automation, healthcare monitoring, and autonomous systems. The growth of fog computing is driven by the need for real-time data processing, reduced network latency, enhanced privacy and security, and the ability to handle the massive data from IoT devices.

Today, fog computing is widely acknowledged as a key technology for supporting real-time processing, low-latency communication, and other services essential to many IoT applications. As technology continues to evolve, fog computing is expected to play a crucial role in enabling the next generation of interconnected and intelligent systems.

Why Fog Computing Better than Cloud Computing?

Compared to cloud computing, fog computing is not superior, rather it is a complementary technology that can expand the possibilities of cloud computing by locating computation and storage closer to the network's edge. Here are some contexts where fog computing is preferable to cloud computing [15–17]:

1. Heterogeneity: It has the capability to process and store data obtained from various sources and collected by different means of communication [18].
2. Low latency: The communication delay caused by the distance between IoT devices and the cloud can be challenging to meet time constraint requirements of applications like smart grid, smart factories, and healthcare [19]. Fog computing can reduce latency by enabling data processing to be distributed across devices located near to data source. This improves the overall performance of the applications that require real-time processing.
3. Reduces bandwidth: Fog computing helps to preserve bandwidth and ease network congestion by processing data locally. Instead of transmitting all data to a remote cloud data centre, fog nodes can filter, aggregate, or preprocess data at the edge, reducing the amount of data that needs to be transferred over the network. This approach optimizes bandwidth utilization, decreases network congestion, and reduces the cost associated with transferring large volumes of data to the cloud.
4. Offline processing: Fog computing is capable of processing the data even when a device is not connected to the cloud, which is useful in systems with limited connectivity. This ensures continuity and resilience, particularly in scenarios where uninterrupted operation is critical, such as remote locations or during network outages.

5. Improved privacy and security: Fog computing allows data to be processed locally on devices, which can enhance the privacy and security of sensitive information. As data remain within the organization's controlled environment and reduce exposure to potential security risks associated with transmitting sensitive information over public networks, applications such as healthcare, finance, and government, which handle sensitive data, can benefit from the enhanced privacy and security provided by fog computing.

6. Geographical distribution: Fog environment comprises fog nodes that are geographically distributed over a large area providing high-quality streaming to devices in motion, thus supporting mobility also.

7. Scalability and flexibility: Fog computing offers a scalable and flexible architecture by distributing computational resources across the network edge. The ability to scale horizontally with distributed fog nodes allows for efficient utilization of resources and enables organizations to handle growing workloads without overburdening the cloud. It also offers flexibility in managing resources, as organizations can adapt the fog infrastructure to suit specific requirements, customize software configurations, and optimize resource allocation based on needs.

8. Cost efficient: By offloading computational tasks and data processing to edge devices and fog nodes, organizations can reduce the volume of data transferred to the cloud and therefore optimize cloud resource usage. This leads to cost savings in terms of reduced bandwidth costs, lower data storage requirements, and optimized cloud service subscriptions.

A well-defined comparison using various other aspects of both the models is done in Table 2.1:

It is important to keep in mind that fog and cloud computing differences are not always well-defined; these can be used together to achieve certain optimal results. Fog computing can provide more effective, real-time, and distributed data processing, but it is not necessarily superior to cloud computing as a whole. By combining the benefits of two technologies, organizations can build an infrastructure that are more reliable, efficient, and well suited to the requirements of their applications and services.

STRUCTURE OF FOG COMPUTING

The structure of fog computing model (Figure 2.3) is designed to bring computation, storage, and networking abilities close to the network's edge. It consists of several layers:

1. Edge devices: Devices like sensors, actuators, or other end devices that help in the generation and collection of data are known as edge devices. These are located at the network's edge and serve as the entry point for data into the fog-computing infrastructure. Edge devices form the foundation of fog computing by capturing and transmitting data to the subsequent layers for processing and analysis.

TABLE 2.1

Comparison between Fog Computing and Cloud Computing

Aspects	Fog Computing	Cloud Computing
Definition	It is a decentralized computing infrastructure where data processing and storage occur at the network's edge	It is centralized computing infrastructure where data processing and storage occur in data centres or cloud
Architecture	It is distributed in nature and has a hierarchical architecture with multiple layers of computing nodes	It has a centralized architecture with a single layer of computing nodes
Latency	Low latency	Higher latency
Bandwidth	Reduced network bandwidth requirements	Higher network bandwidth requirements
Reliability	Reliability higher as fog nodes operate autonomously	Reliability less as dependency is on cloud and data centres
Proximity	Computing resources are located at the network's edge, providing low latency and fast response times	Computing resources are located in data centres, which results in higher latency and longer response times as compared to fog computing
Scalability	Scalability is achieved using multiple layers of computing nodes to handle a large amount of increasing data and its processing	Scalability is achieved using multiple servers in data centres to handle a large amount of increasing data and its processing
Security	Reduced risk of data breaches during transit	It is vulnerable to threats as data are transmitted over the internet
Privacy	Improved privacy as data are processed locally and sensitive information is kept within the network	Privacy concern is high as data transfer and storage are managed by third-party cloud environments
Can be used in	IoT applications, analysis in real-time, and low-latency applications	Web applications, data storage and retrieval, and applications that require high availability and scalability

2. Fog nodes: The next layer consists of fog nodes, also referred to as fog servers or fog gateways. These nodes serve as intermediaries between the edge devices and the cloud infrastructure. Fog nodes act as computing and storage resources like servers, routers, gateways, and other devices. Fog nodes enable the processing and analysis of data to be performed at the network's edge, which reduces latency and improves performance. They provide services to applications that require quick response [20].

3. Cloud: The cloud infrastructure layer represents the traditional cloud computing resources, which can be public, private, or hybrid clouds. It serves as the central repository for data and consists of high-performance servers [21]. Cloud also helps to provide additional computing and storage capabilities as needed. The cloud infrastructure is responsible for handling tasks that require significant computational resources, long-term storage, or complex data analytics. It serves as a scalable and centralized resource for offloading intensive computational tasks from the fog infrastructure when necessary.

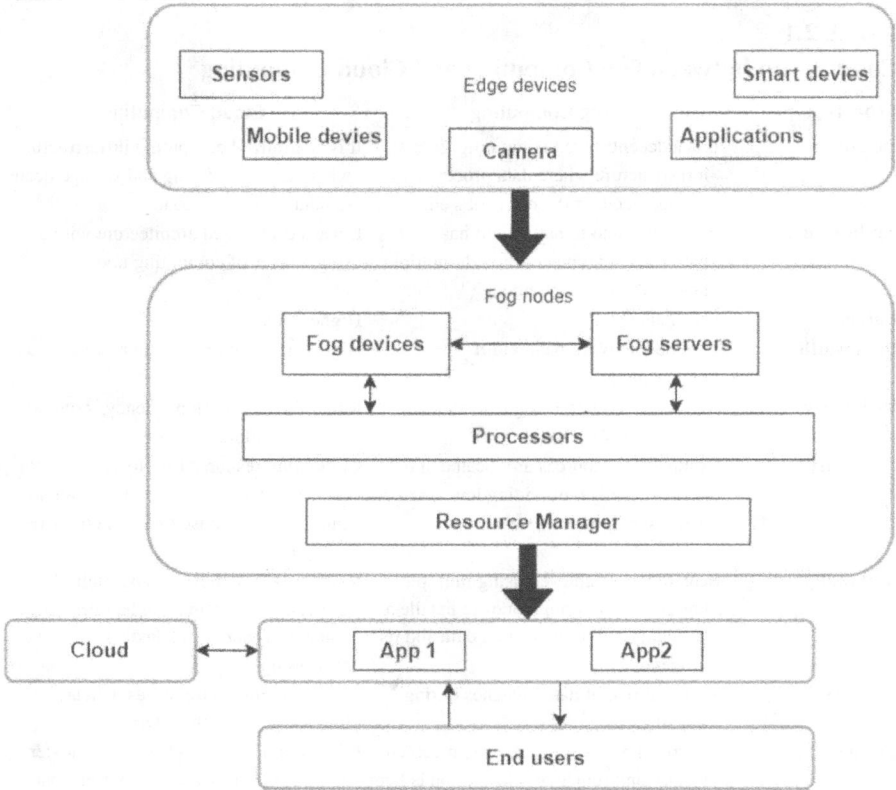

FIGURE 2.3 Fog computing structure [22].

4. Networking: Networking is a critical component of fog computing structure that enables data to be transmitted and processed across different layers of architecture. It connects edge devices, fog nodes, and cloud infrastructure, enabling seamless data transfer and communication. Fog computing is a combination of wired and wireless networking models like Wi-Fi, Bluetooth, 5G, Zigbee, and others. Network also ensures reliable and efficient data transmission between different layers of the fog computing architecture.

5. Applications and services: Applications and services are installed on fog nodes and the cloud and can include real-time processing, analytics, monitoring, and other functions. Depending on the particular needs, these applications and services can be created to operate on various layers of the fog computing architecture.

6. Management and orchestration: The management and orchestration are the responsibility of the top layer of the fog structure. It manages the overall fog infrastructure, including monitoring fog nodes, resource allocation, load balancing, security management, and coordination of data processing tasks. It ensures the efficient utilization of resources and facilitates seamless integration between the edge devices, fog nodes, and cloud infrastructure.

Overall, the structure of fog computing makes it possible to process and analyse data effectively in real-time by moving storage and processing resources to the network's edge, while still utilising the cloud's capabilities for more difficult tasks.

EVOLUTION OF FOG COMPUTING IN SMART CITIES

The evolution of fog computing in smart cities has followed a progressive path, driven by the growing need for real-time data processing and low-latency applications in urban environments. Initially, early smart city concepts focused on centralizing data processing in cloud data centres, but its limitations became evident. Fog computing emerged as a solution, championed by Cisco Systems in 2012, advocating for localized data processing at the network edge. This shift allowed for reduced latency and improved response times, critical for real-time applications in smart cities. As fog computing gained momentum, it integrated with IoT devices and edge computing technologies, becoming an essential component of smart city deployments. IoT devices such as sensors, surveillance cameras, and actuators began using fog nodes to process and analyse data locally, minimizing the need for transmitting large volumes of data to the cloud. This integration resulted in faster understandings, reduced network bandwidth requirements, and improved overall system performance. With time, fog computing evolved to incorporate edge intelligence, enabling edge devices and fog nodes to make localized decisions and perform advanced analysis. This empowered smart city infrastructure to respond quickly to changing conditions, optimize resource utilization, and deliver context-aware services to citizens and urban investors.

To balance local data processing and cloud computing benefits, hybrid fog-cloud architectures emerged in smart cities. These architectures combine fog computing at the network edge for real-time processing and decision-making with cloud infrastructure for resource-intensive tasks and long-term storage. They are meant to provide scalability, flexibility, and cost-effectiveness to smart city solutions. Fog computing when combined with technologies like AI, machine learning, and 5G networks, further enhances its capabilities of smart city applications. Real-time data analysis, predictive modelling, autonomous systems, and intelligent decision-making have become feasible, empowering smart city services.

A smart city can be defined as a concept that makes use of advance technology and data analysis to increase both the sustainability and the quality of life for its residents. Smart city applications take help of IoT devices to collect data, send it for processing, and then act as per decision [23]. Its goal is to improve the economic, social, and environmental well-being of its residents while also making urban areas more efficient, responsive, and sustainable. Smart city infrastructure consists of numerous sensors, entities, and functionalities that must communicate with one another in order to be managed, controlled, and functionally automated [24]. Some of the essential components of a smart city include:

1. Intelligent transportation systems: Technologies like real-time traffic monitoring, intelligent traffic lights, and smart public transport system that offer time-to-time updates on routes, schedules, and availability can be used in smart cities to ease traffic congestion and improve traffic flow.

2. Energy efficiency: Different technologies can help smart cities become more energy efficient and cut back on carbon emissions. Smart buildings with sensors, automatic energy management systems, and energy-efficient lights and electronics are a few examples of how smart cities can optimize energy usage and its distribution.
3. Public safety: Smart cities can use technology to enhance public safety and security by real-time monitoring of public places, intelligent surveillance systems, and implementing quick response systems.
4. Environmental sustainability: Smart cities can help to improve environmental sustainability and reduce waste using smart waste management systems, urban agriculture systems, and maintaining green spaces and parks.
5. Citizen engagement: Citizen engagement and participation in urban governance can be increased in smart cities by including online platforms for public participation and feedback, mobile apps to provide real-time information on events and services, and making government data more accessible and transparent.

In general, a smart city aims to use data and technology to develop more liveable, fair, and sustainable urban environments that improve the well-being of all its residents. In the development of smart cities, fog computing plays a big part, as it enables more effective, real-time, and distributed processing of data using sensors, cameras, and other IoT devices. The growth of fog computing in relation to smart cities can be seen in the following ways:

1. Real-time processing of data: Real-time processing using sensors, cameras, and other IoT devices is made possible by fog computing, which can be crucial for applications like traffic control, emergency response, and public safety. By processing data locally on edge devices or fog nodes, smart city applications can respond more quickly to changing conditions and events.
2. Low-latency communication: Fog computing can reduce latency in smart city applications by enabling data processing to be distributed across devices located close to data source. This results in improved performance of applications and services that require real-time processing.
3. Edge analysis: Fog computing enables edge devices to perform basic analysis of data before transmitting it to cloud, reducing the amount of data to be transmitted over the network resulting in improved system performance. For example, in smart parking system, sensors can be used to do an analysis on where a free parking lot is available.
4. Data privacy and security: Fog computing can enhance data privacy and security in smart city applications by enabling data to be processed locally on edge devices or fog nodes, rather than being transmitted to Cloud. This reduces data breaches and ensures the security of sensitive data. However, new interfaces and protocols may be needed in order for fog software to interact with different hardware platforms and automatically detect any data security breaches if that happen [25].

5. Resource efficiency: Fog computing enables edge devices to carry out processing and analysis; this results in reducing the effort required to transmit data over the network and increases resource efficiency of smart city applications. This helps to solve the problem of network congestion and reduces energy consumption.

As a result, implementing fog computing in smart cities has made it possible to design more efficient, responsive, and secure systems that will improve the quality of life for its residents and the sustainability of the environment.

Throughout this evolution, standardization bodies and industry consortia like the OpenFog Consortium and the Industrial Internet Consortium have played a crucial role. Their efforts have focused on defining common architectures, interoperability standards, and best practices for deploying fog computing solutions in smart cities, encouraging collaborations and driving the adoption of fog computing technologies. The ongoing research and development efforts continue to refine fog computing solutions, optimizing performance, scalability, security, and interoperability. This continuous evolution aims to enable smart cities to address the complex challenges of urbanization and sustainability, facilitating interconnectedness and data-driven decision-making for the benefit of citizens and urban environments.

CONCEPT OF FOG COMPUTING IN SMART CITIES

Here, we will discuss the concept of fog computing in smart cities in both diagrammatical (Figure 2.4) and descriptive ways. A layered architecture showcasing the components involved in fog computing within a smart city context is explained. At

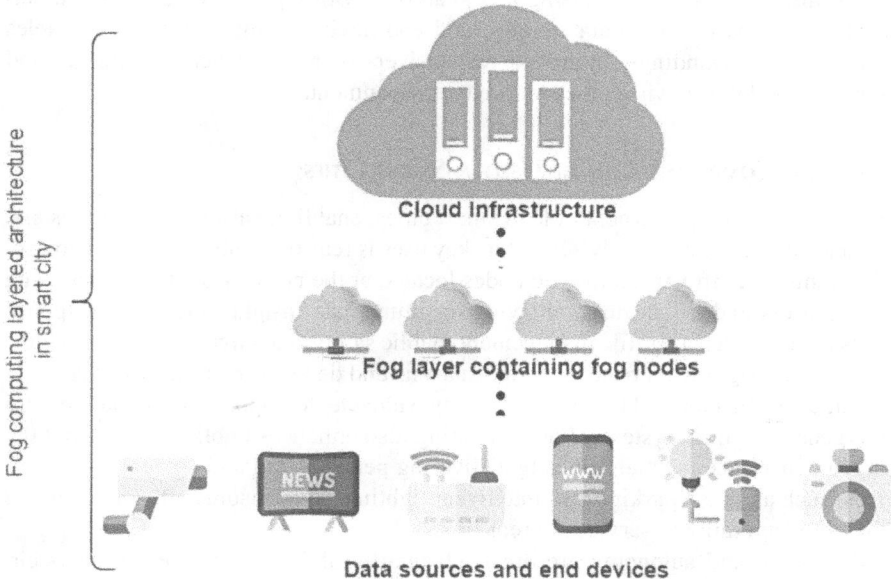

FIGURE 2.4 Fog computing layered architecture in the smart city.

the top layer, we have the cloud infrastructure, which represents the centralized data centres and computing resources typically located outside the smart city. The cloud serves as a repository for large-scale data storage, complex analytics, and long-term data processing. Next to Cloud layer, we have the fog layer, which consists of distributed fog nodes strategically placed throughout the smart city. These fog nodes are typically located at the network edge, closer to the data sources and end devices. They act as intermediary processing units between the cloud and the devices within the city. Connected to the fog nodes, there is a diverse range of IoT devices and sensors spread across the smart city infrastructure. These devices collect data from various sources such as traffic sensors, environmental sensors, surveillance cameras, and smart meters. The data collected by these devices form the foundation for smart city operations.

The fog nodes receive and process these data locally, bringing computation, storage, and networking capabilities closer to the data sources. This localized processing allows for real-time analysis, immediate decision-making, and reduced latency. The fog nodes can perform data filtering, aggregation, data fusion, or even run edge analysis algorithms to extract valuable insights from the collected data. Additionally, the fog nodes may also facilitate direct device-to-device communication within the smart city, enabling local interactions and collaborations among IoT devices without relying on cloud connectivity. The fog layer may integrate with the cloud layer through secure connections, allowing selective data transfer and synchronization between the fog nodes and the cloud infrastructure. This integration ensures that data requiring long-term storage, advanced analytics, or cross-city insights can be efficiently processed and stored in the cloud.

The diagrammatic representation of fog computing in smart cities illustrates the decentralized nature of processing and analytics, with fog nodes placed at the network edge, closer to the data sources, and end devices. This architecture enables real-time understandings, improved responsiveness, reduced network traffic, and enhanced scalability within the smart city environment.

How Fog Computing Can Be Used in Smart Cities

Fog computing plays a crucial role in smart cities, enabling various applications and services to operate efficiently. One of its key uses is real-time monitoring and control of systems and infrastructure. Fog nodes located at the network's edge process data from sensors and IoT devices, allowing for immediate insights and quick response times in areas such as traffic management, public safety, and environmental monitoring. Another significant benefit is edge analysis and decision-making which reduces dependency on cloud. This is particularly valuable for applications that require emergency response systems. Fog computing also enhances public services and the citizen experience in smart cities by delivering personalized and context-aware services, such as smart parking and intelligent lighting. This ensures efficient resource allocation and tailored services in real-time.

Resilience and autonomy are other advantages of fog computing in smart cities. Fog nodes deployed at the edge enable critical services to operate even during network disruptions. This is helpful in empowering autonomous systems like self-driving vehicles, enhancing operational continuity and reliability. Addressing privacy and security concerns, fog computing offers stricter control over data sharing and aligns with regulatory requirements. Scalability and cost-effectiveness are further benefits of fog computing. By distributing computational resources across the network edge, fog nodes can handle increasing workloads efficiently. This reduces reliance on centralized cloud infrastructure and optimizes resource utilization, resulting in cost savings and scalability.

Further, fog computing can be used in smart cities in various ways to improve the efficiency, responsiveness, and sustainability of the environment. Here are some examples of smart city applications using fog computing (Figure 2.5):

1. Intelligent traffic management: Fog computing can be used to analyse real-time traffic using sensors and cameras located throughout the city. By processing these data locally on edge devices or fog nodes, traffic management systems can respond more quickly to changing conditions and optimize traffic flow.
2. Smart parking: Fog computing can analyse real-time data from parking sensors located throughout the city. Processing this data, smart parking systems can provide information on parking availability reducing search times for drivers and optimizing parking utilization.

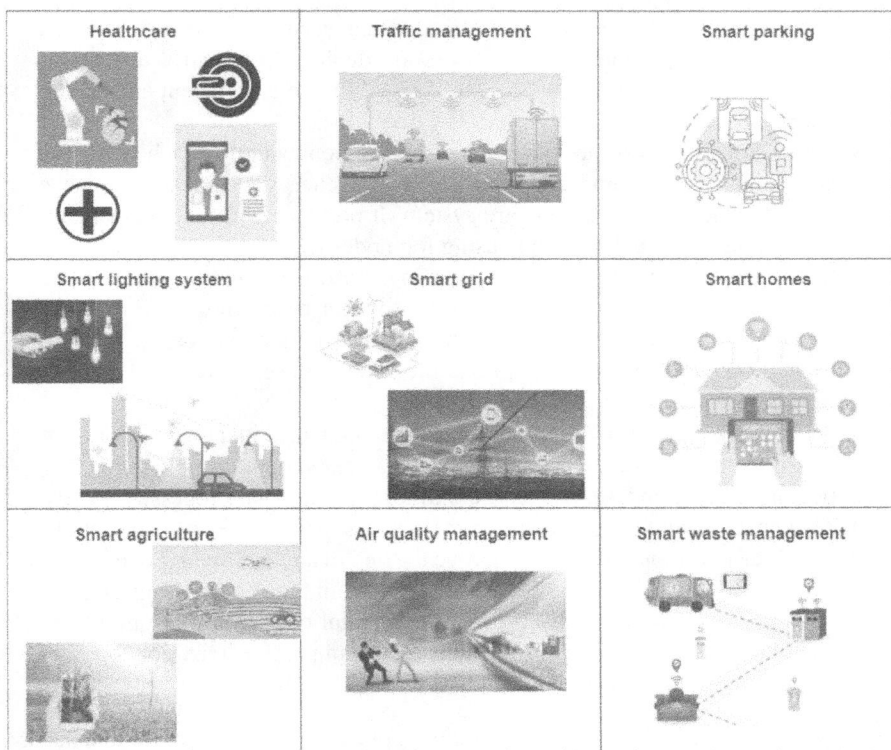

FIGURE 2.5 Various applications of fog computing in the smart city.

3. Environmental monitoring: Environmental conditions such as air quality, noise pollution, and water quality can be analysed in real-time. The information collected can be used to alert city officials about potential problems.
4. Public safety: Fog computing can analyse real-time data from surveillance cameras and various sensors located throughout the city. By processing these data locally on edge devices or fog nodes, public safety systems can identify potential threats in real-time and respond more quickly to emergencies.
5. Smart lighting: Fog computing can be used to optimize the operation of street lights and other lighting systems in the city. By processing real-time data on lighting conditions and energy consumption locally on edge devices or fog nodes, smart lighting systems help lower energy use, improve safety and security, and enhance the overall aesthetic of the city.
6. Healthcare: Cloud computing environment is not suitable for real-time healthcare applications in case of remote patient monitoring as there may be a need for instant actions to be taken in case of unexpected situations. In fog computing, fog nodes help to perform these processing improving network conditions and latency that provide real-time information on medical equipment and patient conditions [26].
7. Smart grid: In it, energy is generated using various stations distributed widely, and smart meters and sensors are deployed to control and monitor the energy consumption process. This results in an efficient and secure energy management.
8. Smart homes: These are where devices are connected to each other via communication channels helping to reduce energy utilization, manage security, and improve healthcare system. It protects from any mishappening by analysing real-time data using fog nodes [27].
9. Smart waste management: By integrating sensors and monitoring devices with fog nodes, waste collection processes can be optimized. Fog nodes process data from sensors to optimize waste collection routes, monitor waste levels in containers, and schedule collection services based on real-time demand, reducing costs and enhancing efficiency.
10. Air-quality management: Air quality can be managed in a fog environment by integrating air-quality sensors with fog nodes, data on pollutants, particulate matter, and atmospheric conditions that can be processed locally. This enables quality monitoring, timely alerts, and localized interventions to reduce pollution levels and improve the health and well-being of citizens.
11. Agriculture: By connecting sensors, actuators, and fog nodes to agricultural systems, real-time monitoring of environmental conditions, soil moisture, and plant health can be achieved. This results in precise irrigation, optimal nutrient delivery, and automated control of the greenhouse.

Overall, fog computing makes smart city applications more efficient and secure by processing data locally rather than transmitting data to Cloud ensuing improvement in the performance of smart city applications and services.

BUILDING FOG COMPUTING MODEL IN IOT FOR SMARTER CITIES

Building a fog computing model for IoT in a smart city involves a systematic approach. First, it is essential to define the specific use cases and applications that would benefit from fog computing. This could include areas such as traffic management, environmental monitoring, or public safety. Once the use cases are identified, infrastructure planning becomes crucial. Assessing the existing infrastructure and determining the optimal locations for deploying fog nodes is necessary. Factors like proximity to data sources, network connectivity, and power supply must be considered during this stage. The integration of IoT sensors and devices into the smart city infrastructure is the next step. Ensuring compatibility and connectivity between the sensors/devices and the fog computing nodes is important for seamless data exchange. Data collection and processing play a significant role in fog computing. Establishing mechanisms to collect data from sensors and devices and implementing data filtering and aggregation techniques at the edge helps reduce the volume of data transmitted to the fog nodes. Processing the collected data locally at the fog nodes enables real-time analytics, data fusion, and the application of machine learning algorithms.

Then comes network connectivity. Reliable and secure connections between the fog nodes, cloud infrastructure, and other components of the smart city ecosystem need to be established. This may involve deploying suitable networking technologies such as wireless networks or wired connections to ensure efficient data transmission and low-latency communication. Robust security measures, including encryption, access controls, authentication mechanisms, and intrusion detection systems, must be implemented to protect the fog computing infrastructure, IoT devices, and the processed data. Privacy measures should be in place to address citizen concerns and comply with data protection regulations. Finally, there is a need to develop applications, where specific applications and services are built on top of the fog computing model. This involves developing software and algorithms tailored to the identified use cases, ensuring seamless integration with the fog computing infrastructure and providing actionable insights and services to enhance the smart city's functionality.

Fog computing is a powerful technology that can be used in IoT systems to make smarter cities. Here are some ways in which IoT for smarter cities can benefit from fog computing:

1. Edge computing: In traditional cloud computing models, IoT devices send data to the cloud for processing and analysis. But in case of fog computing, edge devices are used to process and analyse data locally, as a result, no or very less amount of data (if required) is forwarded to the cloud. This reduces latency and enables real-time processing, making IoT systems more responsive and efficient.

2. Distributed data storage: Fog computing also enables distributed data storage, which means that data can be stored closer to where it is generated, reducing the amount of data needed to be transmitted over the network. This results in less network congestion, improved data security, and reduced data storage costs.

3. Data processing: With fog computing, data can be processed in real-time, which means that IoT systems can respond more quickly to changes in the environment making smart cities safe and secure.
4. Predictive analysis: Fog computing can perform predictive analysis, which means that data can be analysed in real-time to identify trends and patterns that lead to predicting future actions and taking wise decisions.
5. Energy efficiency: Fog computing helps IoT devices use less energy by lowering the amount of data that needs to be sent to the cloud.

This leads to the conclusion that fog computing is a powerful technology that can be used to make IoT systems in smart cities more efficient, responsive, and secure. Smart cities can create IoT systems that are better suited to meet the requirements of their citizen and raise the overall quality of life by utilising the potential of edge computing and distributed data storage. Building a fog computing model in IoT for smarter cities involves several steps.

Here are some key steps (Figure 2.6) that can be followed to build a fog computing model for IoT in smart cities:

1. Identify the requirements: The first step is to identify the requirements for the IoT system in the smart city like the type of sensors and devices that need to be connected, and the kind of data that need to be collected and analysed.
2. Determine the edge devices: The next step determines the edge devices to collect and process data.
3. Design the fog architecture: The fog architecture should be designed such that it optimizes the performance and efficiency of the system. It should be

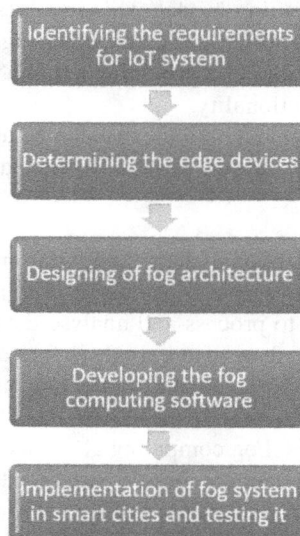

FIGURE 2.6   Steps to build a fog computing model for IoT in smart cities.

able to determine the location and number of fog nodes used, as well as the communication and data processing protocols that will be used.

4. Developing the fog computing software: The fog computing software should be developed to manage the data processing and communication between the edge devices and the fog nodes. This includes developing software that can handle data breakdown, processing, storage, and analysis, and also the software that can manage the communication between the edge devices and the fog nodes.

5. Implementing the system: The fog computing system should be implemented in the smart city. It includes steps from deploying the sensors and devices to installing the fog nodes, and finally configuring the communication protocols and data processing algorithms.

6. Test and optimize the system: Test and optimize the system: The fog computing system should be tested and optimized to ensure that it is working as expected. This can be done by monitoring the performance of the system, identifying issues, and making modifications that will boost the system's functionality and effectiveness.

Building a fog computing model in IoT for smarter cities involves careful planning, design, and implementation in order to guarantee that the system is efficient, responsive, and secure leading to overall sustainability and liveability of urban environments.

APPLICATION OF FOG MODEL IN IOT FOR SMART CITIES

A city that makes use of IoT technology to raise sustainability, optimize services, and better the quality of life for its citizens is known as an IoT-based smart city. It manages resources and services efficiently by using interconnected devices, sensors, and systems that collect and analyse data in real-time. IoT technology enables city administrators to make well-informed decisions on a variety of issues, like energy management, waste management, public safety, and traffic flow. The information gathered from IoT devices can be utilized to improve the efficiency of city services and help in cost saving.

IoT-based smart cities offer a range of benefits to both city officials and citizens. For example, smart traffic management systems can reduce congestion and improve air quality. It can also be used to adjust the timings of traffic lights based on traffic conditions, weather or other factors. While smart waste management system describes the procedures for collecting, dumping, recycling, and landfilling municipal waste [28], fog computing can optimize this collection schedule and reduce costs along with keeping human health in mind. Furthermore, public safety can be enhanced by providing real-time alerts and monitoring of public areas. Fog nodes can process the video data from cameras in real-time to detect suspicious behaviour and alert the authorities. For citizens also, IoT-based smart cities offer a range of benefits as well. For example, smart lighting systems can provide better lighting in public areas, improving safety and security at night. Smart cities based on the IoT can also improve access to information and services by sending out alerts about local events or public transport. It also helps in reducing energy consumption and the greenhouse effect. In all, smart cities having IoT infrastructure act as a powerful tool for enhancing livelihood and improving the quality of life. Smart grids can help to monitor and optimize energy consumption by integrating renewable energy resources. Further, environment monitoring can also be done by collecting data on air quality, temperature, and humidity using sensors. Healthcare industry also had a big relief using fog computing environment as health parameters of remote patients can be monitored at all times. Moreover, in case of emergency, immediate consultation can be given by the doctor available [29]. Patients' reports can be checked and diagnosis can be given in real-time. In the years to come, we could be seeing more advanced IoT-based applications and their advantages emerging as the technology develops.

The best example to understand that how fog computing model manages real-time data in smart cities is the smart grid management system. A smart grid is a modern electrical power distribution system that incorporates communication, energy storage, and renewable energy resources. In order to gather and communicate information on energy production, consumption, and storage, smart grids rely on IoT devices and sensors. Various ways in which fog computing model can be used to manage smart grids are discussed below:

1. Real-time data processing: Fog computing provides a distributed infrastructure to process data in real-time; this allows for immediate response to changes in energy production and consumption in smart grid systems. To reduce latency and enhance performance, fog nodes are placed close to the devices and sensors that produce data.
2. Energy optimization: Fog computing system can optimize the distribution of energy in the smart grid. By collecting and analysing data from IoT devices and sensors, fog nodes can make decisions about where and when to distribute energy to reduce waste and increase efficiency. For example, fog nodes can adjust the voltage of the electrical grid to match the demand for electricity.
3. Demand response management: These programs are designed to encourage consumers to reduce their energy consumption during peak times. Here, fog nodes can collect data from smart meters and other IoT devices to determine when energy demand is likely to be high and inspire them accordingly.
4. Security and reliability: By placing fog nodes close to the network's edge, the model can ensure that critical data are protected from cyberattacks. In addition, fog nodes can provide backup power in case of an outage or other disruption.

ANALYSIS AND RESULTS

Thanks to the developing computing architecture known as fog computing, the processing and analysis of data can now be done at the network's edge, much closer to where it is generated. This helps in making real-time decisions for smart cities using data from sensors, cameras, and other sources. Among many advantages of fog

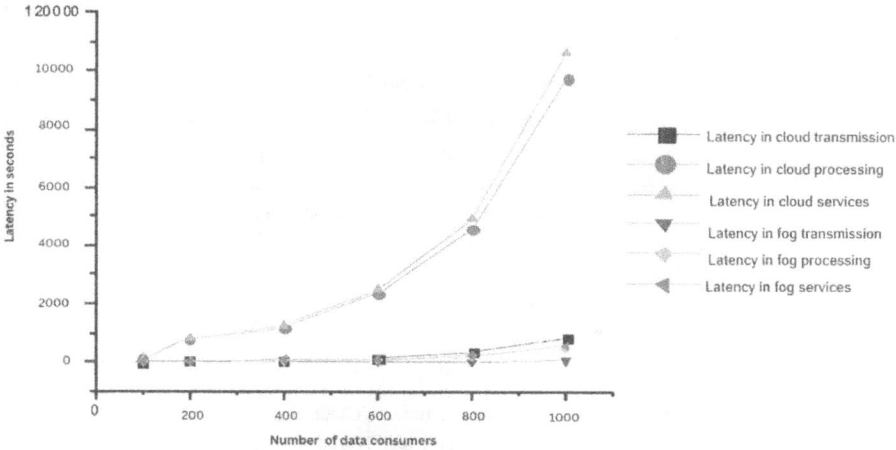

FIGURE 2.7 Comparison of latency in transmission, processing, and services between cloud infrastructure and fog infrastructure.

computing, the important ones are its ability to reduce latency and improve response times which is crucial for smart city applications like traffic management, public safety, and environment monitoring.

After analysis, we find that the latency faced by Cloud environment while transmission of data, processing of data, and providing services of various applications is more than that provided by a fog computing environment. A graph displaying this difference is shown in Figure 2.7. It makes it very clear to us how implementing fog computing infrastructure in different fields, especially in smart cities, is beneficial to us. It will help us to reduce latency while transmitting data, processing, and providing several services which may be required during actual implementation.

Other benefits of fog computing are its capability to reduce bandwidth usage and lessen network congestion. As data are processed locally, less data are transmitted over the network, which improves the network performance and reduces latency. In addition, fog computing can help to improve data security and privacy. Protection from unauthorized access and data breaches is another important advantage of fog computing. Fog computing when implemented in smart cities results in an efficient system as compared to cloud computing if implemented. A comparative analysis of both is done in Table 2.2.

It is well understood from the above comparison that fog computing in smart cities offers more advantages over cloud computing in terms of lower latency, better scalability, reduced network bandwidth usage, increased data privacy and security, and improved energy efficiency. However, fog computing in smart cities is not without difficulties it has to face. One of the major challenges is the requirement for efficient data management and integration. As data are generated using various sources, it becomes difficult to ensure that it is collected, processed, and analysed in a reliable manner. Another challenge is the need for effective resource management as fog computing requires a significant number of computational resources.

TABLE 2.2

Comparison of Fog Computing and Cloud Computing in Smart Cities

	Fog Computing in Smart Cities	Cloud Computing in Smart Cities
Latency	The processing and analysis of data occur at the edge of the network; therefore, less time is required by data to travel to a remote data centre for processing	The processing and analysis of data occur at remote data centre; this increases the amount of time taken by data to travel to and from the data centre
Scalability	For better scalability, the processing and storage of data is distributed across multiple nodes known as fog nodes at the edge of the network	Scalability is attained using multiple servers at the data centres that can handle large data and processing tasks
Bandwidth usage	Fog environment reduces the amount of data that needs to be transmitted over the network to the data centre resulting in reduced network congestion and bandwidth usage	Cloud computing requires a large amount of data to be transmitted over the network to remote data centre; this leads to high network congestion and bandwidth usage
Privacy and security	As data are processed and analysed at the network's edge, it is less vulnerable to interference and data breaches during transit	As data are transmitted over the internet, it is vulnerable to interference, making it prone to security breaches
Energy efficiency	As less amount of data is needed to be transferred to and from data centres, there is reduced energy consumption	As a large amount of data is needed to be transferred to and from remote data centres, energy consumption is high

In all, fog computing offers tremendous opportunities for smart cities; it is important to consider all its prospects in such a way that it maximizes its benefits and reduces its risks.

CONCLUSION

Building and developing smart cities is essential for a country's economy to grow and for its citizens to live better lives. However, to construct a smart city, it is necessary to consider the practical ways to use the available resources that are feasible to create a long-term plan that can be implemented without any obstacles [30]. To satisfy these requirements is necessary to use IoT along with fog computing. Fog computing is an emerging computing standard that has the power to change the way we design and operate smart cities. In the context of smart city, fog computing processes and analysis data at the network's edge close to its source. This makes the fog computing model more advantageous over traditional cloud computing, which relies on centralized data centres. Its tendency to reduce latency and fast response time makes it more important for real-time applications like traffic management, quick response in case of emergency, and environmental monitoring. Another advantage of fog computing is improved scalability. Fog computing has capability to process and store data across multiple nodes known as fog nodes which are distributed over network's edge, allowing for better scalability and improved resource utilization. This can enable smart cities to handle increasing data and processing demands without overburdening central data centres.

Fog computing can help lessen network traffic and congestion, improving the performance and efficiency of the smart city infrastructure. Energy efficiency in smart cities can also be improved using fog computing infrastructure, by reducing energy consumption and carbon emission. In data privacy and security also, it plays an important role as it reduces the risk of any outside interference and data breaches during data transfer.

As we conclude, we find that fog computing is a promising technology that can assist smart cities in overcoming difficulties that arise with using traditional cloud computing. It offers faster response times, better scalability, reduced network bandwidth usage and congestion, increased data privacy and security, and improved energy efficiency. As smart city infrastructure continues to expand, fog computing is likely to become more crucial in supporting the development of more efficient, sustainable, and secure environments.

REFERENCES

1. Yi, S., Li, C. and Li, Q., 2015. A survey of fog computing: concepts, applications and issues. In *Proceedings of the 2015 Workshop on Mobile Big Data* (pp. 37–42).
2. Iorga, M., Feldman, L., Barton, R., Martin, M.J., Goren, N.S. and Mahmoudi, C., 2018. *Fog Computing Conceptual Model, Special Publication (NIST SP)*, National Institute of Standards and Technology, Gaithersburg, MD [online]. https://doi.org/10.6028/NIST.SP.500-325 (accessed 23 November 2023).
3. Kaur, M. and Aron, R., 2022. A novel load balancing technique for smart application in a fog computing environment. *International Journal of Grid and High-performance Computing (IJGHPC)*, *14*(1), pp. 1–19.
4. Bajaj, K., Sharma, B. and Singh, R., 2021. UWS academic portal. In *Conference on Artificial Intelligence and Machine Vision (IEEE AIMV 2021)* (Vol. 24, p. 26th).
5. Bonomi, F., Milito, R., Zhu, J. and Addepalli, S., 2012. Fog computing and its role in the internet of things. In *Proceedings of the first edition of the MCC workshop on Mobile cloud computing* (pp. 13–16).
6. Stojmenovic, I., Wen, S., Huang, X. and Luan, H., 2016. An overview of fog computing and its security issues. *Concurrency and Computation: Practice and Experience*, *28*(10), pp. 2991–3005.
7. Satyanarayanan, M., Bahl, P., Caceres, R. and Davies, N., 2009. "The case for VM-based cloudlets in mobile computing," *IEEE Pervasive Computing*, *8*(4), pp. 14–23. doi: 10.1109/MPRV.2009.82.
8. Stojmenovic, I. and Wen, S., 2014. The fog computing paradigm: Scenarios and security issues. In *2014 Federated Conference on Computer Science and Information Systems* (pp. 1–8). IEEE.
9. Baccarelli, E., Naranjo, P.G.V., Scarpiniti, M., Shojafar, M. and Abawajy, J.H., 2017. Fog of everything: Energy-efficient networked computing architectures, research challenges, and a case study. *IEEE Access*, *5*, pp. 9882–9910.
10. Perera, C., Qin, Y., Estrella, J.C., Reiff-Marganiec, S. and Vasilakos, A.V., 2017. Fog computing for sustainable smart cities: A survey. *ACM Computing Surveys (CSUR)*, *50*(3), pp. 1–43.
11. Hu, P., Dhelim, S., Ning, H. and Qiu, T., 2017. Survey on fog computing: Architecture, key technologies, applications and open issues. *Journal of Network and Computer Applications*, *98*, pp. 27–42.

12. Varshney, P. and Simmhan, Y., 2017. Demystifying fog computing: Characterizing architectures, applications and abstractions. In *2017 IEEE 1st international conference on fog and edge computing (ICFEC)* (pp. 115–124). IEEE.
13. Mouradian, C., Naboulsi, D., Yangui, S., Glitho, R.H., Morrow, M.J. and Polakos, P.A., 2017. A comprehensive survey on fog computing: State-of-the-art and research challenges. *IEEE Communications Surveys & Tutorials*, 20(1), pp. 416–464.
14. Mahmud, R., Kotagiri, R. and Buyya, R., 2018. Fog computing: A taxonomy, survey and future directions. *Internet of Everything: Algorithms, Methodologies, Technologies and Perspectives* (pp. 103–130). Springer.
15. Yi, S., Hao, Z., Qin, Z. and Li, Q., 2015. Fog computing: Platform and applications. In *2015 Third IEEE Workshop on Hot Topics in Web Systems and Technologies (HotWeb)* (pp. 73–78). IEEE.
16. Ai, Y., Peng, M. and Zhang, K., 2018. Edge computing technologies for Internet of Things: A primer. *Digital Communications and Networks*, 4(2), pp. 77–86.
17. Sabireen, H. and Neelanarayanan, V.J.I.E., 2021. A review on fog computing: Architecture, fog with IoT, algorithms and research challenges. *ICT Express*, 7(2), pp. 162–176.
18. Costa, B., Bachiega, J. Jr, de Carvalho, L.R. and Araujo, A.P., 2022. Orchestration in fog computing: A comprehensive survey. *ACM Computing Surveys (CSUR)*, 55(2), pp. 1–34.
19. Puliafito, C., Mingozzi, E., Longo, F., Puliafito, A. and Rana, O., 2019. Fog computing for the Internet of Things: A survey. *ACM Transactions on Internet Technology (TOIT)*, 19(2), pp. 1–41.
20. Malik, S. and Gupta, K., 2021. Smart city: A new phase of sustainable development using fog computing and IoT. In *IOP Conference Series: Materials Science and Engineering* (Vol. 1022, No. 1, p. 012093). IOP Publishing.
21. Babbar, H. and Rani, S., 2022. Security architecture and its methodology for fog computing. *ECS Transactions*, 107(1), p. 4549.
22. Mohanakrishnan, R., 2022. *What is fog computing? Components, examples, and best practices.* Available online: https://www.spiceworks.com/tech/edge-computing/articles/what-is-fog-computing/
23. Wen, W., Demirbaga, U., Singh, A., Jindal, A., Batth, R.S., Zhang, P. and Aujla, G.S., 2023. Health monitoring and diagnosis for geo-distributed edge ecosystem in smart city. *IEEE Internet of Things Journal*, 10(21), pp. 18571–18578.
24. Bhardwaj, K.K., Banyal, S., Sharma, D.K. and Al-Numay, W., 2022. Internet of Things based smart city design using fog computing and fuzzy logic. *Sustainable Cities and Society*, 79, p. 103712.
25. Zahmatkesh, H. and Al-Turjman, F., 2020. Fog computing for sustainable smart cities in the IoT era: Caching techniques and enabling technologies—An overview. *Sustainable Cities and Society*, 59, p. 102139.
26. Badidi, E., Mahrez, Z. and Sabir, E., 2020. Fog computing for smart cities' big data management and analytics: A review. *Future Internet*, 12(11), p. 190.
27. Rahimi, M., Songhorabadi, M. and Kashani, M.H., 2020. Fog-based smart homes: A systematic review. *Journal of Network and Computer Applications*, 153, p. 102531.
28. Javadzadeh, G. and Rahmani, A.M., 2020. Fog computing applications in smart cities: A systematic survey. *Wireless Networks*, 26(2), pp. 1433–1457.
29. Tiwari, D., Prasad, D., Guleria, K. and Ghosh, P., 2021. IoT based smart healthcare monitoring systems: A review. In *2021 6th International Conference on Signal Processing, Computing and Control (ISPCC)* (pp. 465–469). IEEE.
30. Lai, K.L., Chen, J.I.Z. and Zong, J.I., 2021. Development of smart cities with fog computing and Internet of Things. *Journal of Ubiquitous Computing and Communication Technologies (UCCT)*, 3(01), pp. 52–60.

3 Named Data Networking

Content-Based Routing— Architecture, Challenges, and Applications

Harshvardhan Singh Chauhan[1], Himanshi Babbar[1], Imran Memon[2], and Gurbinder Singh Brar[3]
[1]Chitkara University Institute of Engineering and Technology, Chitkara University, Punjab, India
[2]Department of Computer Science, Shah Abdul Latif University, Shahdadkot Campus, Shahdadkot, Pakistan
[3]Departmet of Computer Science and Engineering, CT University, Ludhiana, Punjab, India

INTRODUCTION

In today's communication, the so-called data network (named data network [NDN]) paradigm emerged to replace the traditional network model [1]. Unlike the traditional Internet, which is often based on an address or IP address, NDN offers a content-centred perspective where information is given a unique name and fear of communication becomes important. This new concept changes the definition of "where" data is based on "what" data is and changes the way we think about the web. The essence of NDN lies in its communication structure in context. NDN refers to storing information by name, rather than identifying and communicating with a particular device. This pull-based communication allows customers to express their interest in certain content by sending Letters of Interest containing the title of the requested document. Routers then send those preferences by searching the network for information [2]. When the information is found, it is sent to the customer according to the area of interest. The main point of NDN lies in its content-centric communication model. NDN refers to storing information by name, rather than identifying and communicating with a particular device. This hierarchical naming scheme enables routers to leverage naming hierarchies for efficient routing and caching, optimizing content delivery, and improving network performance. In addition to its communication-centred structure and hierarchical naming, NDN also attaches importance to security and privacy. All data packets in NDN are digitally signed by the manufacturer to ensure the integrity and authenticity of the data. This trust model enables customers to verify the authenticity of the information received, reducing the risks associated with unauthorized content and mitigating attacks [1]. NDN has many advantages, by focusing on content rather than data sources, NDN enables improved content

DOI: 10.1201/9781003438205-3

43

distribution, increased productivity, and retention of popular content. This caching technique reduces latency, makes better use of network resources, and increases scalability [1]. In addition, NDN inherently enhances security and privacy as the pull-based communication model blocks user data and the data integrity mechanism ensures the accuracy and integrity of the information received.

INTEGRATING WSN IN NDN

The integration of wireless sensor networks (WSNs) plays an important role in improving the capacity and performance of data networks (NDNs). As shown in Figure 3.1, WSNs consist of many small, low-power sensor nodes capable of collecting, processing, and transmitting data from the physical world [1]. When integrated into the NDN framework, WSNs can provide real-time information, improve data collection, and contribute to the overall success of NDN [1]. This section examines the important role of WSN in NDN and its impact on various aspects of the network.

Data Analysis and Collection

The WSN is good at understanding and collecting data from various sources. WSN nodes equipped with sensors can monitor temperature, humidity, light intensity, sound, and movement. Sensor nodes can capture data in real-time and send it wirelessly to NDN infrastructure [2]. By integrating WSNs with NDNs, the network can access data produced by multiple sensors, providing a better understanding of the physical world.

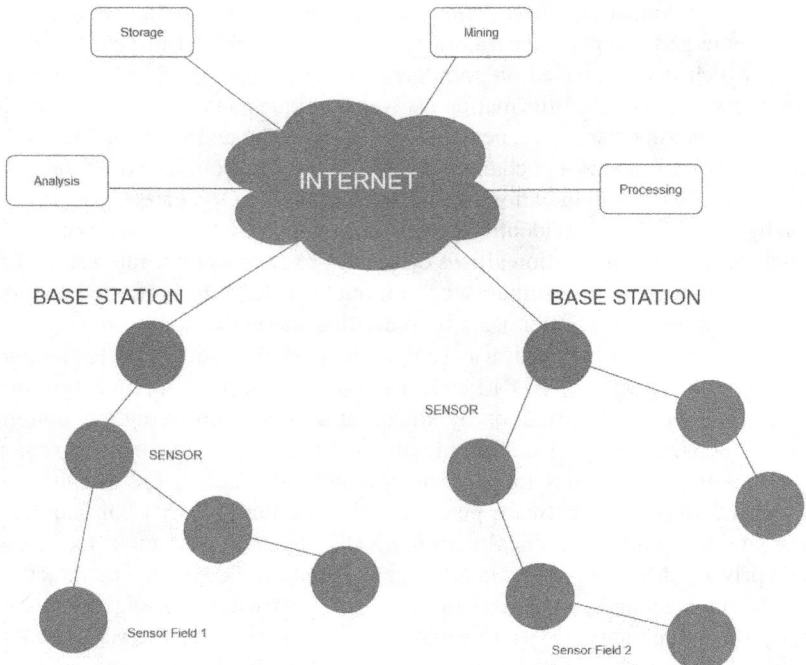

FIGURE 3.1   Structure of wireless sensor networks.

Efficient Data Collection

WSN enhances NDN by providing an efficient data collection process. Sensor nodes can process data generated on the network, create content with unique names, and publish it to NDN. Customers can express their interest in certain information by issuing letters of interest containing the name of the relevant content. NDN infrastructure with WSN can route and forward data requests to clients. This process helps to improve the objectives and quality of data collection, reduce redundant data, and save network resources.

Real-time Data Transmission

WSNs are unique in providing real-time data transmission in the NDN framework. Since the sensor nodes are constantly collecting data from the environment, they can broadcast the received data to the network together with the important list. Consumers interested in real-time data from a particular sensor can provide the data they are interested in to keep the information up to date. WSNs provide timely and accurate information updates, enabling applications such as environmental monitoring, smart cities, and business automation access to the most up-to-date information for decision-making and review.

Data Content Caching

Caching is an important feature of NDN, and WSN can facilitate data caching in the network. WSN nodes can function as caching facilities by storing incoming data in their memory. This approach reduces the need to resend old data as customers can get what they need from a nearby location instead of returning to the original location. Enabling WSN's caching mechanisms improves content delivery, reduces network congestion, and improves overall network performance.

Energy Efficiency and Resource Management

The WSN usually runs on low battery power, so energy-saving strategies are required. When integrated into NDNs, WSNs can benefit from a network communication model that transmits data only when there is a certain interest, thus reducing power consumption [1]. Additionally, WSN nodes can use intelligent techniques such as data aggregation, where data from multiple sensors are combined and sent as a data packet, reducing the number of transmissions and saving energy. These energy savings keep WSNs running longer and help extend the life of NDN infrastructure.

APPLICATIONS OF NDN

NDN has various applications in the computer science field as shown in Figure 3.2.

CONTENT DISTRIBUTION

NDN offers a novel approach to content distribution by focusing on named data rather than named hosts. NDN routers can cache and store content closer to the requesting nodes, reducing network congestion and improving content delivery efficiency. This application is particularly relevant in scenarios where popular content needs to be

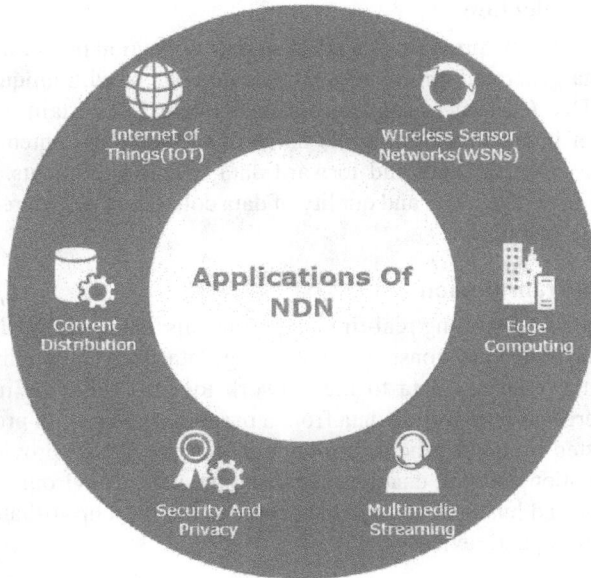

FIGURE 3.2   Applications of NDN in various fields.

distributed to a large number of users simultaneously, such as video streaming, software updates, or news dissemination [2]. Research in this area can explore the performance benefits of NDN-based content distribution networks (CDNs), including reduced bandwidth consumption, improved latency, and enhanced scalability. Additionally, investigating caching strategies, content naming schemes, and content retrieval mechanisms in NDN-based CDNs can provide valuable insights for optimizing content delivery.

INTERNET OF THINGS

The Internet of Things (IoT) encompasses a vast network of interconnected devices that generate and exchange data. NDN's data-centric paradigm aligns well with the IoT concept, where data is of primary importance [2]. NDN enables efficient data retrieval by name, facilitating scalability and reducing the need for maintaining end-to-end connections between devices. Research in this area can explore the application of NDN in IoT environments, focusing on aspects such as data synchronization, security, energy efficiency, and mobility support. Additionally, investigating how NDN can enhance data dissemination, device discovery, and resource management in large-scale IoT deployments can provide valuable insights for IoT architecture design and optimization.

SECURITY AND PRIVACY

NDN's data-centric approach offers inherent security and privacy advantages compared to IP-based networking. Research in this area can investigate the security mechanisms and privacy features provided by NDN, such as data authentication, encryption, and access control. Additionally, exploring the challenges and trade-offs associated with securing NDN-based networks, including key management, trust models, and secure routing protocols, can be an important research direction [3]. Moreover, examining how NDN can address security and privacy concerns in specific applications, such as healthcare, smart cities, or financial systems, can provide insights into the practical implications of adopting NDN in real-world scenarios.

NAMED DATA NETWORKING IN WIRELESS SENSOR NETWORKS

Wireless sensor networks (WSNs) are widely used for various applications, including environmental monitoring, industrial automation, and smart grids. NDN can be applied to WSNs to improve scalability, energy efficiency, and data-centric querying. Research in this area can explore the benefits of NDN in WSNs, focusing on aspects such as data naming and dissemination, routing protocols, energy optimization, and network lifetime extension [4]. Investigating how NDN can facilitate efficient data collection, aggregation, and query processing in resource-constrained WSNs can contribute to the development of more efficient and sustainable WSN architectures.

EDGE COMPUTING

Edge computing is a paradigm where computation and data processing are performed closer to the edge of the network, near the data source or the end users. NDN's data-centric approach aligns well with the principles of edge computing, as it enables efficient data retrieval and caching based on content names. Research in this area can explore how NDN can be leveraged in edge computing scenarios, focusing on aspects such as edge caching, distributed processing, latency reduction, and resource management. Investigating the benefits and challenges of adopting NDN in edge computing architectures can provide valuable insights into improving the performance, scalability, and reliability of edge computing environments.

MULTIMEDIA STREAMING

Multimedia streaming is a demanding application that requires efficient content delivery to many users [1]. NDN's content-centric nature makes it well-suited for multimedia streaming, as it enables efficient content retrieval and caching. Research in this area can focus on optimizing NDN for multimedia applications, including adaptive bitrate streaming, real-time video conferencing, and distributed media delivery [5]. Investigating strategies for content naming, caching policies, multicast support, and congestion control in NDN-based multimedia streaming can contribute to the development of more efficient and scalable multimedia delivery systems. Moreover, exploring the performance benefits of NDN compared to traditional IP-based streaming solutions can provide insights into the advantages of adopting NDN in multimedia applications.

ARCHITECTURE OF CBR IN NDN-WSN

In the NDN, context-based protocols are essential for sending and forwarding context-based packets rather than addresses or addresses [3, 6, 7]. Content-based routing allows NDN routers to make routing decisions by examining packet names [8, 9]. As shown in Figure 3.3, NDN architecture is composed of the following components:

DATA SOURCE

In NDN, a document can be a place that creates or publishes information.

PRODUCERS

These are individuals, organizations, or systems that create information and make it available on the NDN network. For example, a creator could be a video streaming service that broadcasts videos, or a weather station that provides real-time weather information.

IoT DEVICES

IoT devices such as sensors, actuators, or smart devices can be used to process data on NDN. These devices can generate information about the environment, healthcare, energy use, and so on.

DATABASES AND SERVERS

Traditional database or server systems can also be used as data sources in NDN. They can store and process data accessible from the NDN domain.

CONTENT AGGREGATORS

These are sites that collect data from various sources and put it into a single stream. For example, a reporter might collect news from multiple publishers and present them via NDN.

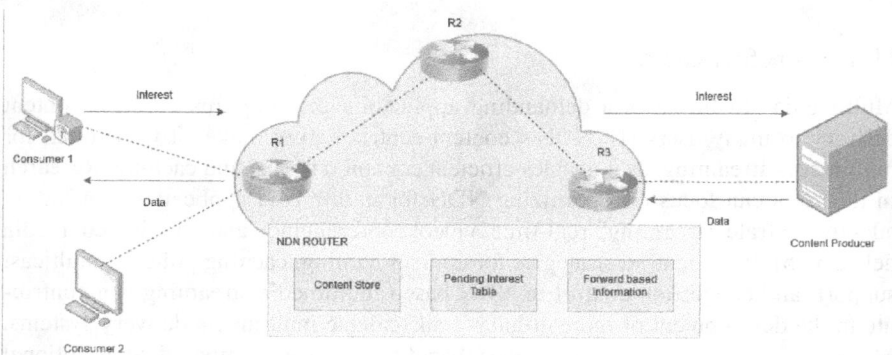

FIGURE 3.3 Architecture of named data networking in a network.

PEER-TO-PEER NETWORKING

NDN supports peer-to-peer (P2P) communication, in which a device can handle data by directly sharing it with other devices used in the network. This leads to knowledge sharing and collaboration.

DATA SINK

In what is called a data network (NDN), the term "repository" refers to a location or system that uses or retrieves data; the point at which the last data is sent in response to the interest rate. In NDN, a repository can be a place where data are stored and used. When a repository wants to store a particular file, it sends a packet containing the name of the requested file to the NDN network. Network routers then use this affinity to send the requested data to the nearest source [10]. When data are found, it is sent back to the repository where it can be processed or used as needed by the application [10].

NDN ROUTERS

NDN routers play an important role in network infrastructure. These routers are responsible for forwarding traffic and data, managing caches, and ensuring the efficiency of the data space [11]. NDN routers form the backbone of the NDN network, facilitating the retrieval and delivery of named data while modifying the caching and routing process to optimize performance and reduce network congestion.

NDN CACHING NODES

A cache node refers to a network node that participates in the caching operation in the NDN network to store and serve packet data [12]. Nodes can contain multiple devices such as routers, servers, and even end users with caching capabilities [13]. Caching of nodes in NDN helps improve the overall efficiency and scalability of the network by bringing data closer to customers and reducing the need to resend data from the original source. Through in-network caching, nodes help increase data retrieval speed, reduce network congestion, and increase the amount of content transmitted.

NDN CONSUMERS

Clients are entities or applications that initially provide data from the network. Users or systems express an interest in receiving certain information by sending interest packets [13–15]. Clients at NDN play an important role in collecting name information from the network. Customers can request specific content or information in advance by expressing interest based on the data list, which can enable data-centric retrieval and increase the value and content delivery by bringing the requested data closer to the customer.

CHALLENGES OF CONTENT-BASED ROUTING

Content-based routing is a messaging pattern used in distributed systems, where messages are routed based on their content rather than relying on fixed destinations or addresses. While content-based routing offers certain advantages, it also comes with its own set of challenges. As shown in Figure 3.4, some of the challenges of content-based routing include the following.

CONTENT COMPLEXITY

The challenge of content complexity in content-based routing refers to the difficulty in extracting meaningful information from complex content. Content can vary in structure, format, and type, ranging from structured data to unstructured text or multimedia. Content complexity poses challenges for content-based routing systems as the routing decisions are based on the analysis and understanding of the content. Different content types may require specialized techniques for accurate content matching and routing decisions. Extracting relevant information from complex content involves tasks such as natural language processing, machine learning (ML), or semantic analysis [16, 17]. These techniques aim to identify key attributes, keywords, or patterns within the content that can guide the routing process. Addressing the challenge of content complexity requires leveraging advanced techniques and algorithms that can analyse and interpret the content effectively [17]. This may involve using domain-specific knowledge, ontologies, or ML models to enhance content understanding and matching capabilities. By overcoming the challenge of content complexity, content-based routing systems can make accurate routing decisions based on the content attributes, ensuring that messages are delivered to the appropriate destinations in the distributed system [17].

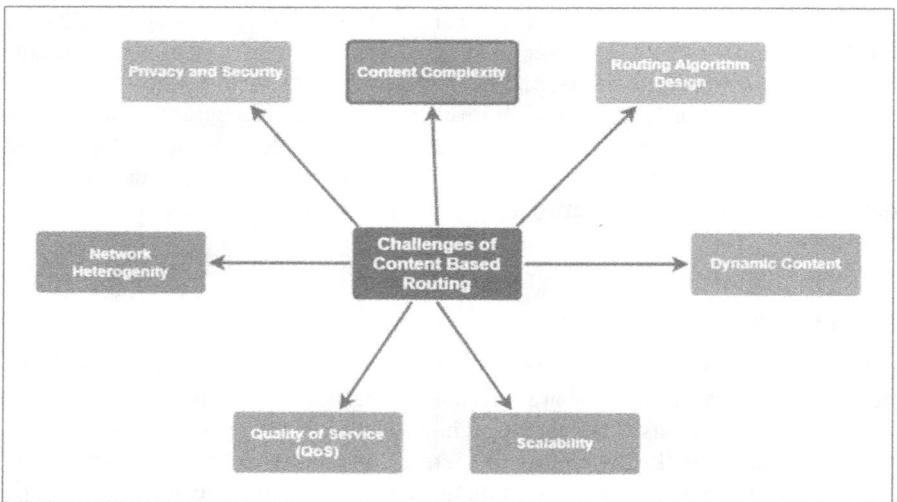

FIGURE 3.4   Challenges of content-based routing.

SOLUTION

The solution to the challenge of content complexity in content-based routing involves employing advanced techniques and algorithms to effectively analyse and extract meaningful information from complex content. Here are some potential solutions:

1. Natural language processing (NLP): Utilize NLP techniques to process and understand textual content. This may involve tasks such as part-of-speech tagging, named entity recognition, sentiment analysis, or topic modelling to extract relevant information for routing decisions.
2. Machine learning: Train ML models to recognize patterns and extract key attributes from complex content. This can involve techniques such as feature extraction, classification, or clustering to identify relevant information that can guide routing decisions.
3. Semantic analysis: Apply semantic analysis techniques to understand the meaning and context of the content. This can involve using ontologies, semantic graphs, or knowledge bases to enrich content understanding and improve the accuracy of routing decisions.
4. Domain-specific knowledge: Incorporate domain-specific knowledge or expert systems that understand the intricacies and specificities of the content. This can help in better interpreting and extracting relevant information from complex content in a contextually appropriate manner [18].
5. Hybrid approaches: Combine multiple techniques such as NLP, ML, and semantic analysis to address different aspects of content complexity. This allows for a more comprehensive analysis of the content and improves the accuracy of routing decisions [18].

ROUTING ALGORITHM DESIGN

The challenge of routing algorithm design in content-based routing pertains to the development of efficient and effective algorithms that determine the routing path for messages based on their content. Designing routing algorithms involves considering various factors such as content matching, network topology, load balancing, and fault tolerance. Routing algorithm design is complex because it requires striking a balance between accuracy and computational efficiency. The algorithms should be capable of analysing the content, comparing it with available routing rules and selecting the appropriate destination(s) for message delivery. Additionally, they need to adapt to changes in network conditions, handle failures, and ensure optimal resource utilization. The challenge lies in devising routing algorithms that can efficiently process a large volume of messages and scale well with increasing network size [18]. These algorithms must consider the diverse range of content types, complex content structures, and dynamic content changes that occur in distributed systems. Addressing the challenge of routing algorithm design involves exploring various techniques and approaches. This may include distributed hash tables (DHTs) for efficient content lookup, adaptive routing algorithms that dynamically adjust routing decisions based on network conditions, or probabilistic routing schemes that balance computational overhead with accuracy. Furthermore, incorporating load balancing mechanisms, fault tolerance strategies, and considering network topology can enhance the efficiency and reliability of the routing algorithms. The design process also involves evaluating the performance and scalability of the algorithms through simulations or real-world deployments [19, 20]. By overcoming the challenge of routing algorithm design, content-based routing systems can achieve accurate and efficient message routing, leading to optimized message delivery and improved overall system performance.

SOLUTION

The solution to the challenge of routing algorithm design in content-based routing involves developing efficient and scalable algorithms that effectively handle content matching, network topology, load balancing, and fault tolerance. Here are some potential solutions:

1. Distributed hash tables: Utilize DHTs to distribute routing information across nodes in a decentralized manner. DHTs provide efficient content lookup by mapping content identifiers to nodes, enabling quick and accurate routing decisions.
2. Adaptive routing: Implement adaptive routing algorithms that dynamically adjust routing decisions based on changing network conditions. These algorithms can consider factors such as network congestion, node availability, or latency to adaptively select the optimal routing path.
3. Probabilistic routing: Employ probabilistic routing schemes that balance computational overhead with accuracy. These algorithms use probabilistic metrics or heuristics to probabilistically select routing paths, allowing for efficient content-based routing decisions.
4. Load balancing: Incorporate load balancing mechanisms to evenly distribute message traffic across multiple nodes. Techniques such as round-robin routing, weighted routing, or dynamic load balancing algorithms can be used to optimize resource utilization and prevent bottlenecks.
5. Fault tolerance strategies: Implement fault tolerance mechanisms to handle failures and ensure reliable message delivery. This may involve backup routing paths, redundancy, or error correction codes to ensure the resilience and fault tolerance of the routing algorithm.
6. Network topology consideration: Take into account the underlying network topology when designing routing algorithms. Consider proximity-based routing or consider the hierarchical structure of the network to optimize routing decisions and minimize message latency [20].
7. Performance evaluation: Evaluate the performance and scalability of routing algorithms through simulations or real-world deployments. This helps identify potential bottlenecks, fine-tune parameters, and optimize the algorithm's efficiency and scalability.

DYNAMIC CONTENT

The challenge of dynamic Content in content-based routing refers to the need to handle content that changes over time. In content-based routing systems, the routing decisions are based on the characteristics and attributes of the content. However, as content evolves, updates occur, and new content is generated; it becomes essential to ensure that routing information remains up to date and accurate [20]. Dynamic content poses challenges for content-based routing systems as they need to track and handle the changes effectively. If routing information becomes stale or outdated, it can result in messages being routed incorrectly or inefficiently, leading to delivery failures or suboptimal performance. The challenge involves developing mechanisms to discover and index dynamic content, propagate updates, and keep the routing information synchronized across the distributed system. It requires addressing issues such as content expiration, content versioning, and efficient content dissemination to ensure that routing decisions are made based on the most recent and relevant content attributes. Furthermore, the challenge extends to efficiently managing the computational and storage resources required to handle the increasing volume of dynamic content and its associated routing information. Addressing the challenge of dynamic content requires implementing dynamic content discovery mechanisms, event-driven architectures, or publish-subscribe models. These mechanisms allow for timely updates and propagation of content changes, ensuring that the routing information remains current and aligned with the evolving content. By effectively handling dynamic content, content-based routing systems can adapt to changes, maintain accurate routing decisions, and ensure the reliable and efficient delivery of messages in distributed systems.

SOLUTION

The solution to the challenge of dynamic content in content-based routing involves implementing mechanisms that enable the discovery, indexing, and synchronization of dynamic content. Here are some potential solutions:

1. Dynamic content discovery: Develop mechanisms to efficiently discover and track changes in dynamic content. This can involve techniques such as content monitoring, event-driven architectures, or subscription-based models to detect updates and new content.
2. Content indexing and versioning: Implement indexing mechanisms that enable efficient lookup and retrieval of content attributes. Maintain versioning information to track changes in content over time, allowing routing algorithms to make informed decisions based on the most up-to-date attributes.
3. Content propagation and dissemination: Establish efficient methods for propagating content updates across the distributed system. This can be achieved through techniques like publish-subscribe models, where content producers notify subscribers of changes, ensuring that routing information remains synchronized.
4. Content expiration handling: Develop mechanisms to handle the expiration of content or time-sensitive routing rules. This ensures that outdated or expired content is not considered for routing decisions, maintaining the accuracy and efficiency of message delivery.
5. Resource management: Efficiently manage computational and storage resources required to handle the increasing volume of dynamic content. This may involve techniques such as distributed caching, distributed processing, or scalable storage systems to handle the growing content and associated routing information.
6. Incremental updates: Optimize the routing algorithm to handle incremental updates instead of reprocessing the entire content. By efficiently

processing only the changes or differences in content, computational overhead can be reduced, enabling faster updates and maintaining system responsiveness.

SCALABILITY

The challenge of scalability in content-based routing refers to the ability of the system to handle an increasing volume of messages and network size while maintaining efficient performance. As the number of messages and nodes in the system grows, scalability becomes crucial to ensure that the routing infrastructure can handle the load without degradation in performance. Scalability challenges arise due to the need to distribute routing information, perform content matching, and maintain efficient routing decisions across a large and dynamic network. Scaling content-based routing systems involves addressing issues such as resource limitations, message overhead, network congestion, and coordination among distributed nodes. As the system scales, it becomes essential to distribute routing information effectively across nodes, ensure load balancing, and optimize resource utilization to prevent bottlenecks and congestion. Additionally, as the message volume increases, the computational complexity of content matching and routing decisions can pose challenges in maintaining real-time responsiveness. To overcome the challenge of scalability, techniques such as sharding, consistent hashing, and distributed caching can be employed to distribute routing information and handle increased message volume. Load balancing mechanisms, dynamic resource allocation, and parallel processing can help distribute the computational load across nodes and improve system scalability. Additionally, employing efficient data structures and indexing techniques can optimize content matching and routing performance. Scalability also involves considering fault tolerance and resilience to handle node failures or network partitions without compromising the overall system's performance. Employing replication, redundancy, and failure detection mechanisms can ensure continuous operation and high availability in a scalable content-based routing system [15, 17, 20]. By addressing the challenge of scalability, content-based routing systems can effectively handle the growing message volume, accommodate a larger network size, and maintain efficient routing decisions, thereby enabling reliable and responsive message delivery in distributed systems.

SOLUTION

The solution to the challenge of scalability in content-based routing involves implementing various techniques and strategies to handle the increasing volume of messages and network size while maintaining efficient performance. Here are some potential solutions:

1. Sharding and partitioning: Divide the routing information and message workload into smaller partitions or shards, distributing them across multiple nodes. This allows for parallel processing and efficient utilization of resources, enabling the system to scale horizontally.
2. Consistent hashing: Use consistent hashing techniques to distribute the routing information and message traffic across nodes in a balanced manner.

Consistent hashing minimizes the impact of adding or removing nodes from the system, ensuring that the load is evenly distributed during scaling.

3. Distributed caching: Implement distributed caching mechanisms to cache frequently accessed routing information or content attributes. Caching reduces the need for repeated computation and database queries, improving response times and relieving the computational load on the system.

4. Load balancing: Employ load balancing techniques to evenly distribute message traffic across nodes. This can involve strategies such as round-robin routing, weighted routing, or dynamic load balancing algorithms to optimize resource utilization and prevent bottlenecks.

5. Dynamic resource allocation: Utilize dynamic resource allocation mechanisms to adjust the allocation of computational resources based on the workload and network conditions. This ensures that resources are efficiently utilized and scaled up or down as needed to handle the changing demands of the system.

6. Distributed processing: Enable distributed processing of routing tasks across multiple nodes. This involves parallelizing content matching and routing decisions, allowing multiple nodes to work collaboratively to handle the increasing workload and improve processing speed.

7. Fault tolerance and resilience: Incorporate fault tolerance mechanisms such as replication, redundancy, and failure detection to ensure high availability and resilience. By detecting and recovering from node failures or network partitions, the system can continue to operate reliably and handle scalability challenges.

QUALITY OF SERVICE

The challenge of quality of service (QoS) in content-based routing refers to ensuring that messages are delivered with specific performance guarantees, such as low latency, high throughput, reliability, and prioritization based on service-level agreements (SLAs) or user requirements. QoS is crucial in content-based routing systems to meet the expectations of users and applications while efficiently utilizing network resources, ensuring QoS involves several challenges [21, 22]. First, content-based routing systems need to prioritize and differentiate messages based on their importance or urgency to meet specific QoS requirements [5, 18]. This requires developing mechanisms for message prioritization and handling varying levels of service. Second, QoS challenges arise due to the dynamic nature of the network and content. As network conditions fluctuate and content changes, maintaining consistent QoS becomes difficult. The system needs to adapt and dynamically adjust routing decisions to optimize QoS parameters in real-time. Third, QoS challenges also involve resource allocation and load balancing. Allocating sufficient computational and network resources to meet QoS requirements while avoiding resource bottlenecks is a complex task. Load balancing strategies and resource management techniques need to be implemented to ensure optimal utilization and QoS adherence. Moreover, QoS challenges extend to fault tolerance and resilience. The system should be able to handle failures, congestion, or network disruptions without compromising the QoS guarantees. This requires fault tolerance mechanisms, redundancy, and efficient recovery strategies to maintain QoS levels during adverse conditions. Addressing the challenge of QoS involves implementing mechanisms for message prioritization, dynamic routing adaptation, resource allocation, load balancing, fault tolerance, and resilience. These solutions ensure that messages are delivered with the desired performance guarantees, meeting the QoS requirements of users and applications [19]. By effectively handling QoS challenges, content-based routing systems can deliver messages in a timely, reliable, and efficient manner, satisfying the expectations of users and enabling the successful operation of distributed systems.

SOLUTION

The solution to the challenge of QoS in content-based routing involves implementing various techniques and strategies to ensure that messages are delivered with the desired performance guarantees. Here are some potential solutions:

1. Message prioritization: Develop mechanisms to prioritize messages based on their importance or urgency. This can involve assigning different levels of priority to messages and ensuring that higher-priority messages receive preferential treatment in the routing and delivery process [23].
2. Dynamic routing adaptation: Implement algorithms that can dynamically adapt routing decisions based on changing network conditions and QoS requirements. This ensures that routing paths are selected to optimize QoS parameters such as latency, throughput, or reliability in real-time.
3. Resource allocation and load balancing: Allocate computational and network resources efficiently to meet QoS requirements while avoiding resource bottlenecks. Load balancing techniques, such as distributing message traffic evenly across nodes or dynamically adjusting resource allocation based on demand, can help optimize resource utilization and improve QoS.
4. QoS monitoring and feedback: Deploy monitoring mechanisms to continuously assess QoS parameters such as latency, throughput, or packet loss. Collecting feedback allows the system to measure and analyse the actual QoS performance and make necessary adjustments to ensure adherence to QoS requirements.
5. Fault tolerance and resilience: Incorporate fault tolerance mechanisms to handle failures, congestion, or network disruptions without compromising QoS guarantees. Redundancy, backup routing paths, and efficient recovery strategies can help maintain QoS levels during adverse conditions.
6. Service-level agreements: Establish clear SLAs with users or applications, defining the specific QoS requirements and expectations. This ensures that the content-based routing system is designed and configured to meet the agreed-upon QoS criteria [24].
7. Traffic engineering: Utilize traffic engineering techniques to optimize network paths, minimize latency, and maximize bandwidth allocation for specific QoS requirements. This involves analysing network conditions, traffic patterns, and dynamically adjusting routing decisions to achieve desired QoS levels.

NETWORK HETEROGENEITY

The challenge of network heterogeneity in content-based routing refers to the presence of diverse network environments, technologies, and capabilities within a distributed system. Network heterogeneity encompasses variations in bandwidth, latency, reliability, topology, and protocol support across different nodes or network segments. Handling network heterogeneity poses challenges in content-based routing systems as they need to adapt to the varying capabilities and constraints of different network components [19]. The challenge lies in designing routing algorithms and mechanisms that can efficiently route messages across heterogeneous networks, ensuring optimal performance and reliable message delivery. Network heterogeneity introduces complexities in content-based routing systems, such as varying communication speeds, network delays, and different levels of reliability or availability. Routing decisions need to consider these factors to select appropriate paths and avoid performance bottlenecks or network congestion. Moreover, network heterogeneity may involve different protocols or communication standards employed by nodes or network segments. This introduces interoperability challenges, requiring translation or adaptation mechanisms to enable seamless communication and routing across diverse network environments. Addressing the challenge of network heterogeneity involves developing adaptive routing algorithms that can dynamically adjust to the characteristics and capabilities of different network components. These algorithms should consider factors like bandwidth, latency, reliability, and protocol compatibility to make informed routing decisions. Additionally, employing techniques such as network monitoring, link quality estimation, or network characterization can provide valuable information about network heterogeneity, allowing the routing system to adapt and optimize routing decisions accordingly. Furthermore, incorporating protocol translation or adaptation mechanisms can facilitate communication across different network environments, enabling interoperability and seamless routing between heterogeneous components. By overcoming the challenge of network heterogeneity, content-based routing systems can effectively handle diverse network environments, ensuring efficient routing, and reliable message delivery across heterogeneous networks in distributed systems.

SOLUTION

The solution to the challenge of network heterogeneity in content-based routing involves implementing various techniques and strategies to handle the diverse capabilities and constraints of different network components. Here are some potential solutions:

1. Adaptive routing algorithms: Develop routing algorithms that can dynamically adapt to the characteristics of heterogeneous networks. These algorithms should consider factors such as bandwidth, latency, reliability, and network conditions to make informed routing decisions. Adaptive algorithms can select paths that optimize performance based on the capabilities of each network segment.
2. Quality-of-service differentiation: Implement mechanisms to differentiate routing decisions based on QoS requirements and network

heterogeneity. This involves assigning different levels of priority or service guarantees to messages based on their content or user/application requirements. QoS differentiation ensures that messages are routed appropriately to meet specific performance expectations across diverse network environments.

3. Protocol translation and adaptation: Incorporate protocol translation or adaptation mechanisms to enable seamless communication and routing between different network environments. These mechanisms can handle protocol conversion, message format transformation, or protocol-specific optimizations, allowing content-based routing systems to bridge the gap between heterogeneous networks.

4. Network monitoring and characterization: Deploy network monitoring tools and techniques to gather information about the characteristics and performance of heterogeneous networks. This includes monitoring bandwidth, latency, reliability, and other network metrics. By characterizing the network heterogeneity, routing decisions can be optimized based on real-time network conditions.

5. Bandwidth management and traffic engineering: Utilize bandwidth management techniques and traffic engineering strategies to optimize network utilization and mitigate performance discrepancies in heterogeneous networks. This may involve traffic shaping, prioritization, load balancing, or adaptive routing mechanisms to ensure efficient resource allocation and minimize congestion.

6. Adaptive message fragmentation: Implement adaptive message fragmentation mechanisms to optimize message delivery across networks with varying packet size limits. This ensures that messages are efficiently divided into smaller packets to accommodate the constraints of heterogeneous networks without compromising content integrity or delivery efficiency.

7. Network-aware routing policies: Develop routing policies that take network heterogeneity into account. These policies can consider factors such as network topology, latency, or available bandwidth to make routing decisions that optimize performance and avoid bottlenecks in the heterogeneous environment.

PRIVACY AND SECURITY

The challenge of privacy and security in content-based routing refers to the protection of sensitive information and ensuring the confidentiality, integrity, and availability of data within the routing infrastructure. Privacy and security are essential aspects of content-based routing systems to safeguard user data, prevent unauthorized access, and mitigate potential security threats [19]. Privacy challenges arise due to the need to handle personal or sensitive information during content matching, routing decisions, and message delivery [24–26]. Ensuring the privacy of user data involves mechanisms for data anonymization, encryption, access control, and compliance with privacy regulations. Security challenges encompass protecting the routing infrastructure from various threats such as unauthorized access, data breaches, network attacks, or malicious manipulation of content. Content-based routing systems need to implement authentication, encryption, intrusion detection, and prevention mechanisms to maintain the integrity and security of the routing infrastructure.

Additionally, privacy and security challenges extend to secure content dissemination, where content publishers need to ensure that their content is securely distributed to authorized subscribers while preventing unauthorized access or content leakage.

SOLUTION

The solution to the challenge of privacy and security in content-based routing involves implementing various techniques and strategies to protect sensitive information and ensure the confidentiality, integrity, and availability of data within the routing infrastructure. Here are some potential solutions:

1. Encryption: Employ strong encryption algorithms to protect data confidentiality during transmission and storage. This ensures that sensitive information remains encrypted and can only be accessed by authorized parties with the appropriate decryption keys.
2. Access control and authentication: Implement robust access control mechanisms and user authentication to prevent unauthorized access to the routing infrastructure. This includes user authentication protocols, role-based access control, and secure login mechanisms to ensure that only authenticated users can perform content-based routing operations.
3. Intrusion detection and prevention: Deploy intrusion detection systems and mechanisms to monitor network traffic and detect potential security breaches or attacks. Intrusion prevention techniques such as firewalls, intrusion prevention systems, and anomaly detection algorithms can help prevent unauthorized access and protect the routing infrastructure from security threats.
4. Privacy-preserving techniques: Utilize privacy-preserving techniques to protect user privacy during content matching and routing decisions. Techniques such as data anonymization, differential privacy, or secure multi-party computation can be employed to perform content-based operations without revealing personally identifiable information.
5. Secure content dissemination: Implement secure mechanisms for content dissemination, such as encryption, digital signatures, and access controls. These measures ensure that content is securely distributed to authorized recipients, preventing unauthorized access or tampering.
6. Security audits and penetration testing: Conduct regular security audits and penetration testing to identify vulnerabilities and assess the effectiveness of security measures. This helps in identifying and addressing potential security weaknesses in the content-based routing system [27].
7. Compliance with privacy regulations: Ensure compliance with relevant privacy regulations and data protection laws. This involves implementing privacy policies, obtaining user consent when required, and providing mechanisms for data subject rights and requests to ensure adherence to privacy regulations such as GDPR or other regional data protection laws.

REAL LIFE CASE STUDY

One real-life case study of NDN is the CCNx (content-centric networking) project, which was an early implementation of NDN principles and concepts. CCNx aimed to demonstrate the feasibility and benefits of a content-centric network architecture

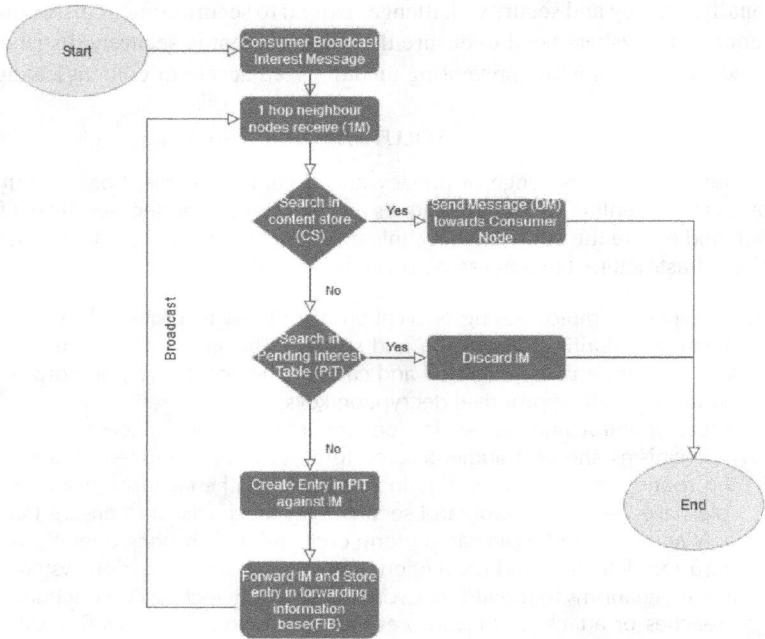

FIGURE 3.5 Case study flow chart of named data networking content-centric networking project.

by implementing NDN in real-world scenarios [2, 5] (refer to Figure 3.5). The project focused on various research areas and applications, including:

1. Content distribution: CCNx was used for efficient content distribution in scenarios such as video streaming. Researchers demonstrated how NDN's content-based approach allows users to retrieve video content directly from nearby caches or even other users, reducing the load on the original content server and improving scalability.

2. IoT and smart environments: CCNx was applied to IoT scenarios, showcasing the benefits of NDN in resource-constrained environments. Researchers illustrated how NDN's content-centric model can simplify IoT communication by leveraging the inherent naming and caching mechanisms of NDN, leading to improved data retrieval and reduced network overhead [28, 29].

3. Collaborative data sharing: CCNx was utilized for collaborative data sharing in distributed environments. The project demonstrated how NDN enables users to share and access content directly, based on content names, rather than relying on traditional host-based addressing. This approach facilitates efficient and secure content sharing among distributed users or organizations.

4. Named-based routing: CCNx explored different routing strategies and algorithms based on content names rather than traditional IP addresses. This research aimed to optimize routing decisions in NDN networks, taking into account content popularity, network conditions, and caching opportunities.

The CCNx project served as a real-life case study that contributed to the understanding and development of NDN [27, 30]. It provided insights into the practical implications, challenges, and potential benefits of adopting a content-centric network architecture. The project demonstrated the feasibility and advantages of NDN in various application domains, highlighting its potential for improving content distribution, IoT communication, collaborative data sharing, and routing efficiency.

CONCLUSION AND FUTURE SCOPE

In conclusion, we have explored the fascinating field of NDNs and WSNs. We explore its architecture, key features, advantages, and challenges. NDN represents a change from traditional data-centric models of data communication and holds promise for the challenges facing the Internet today. By focusing on content rather than space, NDN exhibits advantages such as high-quality content delivery, higher security, and robustness to handle large-scale data and IoT environments. Our research of NDN architectures demonstrates the ability to transform content delivery, enable intranet caching and multicast to reduce network traffic, and improve user experience. In addition, NDN's security features, including data signature and name management, protect against a variety of threats, increasing trust in the network and its users. While the NDN shows great promise, it's not that hard. Migrating from traditional IP-based connections to NDN requires careful planning and consideration in terms of migration strategies and integration with existing systems. Additionally, the complexity and signature verification overhead of content-based applications require optimization to be effective in the export world. As NDN research continues, there are many important issues that deserve future research:

SCALABILITY AND PERFORMANCE

Large-scale experiments and simulations should be conducted to evaluate the scalability and performance of NDN in various network scenarios. Identifying and optimizing potential vulnerabilities can lead to more efficient use of content and better use of network resources.

SECURITY AND PRIVACY

While NDNs provide security features, more research is needed to assess their protection against threats and attacks. Improving security systems and addressing potential issues will increase the reliability of NDN as a secure communication network.

INTEROPERABILITY AND COEXISTENCE

Exploring ways to be compatible with existing IP-based networks and improve interoperability between NDN and other data-centric networking methods will facilitate the transition to NDN and support real-world adoption.

USE CASE VERIFICATION

Real tests and case studies in various field applications will demonstrate the validity and effectiveness of NDN. Exploring how NDN can enhance specific applications such as IoT, edge computing, and content delivery will provide a better understanding of its impact.

POLICY AND MANAGEMENT

Reviewing NDN network management standards and policies is important to ensure fair distribution, accurate content, and access control. Exploring how NDN's data-centric approach relates to data management and governance is critical to improving its social impact.

STANDARDIZATION AND IMPLEMENTATION

Collaborating to standardize the NDN protocol and improve usability will facilitate wider use and integration of NDN in many different environments.

REFERENCES

1. Badotra, S., and S. N. Panda. "SNORT-Based Early DDoS Detection System Using Open Daylight and Open Networking Operating System in Software Defined Networking," *Cluster Computing*, 24, 501–513, 2021.
2. Rathore, P. S., J. M. Chatterjee, A. Kumar, and R. Sujatha. "Energy-efficient Cluster Head Selection through Relay Approach for WSN," *The Journal of Supercomputing*, 77, 7649–7675, 2021.
3. Bill Segall, David Arnold, Julian Boot, Michael Henderson, and Ted Phelps. Content Based Routing with Elvin4, CRC for Enterprise Distributed Systems Technology (DSTC) The University of Queensland, St Lucia, Australia.
4. Muhammad Atif Ur Rehman, Rehmat Ullah, and Byung-Seo Kim. "A Compact NDN Architecture for Cluster-based Information Centric Wireless Sensor Networks," in ICN '19, September 24–26, 2019, Macao, China.
5. Pedro Bizarro, Shivnath Babu, David DeWitt, and Jennifer Widom. "Content-Based Routing: Different Plans for Different Data," in Proceedings of the 31st VLDB Conference, Trondheim, Norway, 2005.
6. Lixia Zhang, Deborah Estrin, Jeffrey Burke, Van Jacobson, James D. Thornton, and Diana K. Smetters. "Named Data Networking (NDN) Project," in NDN-0001, October 31, 2010.
7. Seemu Koponen, Mohit Chawla, Byung-Gon Chun, Andrey Ermolinskiy, Kye Hyun Kim, Scott Shenker, and Ion Stoica. "A Data-Oriented (and Beyond) Network Architecture," in SIGCOMM, 2007.

8. Lasse Overlier, and Paul Syverson. "Locating Hidden Servers," in IEEE Symposium on Security and Privacy, 100–114, 2006.
9. Martin Abadi. "On SDSI's Linked Local Name Spaces," Journal of Computer Security, 6(1–2), 3–21, October 1998.
10. Zuhua Ding, Gene Tsudik, and Shouhuai Xu. "Leak-free Group Signatures with Immediate Revocation," In IEEE ICDCS, 608–615, 2004.
11. M. Caesar, T. Condie, J. Kannan, K. Lakshminarayanan, I. Stoica, and S. Shenker. "ROFL: Routing on Flat Labels," in ACM SIGCOMM, 374, 2006.
12. David D. Clark, John Wroclawski, Karen R. Sollins, and Robert Braden. "Tussle in Cyberspace: Defining Tomorrow's Internet," in SIGCOMM '02: Proceedings of the 2002 Conference on Applications, Technologies, Architectures, and Protocols for Computer Communications, 347–356, ACM, New York, NY, 2002.
13. S. Dharmapurikar, P. Krishnamurthy, and D. E. Taylor. "Longest Prefix Matching Using Bloom Filters," 2003.
14. Kevin Fall. "DTN: An Architectural Retrospective," IEEE Journal on Selected Areas in Communications, June 2008.
15. Paul Laskowski, and John Chuang. "Network Monitors and Contracting Systems: Competition and Innovation. In SIGCOMM, 183–194, ACM, New York, NY, 2006.
16. A. Mtibaa, M. May, M. Ammar, and C. Diot. "PeopleRank: Combining Social and Contact Information for Opportunistic Forwarding," in INFOCOM, 2010.
17. F. Papadopoulos, D. Krioukov, M. Boguñá, and A. Vahdat. "Greedy Forwarding in Dynamic Scale-free Networks Embedded in Hyperbolic Metric Spaces," in INFOCOM, 2010.
18. L. Zhang, S. Deering, D. Estrin, S. Shenker, and D. Zappala. "RSVP: A New Resource Reservation Protocol," in IEEE Network, September 1993.
19. Philip R. Zimmermann. The Official PGP User's Guide. MIT Press, Cambridge, MA, 1995. ISBN: 0-262-74017-6.
20. Dan Massey, Lan Wang, Beichuan Zhang, and Lixia Zhang. "A Scalable Routing System Design for Future Internet," in Proc. of ACM SIGCOMM Workshop on IPv6, 2007.
21. Gerber, A., Lawley, M., Raymond, K., Steel, J., and Wood. Transformation: The missing link of MDA. In International Conference on Graph Transformation. Berlin, Heidelberg: Springer Berlin Heidelberg, October 2002, 90–105.
22. V. Jacobson, D.K. Smetters, J.D. Thornton, M.F. Plass, N.H. Briggs, and R.L. Braynard. "Networking Named Content," in 5th International Conference on Emerging Networking Experiments and Technologies, Rome, Italy, December 2009, 1–12.
23. C.-Y. Chan, P. Felber, M. Garofalakis, and R. Rastogi. "Efficient Filtering of XML Documents with XPath Expressions," VLDB Journal, 11(4), 354–379, 2002.
24. I. Stoica, D. Adkins, S. Zhuang, S. Shaker, and S. Sumna. "Internet indiscretion infrastructure," in Proceeding of DCM SIGCOMM 2W, Pennsylvania. August 2002, 7348.
25. (a) Engineering, 27(9), 827–850, September 2001. (b) Y. Diao, P. Fischer, M. Franklin, and R. To. "YFilter: Efficient and Scalable Filtering of XML Documents," in Proceedings of the 18th International Conference on Data Engineering (ICDE 2002) (pp. 341–342), IEEE, San Jose, CA, February 2002.
26. G. Banavar, T. Chandra, B. Mukherjee, J. Nagarajarao, R. Strom, and D. Sturman. "An Efficient Multicast Protocol for Content-based Publish-Subscribe Systems," in Proceedings of the 19th International Conference on Distributed Computing Systems (ICDCS'99), 1999.
27. S. Kumar, and P. Crowley. Segmented Hash: An Efficient Hash Table Implementation for High-performance Networking Subsystems, In Proceedings of the 2005 ACM Symposium on Architecture for Networking and Communications Systems (pp. 91–103), 2005.

28. Ralph Charles Merkle. "Secrecy, Authentication, and Public Key Systems," PhD Thesis, 1979.
29. W3C. XML Path Language (XPath) 1.0, Nov. 1999.
30. R. Chand, and P. A. Felber. "A Scalable Protocol for Content-Based Routing in Overlay Networks," in Proceedings of the Second IEEE International Symposium on Network Computing and Applications (NCA'03), 2000.

4 Advanced Computing Technologies for Energy-Efficient and Secure IoT Network in Smart Cities
Green IoT Perspective

Shailja Kumari[1], Divya Gupta[1], and Ali Kashif Bashir[2]
[1]Department of Computer Science and Engineering, Chandigarh University, Mohali, Punjab, India
[2]Department of Computing and Mathematics, Manchester Metropolitan University, Manchester, UK

INTRODUCTION

SMART CITY

The term "smart city" first appeared in 1994. According to a prevalent opinion, smart cities are cutting-edge urbanization and smart concepts to address urban challenges, particularly those related to the environment, homes, people, and their well-being. As governments from various nations have a vision for smart cities, technology adoption and implementation are at their pinnacle. The development of smart technology to improve life quality places a heavier load on the environment due to the high energy consumption and increasing carbon footprint. For the sustainability of the environment, there is an urgent need to address energy management and the global carbon footprint.

ENABLING TECHNOLOGIES FOR SMART CITY

A smart city is a collaborative framework that integrates information and communication technologies (ICT), artificial intelligence (AI), edge computing (EC), the Internet of Things (IoT), and the cloud to facilitate easy interactivity. Data from people, devices, and structures can be collected and analysed in a "smart city" to improve control over things like infrastructure, traffic, and energy. Smart energy infrastructure helps to monitor energy utilization and reduce carbon emissions and costs in the city. Everyday processes like governance, transportation, agriculture, logistics, maintenance, education, and healthcare are all automated in some way as a result of technological interventions, and these automated processes can all be controlled, managed, and accessed remotely with the aid of smart devices. These smart ecosystems are emerging as a result. Information, communication, and technology are combined with the traditional infrastructure already in place in "smart cities", which are then coordinated and governed by means of digital technology. Several ICT, including AI, protocols, the IoT, and cloud/EC, must be integrated in order to enable the city to be smarter. Smart city–enabling technologies are described in the following section.

CLOUD COMPUTING

The concept of cloud computing emerged as a distributed, large-scale, service-based computing paradigm in the first decade of the twenty-first century (Jain and Mohapatra, 2019). A cloud platform employs a service paradigm and can provide the end user with a variety of services, including workspaces, data, platforms (hardware + software), and security solutions. Since there are more people, the number of cloud-connected devices, the amount of data to be processed, and the expansion of the IoT and cloud computing are constrained by the high bandwidth needs for data transmission to centralized cloud architectures and the high processing power requirements (d'Oro et al., 2019). The cloud-based computing technology has made significant contributions to the provision of cheaper and faster operating systems, software, and infrastructure solutions. Users do not need to spend a lot of money setting up physical infrastructure; instead, they can rent these from any cloud services provider like Amazon, Google, or Microsoft.

CURRENT CLOUD COMPUTING MODEL

The "Good and Bad" (Pan and McElhannon, 2017) are the following: The success can be attributed to a number of factors, including the following:

- The on-demand pay-as-service it offers users lowers the ownership cost for average customers;
- The elastic computing, storage, and networking resources it offers are flexible and scalable;
- The highly centralized colocation of intensive computation and data enables big-data analytics using machine learning technologies.

CHALLENGES IN CLOUD COMPUTING

- The amount and rate at which IoT devices accumulate data: the delivery of new applications in the current model heavily relies on the proprietary overlays and tools of large companies, and they typically have to transfer all of the data from the edge devices to the remote data centres, which will not be feasible in the future given the volume and velocity of the data generated by IoT devices.

- Latency brought on by the separation between data centres and edge IoT devices; the centralized cloud model also contributes to the fact that edge devices, which are frequently mobile, are typically located some distance from the data centres. High latency could be a significant problem for many applications that need end-to-end communications in the future as the number of edge devices grows exponentially.

EDGE COMPUTING

EC is intended to reduce the computing-related burden on the cloud by executing some preprocessing and/or computation operations at the network edge in order to process device data. EC is hence appropriate for supporting big data analytics and situations where a real-time response is needed for the user or if the end device application has time-criticality restrictions (Ray et al., 2019).

By constructing the essential infrastructure required to utilize the technology, the majority of nations worldwide are attempting to make the concept of "smart cities" a reality. In this situation, EC is crucial for quicker data processing and quicker responses at the network's edges. In recent times, smart cities have been utilizing EC to enhance urban planning, home automation, building security, municipal management, traffic, and parking systems. Traditional IoT networks gather and transmit data and forward it to the cloud for processing. Contrarily, EC devices can process and analyse the data which reduces the burden on the network and improves security (Chavhan et al., 2022).

ARTIFICIAL INTELLIGENCE

A smart city generates tremendous amounts of data, but these data are useless unless it is analysed to extract meaningful and useful information. The processing and analysis of data produced by machine-to-machine communication in a smart city setup is facilitated by AI.

EDGE COMPUTING WITH INTELLIGENCE

The phrase edge intelligence, also referred to as edge AI (Gennaro et al., 2010), has just recently come to be used to describe the fusion of EC and machine learning or AI in general (Mendez et al., 2022).

The smart city uses edge intelligence to combine information from multiple IoT devices by running algorithms at the edge for analysis. As a result, networked devices can make judgments more quickly and with higher quality data. In order to connect things to people and the IoT, smart cities are utilizing human-enabled edge computing and next-generation wireless technologies. As a result, powerful services and automation in the creation of dense and dynamic data sets are produced. A local server, AI, and connectivity to systems for computing in mobile devices, automobiles, and the IoT are necessary for a successful EC architecture (Chavhan et al., 2022).

CHALLENGES OF EDGE-BASED IoT

- **System integration**: EC integrates a variety of numerous servers, networks, and platforms. so, it is a heterogeneous system fundamentally. Thus, it will be challenging to plan, manage, and program resources and data for various applications cations operating on numerous, diverse platforms (Agiwal et al., 2016).
- **Resource management**: Resource management needs to be completely understood and optimized in order for IoT and EC to work together. IoT devices, which frequently lack resources and processing capacity, will be significantly impacted by network delay and congestion, requiring more power to retransmit data in crowded environments (Yu et al., 2017).

INTERNET OF THINGS

The IoT refers to a grouping of sensor-embedded devices that are connected and may communicate with one another as well as the external environment. A smart city's central components are sensory devices. These are tiny chips that have been inserted into the equipment and supplies of the participating entities. These sensors have the capacity to detect their environment and transmit that data to the network gateways for additional processing. Their main objective is to gather information about a phenomenon (such as temperature, pressure, humidity, stress, strain, etc.) from the environment or an event.

INTERNET OF THINGS: OVERVIEW

The concept of the IoT aims to provide interconnectivity of various things in the physical world by using sensor devices. IoT is a heterogeneous environment containing different nodes such as radio-frequency identification (RFID) readers, RFID tags, and mobile devices. Several additional technologies, besides RFID, are employed in the IoT, such as wireless sensor networks (WSN), bar code, ZigBee, and Bluetooth, among others (Balaji et al., 2019).

There are various IoT applications in many aspects of our lives such as agriculture, health, transportation, and environmental monitoring.

IoT was conceptualized by Kevin Ashton first time in 1995. IoT connects heterogeneous devices, which can communicate by information sharing through the internet. IoT is used in various applications in the real world such as smart homes, smart agriculture, smart transportation, and smart grid. The evolution of IoT increases opportunities in the information and communications technology (ICT) sector. IoT devices communicate by using smart sensors with actuators by collecting and transmitting data over the internet. Energy is consumed when a lot of data is created and transported for communication. IoT's primary components include identification, sensing, transmission, computing, service, and semantics. Identification is done by using RFID tags within the network. The sensing process is accomplished by smart devices that can sense and gather information from surroundings with the help of sensors such as temperature, traffic, and humidity. The sensed information is transmitted over the network in order to perform computation. There are a number of communication protocols that can be used for IoT such as Constrained Application Protocol, constrained restful environment, and Message Queue Telemetry Transport. Computational tasks are carried out by hardware and software. Software platforms like Lite OS and Tiny OS are used, as well as hardware platforms like Raspberry PI and Arduino. The cloud platform also performs high-speed computational tasks of IoT. With various IoT applications, there are many challenges such as energy consumption, security, cost, and quality of service (QoS) (Malik and Kushwah, 2022).

High energy consumption further radiates heat into the environment which leads to the emission of CO_2 and the depletion of natural resources. Also due to the increasing production and use of IoT devices, electrical and electronic waste (E-waste) is also increased. To handle these issues, IoT technology needs to be greener by implementing strategies (Arshad et al., 2017).

SMART CITIES AND GREEN IoT FOR ENVIRONMENTAL SUSTAINABILITY

Urbanization and population growth in cities are growing exponentially, posing several challenges to their ability to sustain their economies, societies, and environments. Traffic congestion, poor infrastructure, health issues, a lack of resources instructive challenges, inadequate infrastructure, increasing crime rates, fewer jobs, outdated buildings, power theft, and supply connection problems are such problems in cities. A sustainable living environment that should consider all stakeholders, businesses, and the government is currently needed due to the way people live and the standard of their everyday activities. People will also need to take part in the development of a sustainable living environment and tackle problems like climate change, resource depletion, and biodiversity loss. The GIoT revolution's primary objective is to improve life quality and protect the environment from these types of issues through technological advancements. Cities become smarter due to the GIoT, which connects infrastructure, vehicles, smart sensors, and computers in every part of the city. Stakeholders can increase the protection of human health and life by reducing chemical emissions, water use, and other waste. Although smart devices that are IoT enabled have altered our world by making living easier, it is important to take into account that they can also have a negative impact on the environment. The solution to these issues is only green IoT (GIoT). GIoT will help reduce emissions and pollutants that take advantage of environmental maintenance and surveillance, as well as the cost of operation and power consumption. Researchers and industry professionals have focused a lot of attention recently on how to use the IoT to enable energy conservation in the development of smart cities, which has opened the way for an emerging field known as the GIoT (Chakraborty, 2021). Citizens can play a crucial role in environmental sustainability by adopting the concept of "living green" with the help of GIoT. GIoT services and tools, particularly GIoT, increased use of eco-friendly goods, green roofs, renewable energy sources, conserving energy at home (smart grid), and awareness of the significance of recycling. GIoT allows the deployment of smart environments while reducing their negative effects on the environment and energy consumption. GIoT enables the achievement of a sustainable smart world by promoting the use of eco-friendly ICT and improving quality of life by creating a smart, healthy environment. There are various IoT applications where efficient energy use is required to support a sustainable environment. This chapter focuses on the fundamental connection between sustainability, the IoT, and smart cities (Mohamed et al., 2017) to adopt GIoT-enabled sustainable practices for efficient energy management. The contributions that were made in this chapter are as follows:

- First, the concept of a smart city with its enabling technology, the background of IoT, with its evolution and working has been discussed.
- Second, GIoT with its goal, strategies, and taxonomy is presented.
- Further, related work on energy efficiency for GIoT is presented.
- Further discussed EC with IoT and the recent concept of edge intelligence.
- Lastly, IoT trends describing their applications and challenges in IoT have been presented in detail.

BACKGROUND

EVOLUTION OF IoT

During the mid-1980s, telephone lines or letters were the means of communication. With time, communication got new opportunities and platforms due to the existence of the Internet. Voice over Internet Protocol has existed over the years. Today's era has come up with a new concept of IoT and has left the internet far behind. IoT combines things with the help of the internet and communicates by sharing information. In the early 1990s, an introductory IoT concept was proposed in Auto-Id labs at the Massachusetts Institute of Technology. The first IoT application Trojan Room coffee pot was developed. IoT interconnects the physical world by synchronizing things over the internet. The evolution of IoT undergoes various phases from the Internet to IoT. The first phase was the pre-Internet phase in which communication is via short message service (SMS) over telephone lines. Later mobile telephone devices came into existence as a communication medium. The second phase was the Internet of content phase. In this phase, large-sized messages can be communicated via the World Wide Web, email with attachments, and focused on entertainment and information. The third phase was the Internet of Service, which provides possibilities for electronic applications like E-commerce and E-productivity. The fourth phase was the Internet of People in which people got associated with social media and many other mediums like YouTube, Facebook, and Skype. The next phase is the IoT which helps in automation, tracking, and monitoring parameters like temperature and pressure. In the transformation phase of IoT, AI can be incorporated so that decision-making can be possible with minimizing human intervention. IoT plays a role in all life activities such as tracking, intelligent decision-making, and improving environmental status. Figure 4.1 depicts the evolution of IoT and its achievement.

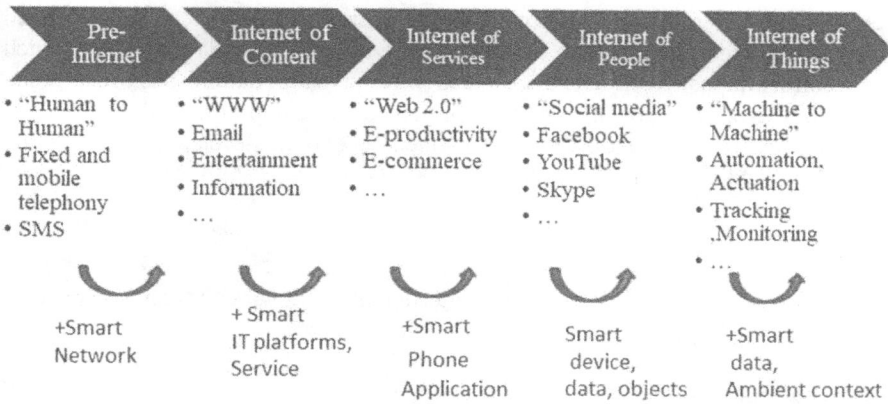

FIGURE 4.1 Evolution of IoT.

WORKING

The IoT mechanism consists of various components such as the identity of source, sensing, communication, computing, service, and semantics. Identification ensures the provision of data or services to the target address.

The information from the surrounding of the identified source/target is collected by sensors such as temperature, heat, and humidity sensors, under the sensing stage and then sent to the data centre by communication. These data are then processed and analysed using various parameters and conditions to provide service. Computation processes can be done by various microprocessors, microcontrollers, and some software applications. Service concerns information aggregation, identity, collaborative work, and ubiquitous. Lastly, semantics incorporates intelligent knowledge-based decision-making.

GREEN IoT: OVERVIEW

GOAL OF GIoT

The GIoT goal is to reduce energy use, reduce costs, and increase security. The IoT revolution is being driven by the GIoT. Some characteristics need to be evaluated in order to produce IoT as green. The primary characteristic is to increase the energy efficiency of devices using network architecture and communication protocols, and smaller network sizes, to provide a safe, cost-effective system with little pollution and emissions to the environment. Six primary enabling technologies, including green RFID, green wireless sensor networks (GWSN), green machine-to-machine communication (GM2M), green data centres (GDC), green cloud computing, and eco-friendly ICT (Malik and Kushwah, 2022).

GIoT techniques and approaches are to create a green and smart environment that uses enabling technologies to reduce pollution and energy consumption, while also promoting a smart and high quality of life. GIoT technologies have numerous benefits, but the highest contribution is the decreased energy consumption and associated cost savings, as well as the decreased carbon dioxide emissions, which have a direct impact on climate change and lower global pollution. The use of green technology, from an economic standpoint, lowers costs and creates the possibility of improving living standards, the environment, economic development, and green global modernity (Memić et al., 2022).

STRATEGIES FOR GIoT

GIoT realization can be done by considering the following design components (Maksimovic, 2018):

- Usage of bio-products and eco-friendly products.
- Reducing energy consumption in the IoT environment.
- Making use of renewable and green sources of energy.
- Distributed processing to overcome latency limitation.
- Reduce network data transmission and processing for energy efficiency.
- Reduce the network path for the data by using an energy-efficient routing approach.
- Adopting security mechanisms.

GREEN IoT TAXONOMY

GIoT taxonomy represents the classification of the technological basis on which the energy-efficient models are employed in them. The taxonomy of GIoT is represented in Figure 4.2 (Arshad et al., 2017).

- **Software-based GIoT techniques:** Data centres can play a key role in GIoT by adopting energy-efficient centres. Selective sensing is used by a context-aware sensing platform to save energy. An energy-efficient scheduling technique is used to reduce unnecessary energy consumption when sensors are powered on but not in use, by changing sensor statuses to on-duty, pre-off-duty, and off-duty in accordance with the demands of the situation.
- **Hardware-based GIoT techniques:** The frequent energy needs of sensor nodes can be reduced by including passive sensors, which would result in even lower energy usage and waste. CoreL is for low-computational operations, while CoreH is for high-computational tasks in an energy-efficient twin-core processor for GIoT. It reduces energy consumption by developing a scheduling algorithm that assigns various tasks to these cores based on the resources they could require.
- **Policy-based GIoT techniques:** The plans according to real-time data from IoT devices. The process of creating policies to achieve energy efficiency involves several stages, including monitoring (various energy consumption conditions), information management, feedback, and automation system.

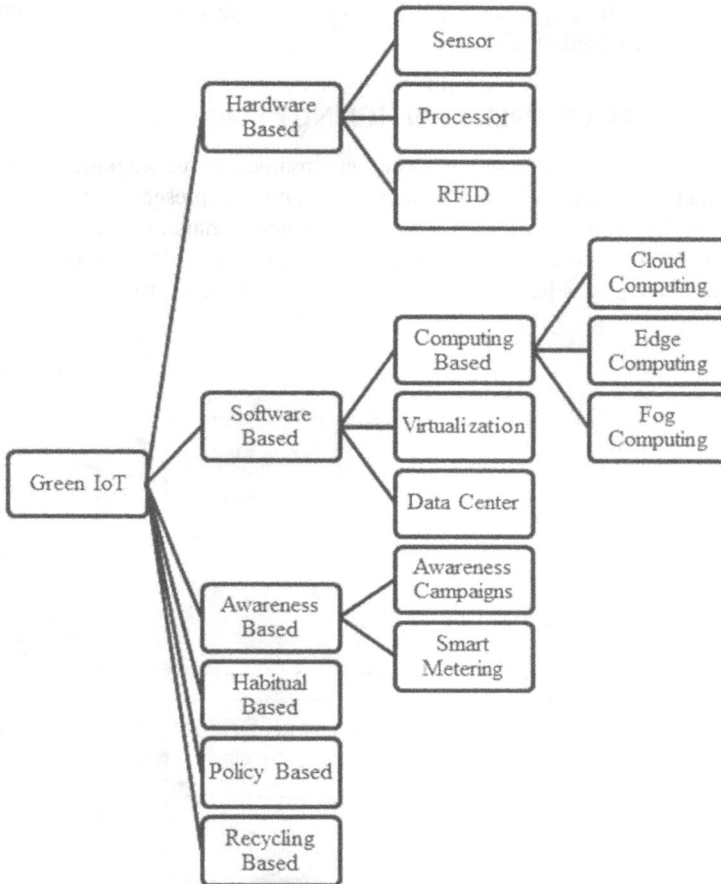

FIGURE 4.2 Green IoT taxonomy.

- **Awareness-based techniques:** The effectiveness of awareness programs in reducing energy usage varies depending on the culture and the country because it is impossible to forecast or estimate the number of individuals who will be involved in the events.
- **Changing habits towards GIoT:** Adopting some of these behaviours will help to become more energy efficient and reduce carbon footprint. Simple routines can help us use less energy during regular tasks. One strategy is to use automation systems to monitor the patterns of energy consumption in workplaces, homes, and industries.
- **Recycling for GIoT:** Making equipment for an IoT network from recyclable materials can contribute to the network being environment-friendly. Mobile phones, for instance, are constructed from some of the most limited natural resources, such as copper and plastic, and include some materials

that are not biodegradable and may increase the effect of greenhouse if properly disposed of after use.

RELATED WORK OF ENERGY EFFICIENCY FOR GREEN IoT

GIoT is an enabler of a sustainable and smart environment to reduce power consumption and carbon emissions. Energy management taxonomy is represented in Figure 4.3 and described in detail in Memić et al. (2022). The energy management scheme (EMS) proposed IoT nodes for heterogeneous energy constraints. EMS adopted three strategies; the first strategy is based on data minimization which is to be transmitted over

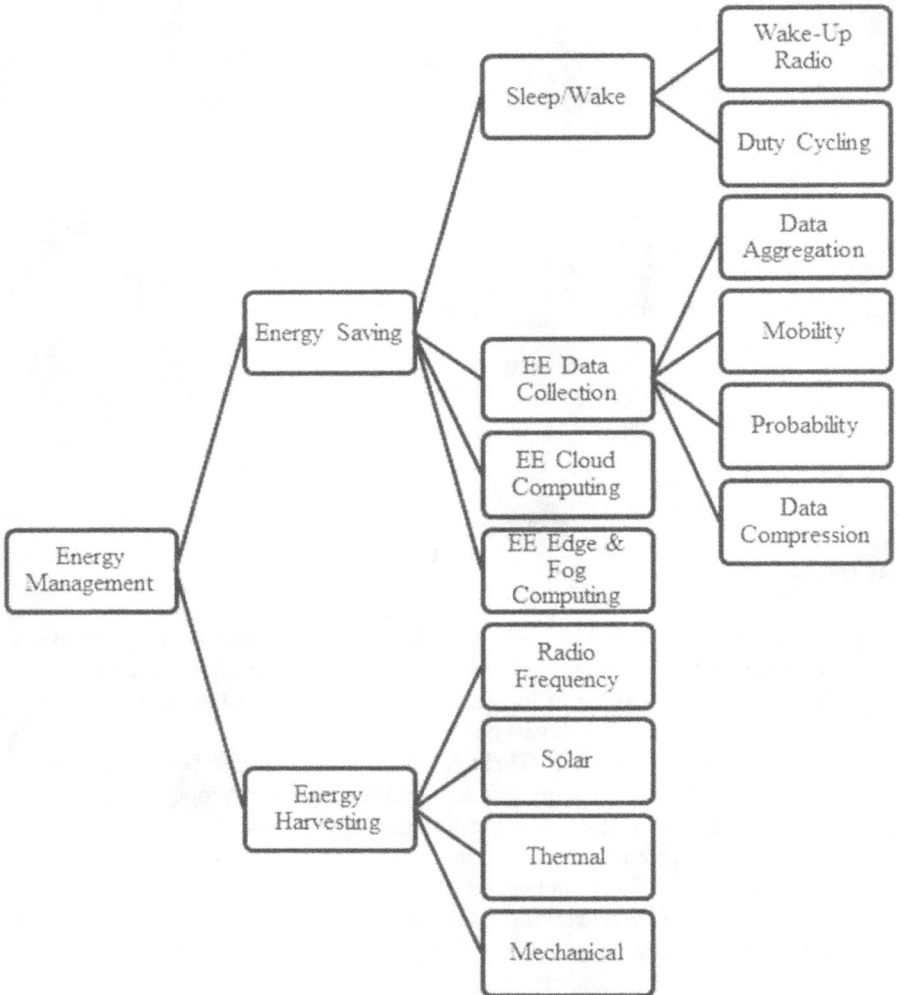

FIGURE 4.3   Energy management taxonomy.

the IoT network. The second strategy is based on scheduling in which work is scheduled at IoT nodes, and finally, fault-tolerant provision of nodes for the IoT.

Data at sensor nodes consume higher energy due to various processes like data gathering, processing, and data transmission. Each node's data reduction is different due to the class of nodes. Associated functions for data are compression, fitting the data, and prioritization. Data compression concerns data processing and transmission, and data fitting processes mean the gathering and transmission of data. Data compression refers to processes and transmission. Network congestion is handled by queuing theory data scheduling which reduces the consumption of energy by delaying processes related to nodes using issues like current energy level, source of energy, alternate nodes, the energy level of heterogeneous nodes, the significance of node, particular task period or location of node finally fault tolerance strategy handles energy failure of a node by replacement with alternate one. On the basis of significance and tolerance of faults based on node levels, there are different levels of IoT sensor nodes. As performance indicators for the proposed EMS to benchmark, average energy consumption rate, data throughput number of failing nodes, and network lifetime were taken into consideration.

In Maksimovic (2018), the authors discussed GIoT as a facilitator of a smart, sustainable, and healthy environment. Key beneficiaries of GIoT are in various sectors through environmental protections, increased profit margins, and customer satisfaction.

Authors have proposed a protocol, Reservation Aloha for No Overhearing (RANO), that avoids the problem of overhearing during RFID communication. However, self-wakeup hardware based on times in said proposal is not accounted for energy consumption.

In Zhu et al. (2015), technologies to make GIoT such as green RFID, GCC, GWSN, GDC, and GM2M are reviewed and discussed reduction in RFID size concern with GRID which results in a reduction of non-biodegradable material, transmitting power adjustment dynamically at nodes for green communication. Using energy-efficient routing techniques, sled mode of low power during in-activity radio optimization is concerned with GWSN. Energy harvesting activity scheduling helpful in GM2M implementation and data centres powered by green energy were discussed in the article. However, these approaches are not supported or proven analytically with data that can validate mentions.

According to Chatterjee and Armentano (2015), the proposed system model focuses on three stages, on-duty, off-duty, and pre-off-duty, as a scheduling algorithm that is energy-efficient for IoT. Energy use as reported by using the proposed algorithm is not compared in the absence of three stages.

In Albreem et al. (2021), authors provide insights on software and hardware approaches for a reduction in energy consumption; FM2M for data exchanges employing data aggregation in an efficient way is discussed. Context-based computing for data reduction under green design also discussed and proposed fog-based energy consumption model.

The authors proposed an offloading scheme based on reinforcement learning for IoT devices with the harvesting of energy. IoT device does offload at the edge node on the basis of the current level of the battery and the predicted amount of harvested energy. However, the learning speed is not so effective. The authors proposed a solution for data compression at edge nodes and then transmitted over the network. Data compression is based on error-bounded lousy mechanisms for application areas where high performance is required. Analysis of compressed data is done to prove the accuracy of the result. Some of the specific EC strategies have been the subject of other researchers' attention, who studied multi-access edge computing (MEC) in 5G and obtained information. Though it tries to provide a comprehensive grasp of MEC rather than a basic summary, its work does not provide us the edge. Additionally, that study does not cover all potential platforms and frameworks that may be used to deploy an AI model at the network edge, but it does offer a general overview of the conditions that would need to be met in order to effectively carry out ML tasks there.

RELATED WORK OF SECURITY ENHANCEMENT IN SMART CITY

IoT also has applications in security solutions. In surveillance systems, the cameras or a device at the network's edge could process the raw data to deduce a conclusion or combine the data (Malik and Kushwah, 2022).

The method is employed in the work of (Albreem et al., 2021) to successfully use cameras to manage the speed of moving vehicles on highways (Lin et al., 2015).

In the context of vehicles, EC in cooperation with security methods such as blockchain technology is to guarantee data privacy and effective energy connections.

IoT TRENDS

APPLICATIONS OF IoT

IoT plays a role in all life activities such as tracking, intelligent decision-making, and improving environmental status. Various applications of IoT are shown in Figure 4.4.

- **Smart environment:** Environment-based IoT services in order to use environmental resources in an efficient and effective way. Activities like monitoring air pollution and forest fire detection are performed in a smart (Malik and Kushwah, 2022).
- **Smart agriculture**: IoT deployment in agriculture helps in the automation process and also in decision-making. IoT sensor nodes help in prediction and decision-making by providing data like soil nutrients and humidity, solar exposure, leaf wetness, temperature level, rainfall pattern with the amount, insect attacks, animal attacks, and so on (Maksimovic, 2018).
- **Smart home**: Smart home is the integration of diverse sensing devices, processing units, and networks with smart things that provide a comfortable, intelligent, secure, and safe home environment (Jabbar et al., 2018).

Access to doors, air conditioning, windows, room ventilation, and so on controlled and monitored through IoT network access. Smart homes reduce energy wastage and provide automation (Albreem et al., 2021). Home equipment like TVs, AC, and fridges can be remotely controlled using IoT technologies. For example, a

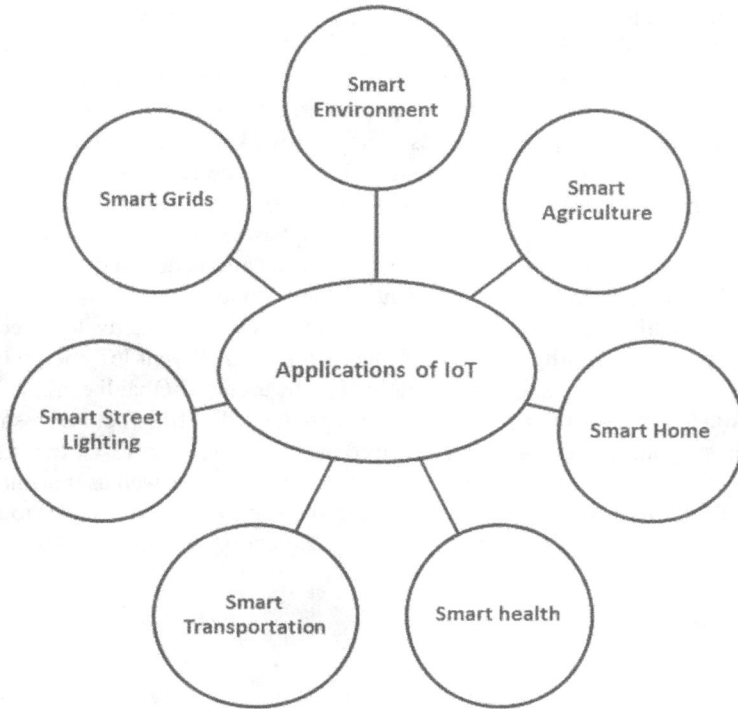

FIGURE 4.4 Applications of IoT.

smart AC can have RFID which can set the temperature according to surrounding conditions on the basis of sensor data and can be remotely controlled by a smartphone using an application. Smart home activities improve resource utilization, energy saving, money, and environment saving (Arshad et al., 2017). With the advancement of IoT technology, traditional homes turned to smart homes. Home appliances, including refrigerators, air conditioners, curtains, geysers, lights, and doors, are connected to the internet with the potential for communication and processing. With the development of new technology like fog computing and EC, the owner can remotely monitor their home while seated a long way from where it is located (Gupta et al., 2022).

- **Smart health:** Health-related services like maintaining health information, making automation results of the diagnosis, health parameter analysis, and processing in real-time. IoT enables paramedics by wearing biosensors by which critical parameters of patients' health can be accessed remotely so that treatment schedules can be planned on the basis of data (Chatterjee and Armentano, 2015).
- **Smart transportation**: IoT technology is used in the logistics and transportation industry. Smart transportation helps in tracking vehicles and products in real-time using RFID and sensors remotely (Arshad et al., 2017).

- **Smart grids:** A power network that allows for the two-way flow of both electricity and information is a" smart grid." In contrast to conventional electric power, a smart grid (Lin et al., 2015) enables for exchange of electricity between power networks and electricity consumers. Additionally, integrating IoT into power grids creates a two-way information flow that enables electricity customers to actively participate in the system's operation and grids to self-heal (Al-Obaidi et al., 2022).
- **Smart street lighting:** This application emphasizes ease as well as a reduction in energy use. In reaction to both vehicles and pedestrian activity, the night's autonomous and intelligent lighting control system is activated. Additionally, it may be set to automatically activate or deactivate based on the quantity of sunlight. The implementation of intelligent IoT energy conservation techniques for street lighting can be used (Al-Obaidi et al., 2022).
- **Smart parking:** One of the core concepts for IoT infrastructure in smart cities is smart parking. Finding a free parking space can assist in cutting down on the quantity of petrol and pollutants created, as well as the amount of electricity used. In smart cities, using smart parking technologies to find open spots would ease traffic and increase energy efficiency (Al-Obaidi et al., 2022).

CHALLENGES OF IoT

IoT revolutionizes the IT industry, but issues such as security require research so that users can trust it. However, due to billions of IoT devices, energy consumption is the most significant challenge. IoT devices such as sensors, actuators, wireless networks, and processors, data centres all need the power to operate. These devices consume energy on a daily basis so energy is required (Arshad et al., 2017). *Cost*: The technology itself includes a large number of interconnected devices; as a result, maintenance and service costs will continue to rise rapidly, producing an imbalance in a nation's economy. This is due to how frequently customer and business demands for IoT change. One solution to this problem is the development of sensors or other equipment that needs little to no maintenance. As a result, the cost of service will be lower, and certain economic circumstances will be avoided (Balaji et al., 2019).

The setup cost includes the cost of the IoT hardware necessary for a smart environment. The cost of operating is determined by IoT platforms that enable data processing, device maintenance, and data exchange, among cloud servers, gateways, and IoT devices, in addition to energy and maintenance (Sinha and Dhanalakshmi, 2022). *Energy consumption*: With the billions of IoT devices currently in use and the rising energy consumption for information technology, this poses a serious problem. This is predicted to increase to 7.335 TWh annually by 2025. The majority of the equipment used in IoT is powered by batteries, and once the sensor is deployed, it is nearly impossible to replace, resulting in high power consumption and ultimately a world energy crisis. Therefore, creating sensors that don't require batteries or batteries with long life cycles is another challenge. Energy consumption can be reduced by using an energy-saving approach or by using renewable sources of energy. IoT devices often have limited battery life, thus applications must rely on infrastructure and policies that encourage wise device use (Silva et al., 2023). *Standardization*: In order to accomplish their objectives, IoT devices should be simpler to operate and have greater compatibility with other connected devices. In order to facilitate the heterogeneity of networks and interoperability, it is necessary to design some standard protocols for communication that are significantly simpler for the user (Sinha and Dhanalakshmi, 2022). *Security and privacy concern*: The handling of personal data, together with security and privacy concerns, is one of the biggest issues facing IoT. Being a networked platform, IoT allows for the uploading of any personal data into the cloud that is quite susceptible. The security and privacy of an individual will obviously be compromised if there is a flaw in IoT security. Utilizing a secure gateway and creating security algorithms can create a safer environment (Balaji et al., 2019). IoT layer security breaches such as data theft, attack, and so on. The primary focus is on physical elements like actuators and sensors. Malware, criminals, or unintentional human activity could cause physical equipment to malfunction. As a result, it is difficult to implement complex and intricate algorithms in IoT devices because of their limited memory and limited connectivity. It is also possible to suffer from routing attacks, DOS threats, and congestion (Sinha and Dhanalakshmi, 2022). Major issues for security and privacy (Pan and McElhannon, 2017) are the following:

- For the edge cloud infrastructure and applications of the future, security continues to be one of the most significant challenges. Security issues will be multiplied because the future edge cloud may incorporate several technologies (such NFV, SDN, and IoT). In the beginning, edge clouds would have the same security issues as the classic cloud computing paradigm due to the deployment of virtualization technology (such as VM security).
- Due to their sparse distribution and proximity to the consumers' premises, edge cloud servers may be more vulnerable to physical assaults.
- There will always be security concerns with regard to particular technologies like NFV, SDN, and IoT in edge clouds. Additional security concerns may arise from the connections or interactions between these technologies since the future edge clouds may be a synergistic effort and each one of them may play distinct responsibilities.

Resource utilization: Resource optimization is necessary in IoT-enabled smart environments in order to determine the optimum number of gateways, IoT devices, transmitted data, and cloud storage needed to increase profitability. The heterogeneous devices and environment make this more challenging. Implementing complex algorithms and statistical models that can determine the optimal resource distribution will be required (Sinha and Dhanalakshmi, 2022). CO_2 *emission*: The major environmental issue is due to the emission of CO_2, which is generated by the carbon footprints of smart devices. The "smart devices" refers to any device that is now connected to the internet and stored on a cloud. According to reports, by 2050, the number of connected devices is expected to increase to 170 billion. As a result, on average, the number of IoT devices grows by 12% annually, which increases the percentage of GHG (greenhouse gas) emissions, and carbon footprints cause depletion of natural resources and pollution on Earth (Sharma and Panwar, 2020). *QoS*: Before being transmitted to the end user, the volume of data sensed from the surrounding adheres to certain quality standards. All of the sensor nodes used in the IoT environment must have their available resources and calculations load balanced in order to provide QoS. However, the changing network topology and constraints such as energy, delay, and processing capabilities render the current QoS solutions ineffective in an IoT setting. Therefore, new methods for delivering QoS solutions need to be built to support quick content delivery with minimum retransmission (Gupta et al., 2022).

To provide a reliable backbone for IoT connectivity, QoS is expected to be able to adapt to changes in network resources and capabilities. However, the IoT is such a complicated and heterogeneous network that connects various network kinds using various communication technologies. Providing QoS in real-time across the network is challenging due to network and communication-related problems such as bit errors, lengthy delays, collisions, packet droppings, and jitter impairments. The provision of QoS while accounting for energy efficiency is significantly more difficult (Tuysuz and Trestian, 2020). *Interoperability*: Due to the lack of IoT standards at this time, solutions published in the literature frequently concentrate on their own designs. However, due to the significant physical and network heterogeneity and the application domain of these solutions, multiple architectural designs with different protocols and components have been made. This inevitably results in a little degree of interoperability among the ideas/projects put out. IoT-based vendor-specific new designs are continuously being proposed, despite the fact that the rapid and balanced development of IoT and the optimization of energy efficiency will be significantly damaged. In this situation, it is difficult to develop multi-layered, interoperability-focused GIoT solutions before any standardization efforts or drafts have been published (Tuysuz and Trestian, 2020) [26]. *Analytics of big data*: Integration of IoT is into current integrated information system architecture; IoT relies primarily on a conventional ICT environment and is influenced by everything connected to the network, and therefore, a significant amount of effort is required. Additionally, billions of Internet-connected objects are anticipated to generate a huge volume of real-time data flows. Actually, without effective comprehension and analysis, these raw data are useless. However, obtaining useful information from the large amounts of data i.e. of IoT objects collected through various networks and communication methods. Technology demands strong big data analytics abilities, which may be difficult for end users (Tuysuz and Trestian, 2020). *Electronic waste (E-waste)*: E-waste is a rapidly developing environmental problem, particularly in the majority of technologically advanced nations. Due to the crucial role that IoT plays in evolving urban areas into smart environments, cities are becoming smarter. However, research predicted that by the end of 2030, 50 billion IoT devices will be connected. IoT devices often have a short lifespan, which makes it evident that there may be problems with E-waste. Therefore, while incorporating IoT technology during the designing stage, built environment experts should take this issue into consideration (Al-Obaidi et al., 2022).

CONCLUSION

With the rapid evolution of IoT technologies and the development of the "smart city", it must be monitored carefully from the point of view of environmental and natural resources to ensure sustainability and efficiency. GIoT is the concept that enables the sustainable development of IoT by reducing energy consumption and harmful impacts on the environment. In this chapter, the background of IoT with its transition phase of evolution has been discussed. The detailed concept of GIoT with its goal and strategies that should be followed for GIoT has been provided. A detailed review of related work on energy efficiency for GIoT has been performed. Furthermore, IoT trends are presented in detail by providing applications and challenges of IoT. To create a smart, safer, and healthy environment, more research to achieve GIoT is required.

REFERENCES

Agiwal, M., Roy, A., and Saxena, N. (2016). Next generation 5G wireless networks: A comprehensive survey. IEEE Communications Surveys & Tutorials, 18(3):1617–1655.

Albreem, M. A., Sheikh, A. M., Alsharif, M. H., Jusoh, M., and Yasin, M. N. M. (2021). Green internet of things (GIoT): Applications, practices, awareness, and challenges. IEEE Access, 9:38833–38858.

Al-Obaidi, K. M., Hossain, M., Alduais, N. A., Al-Duais, H. S., Omrany, H., and Ghaffarianhoseini, A. (2022). A review of using IoT for energy efficient buildings and cities: A built environment perspective. Energies, 15(16):5991.

Arshad, R., Zahoor, S., Shah, M. A., Wahid, A., and Yu, H. (2017). Green IoT: An investigation on energy saving practices for 2020 and beyond. IEEE Access, 5:15667–15681.

Balaji, S., Nathani, K., and Santhakumar, R. (2019). IoT technology, applications and challenges: A contemporary survey. Wireless Personal Communications, 108:363–388.

Chatterjee, P., and Armentano, R. L. (2015). Internet of things for a smart and ubiquitous eHealth system. In 2015 International Conference on Computational Intelligence and Communication Networks (CICN), pp. 903–907. IEEE.

Chavhan, S., Gupta, D., Gochhayat, S. P., Khanna, A., Shankar, K., and Rodrigues, J. J. (2022). Edge computing AI-IoT integrated energy-efficient intelligent transportation system for smart cities. ACM Transactions on Internet Technology, 22(4):1–18.

d'Oro, E. C., Colombo, S., Gribaudo, M., Iacono, M., Manca, D., and Piazzolla, P. (2019). Modeling and evaluating a complex edge computing based systems: An emergency management support system case study. Internet of Things, 6:100054.

Gennaro, R., Gentry, C., and Parno, B. (2010). Non-interactive verifiable computing: Outsourcing computation to untrusted workers. In Advances in Cryptology—CRYPTO 2010: 30th Annual Cryptology Conference, Santa Barbara, CA, USA, August 15–19, 2010. Proceedings 30, pp. 465–482. Springer.

Gupta, D., Rani, S., and Shah, S. H. A. (2022). Utilizing ICN caching for IoT big data management in WSN-based vehicular networks. IoT and WSN Based Smart Cities: A Machine Learning Perspective, pages 225–241. Springer.

Jabbar, W. A., Alsibai, M. H., Amran, N. S. S., and Mahayadin, S. K. (2018). Design and implementation of IoT-based automation system for smart home. In 2018 International Symposium on Networks, Computers and Communications (ISNCC), pp. 1–6. IEEE.

Jain, K., and Mohapatra, S. (2019). Taxonomy of edge computing: Challenges, opportunities, and data reduction methods. *Edge Computing: From Hype to Reality*, pp. 51–69. Springer.

Lin, J., Yu, W., and Yang, X. (2015). Towards multistep electricity prices in smart grid electricity markets. IEEE Transactions on Parallel and Distributed Systems, 27(1):286–302.

Maksimovic, M. (2018). Greening the future: Green Internet of Things (G-IoT) as a key technological enabler of sustainable development. Internet of Things and Big Data Analytics toward Next-Generation Intelligence, pp. 283–313.

Malik, A., and Kushwah, R. (2022). A survey on next generation IoT networks from green IoT perspective. International Journal of Wireless Information Networks, 29(1):36–57.

Mendez, J., Bierzynski, K., Cuéllar, M., and Morales, D. P. (2022). Edge intelligence: Concepts, architectures, applications, and future directions. ACM Transactions on Embedded Computing Systems (TECS), 21(5):1–41.

Mohamed, N., Al-Jaroodi, J., Jawhar, I., Lazarova-Molnar, S., and Mahmoud, S. (2017). SmartCityWare: A service-oriented middleware for cloud and fog enabled smart city services. IEEE Access, 5:17576–17588.

Pan, J., and McElhannon, J. (2017). Future edge cloud and edge computing for Internet of Things applications. IEEE Internet of Things Journal, 5(1):439–449.

Ray, P. P., Dash, D., and De, D. (2019). Edge computing for internet of things: A survey, e-healthcare case study and future direction. Journal of Network and Computer Applications, 140:1–22.

Sharma, N., and Panwar, D. (2020). Green IoT: Advancements and sustainability with environment by 2050. In 2020 8th International Conference on Reliability, Infocom Technologies and Optimization (Trends and Future Directions) (ICRITO), pp. 1127–1132. IEEE.

Silva, P. V. B. C., Taconet, C., Chabridon, S., Conan, D., Cavalcante, E., and Batista, T. (2023). Energy awareness and energy efficiency in internet of things middleware: A systematic literature review. Annals of Telecommunications, 78(1–2):115–131.

Sinha, B. B., and Dhanalakshmi, R. (2022). Recent advancements and challenges of internet of things in smart agriculture: A survey. Future Generation Computer Systems, 126:169–184.

Tuysuz, M. F., and Trestian, R. (2020). From serendipity to sustainable green IoT: Technical, industrial and political perspective. Computer Networks, 182:107469.

Yu, W., Liang, F., He, X., Hatcher, W. G., Lu, C., Lin, J., and Yang, X. (2017). A survey on the edge computing for the in Internet of Things. IEEE Access, 6:6900–6919.

5 Amalgamation of 5G and Edge Computing for Latency Reduction in Metaverse Using WSN and IoT

Kanwal Preet Kour[1], *S. Kanwal Deep Singh*[2],
Deepali Gupta[3], *and Kamali Gupta*[3]
[1]Department of Computer Science and Engineering,
Lovely Professional University Phagwara, Punjab, India
[2]Department of Computer Science,
Federation University, Australia
[3]Chitkara University Institute of Engineering &
Technology, Chitkara University, Punjab, India

INTRODUCTION

It was in 1992, in a novel called "Snow Crash", by Neal Stephenson, where users interacted with each other through avatars [1]. Despite being around for over a decade, the concept of metaverse gained widespread attention in 2021 when the company Facebook changed its name to "Meta". Metaverse refers to a simulated virtual environment created by using on web-based technologies. Immersive technologies which are an extension of VR (virtual reality) in collaboration with IoT (Internet of Things) are making it the major technological wave, after internet in the present era [2]. The gaming industry in metaverse generates billions of dollars, and it is predicted to have a market value of USD 126 billion based on people's interests [3, 4]. The pandemic has highlighted the importance of metaverse as it offers real-like immersive experiences and a new concept of social networking. However, there are challenges associated with communication, scalability, authentication, and reliability that need to be taken care of by the introduction of new protocols and standards for communication. Metaverse's full potential can only be realised through high-speed internet, such as 5G and other future standards, which minimises latency [5].

As seen in Figure 5.1, the idea of the metaverse has developed over many years. MIT developed a Movie Map in 1970 that allowed users to have an immersive virtual tour of the Colorado town of Aspen. The metaverse was born out of this VR experience. Many science fiction stories written before this depict a failed economy giving rise to

DOI: 10.1201/9781003438205-5

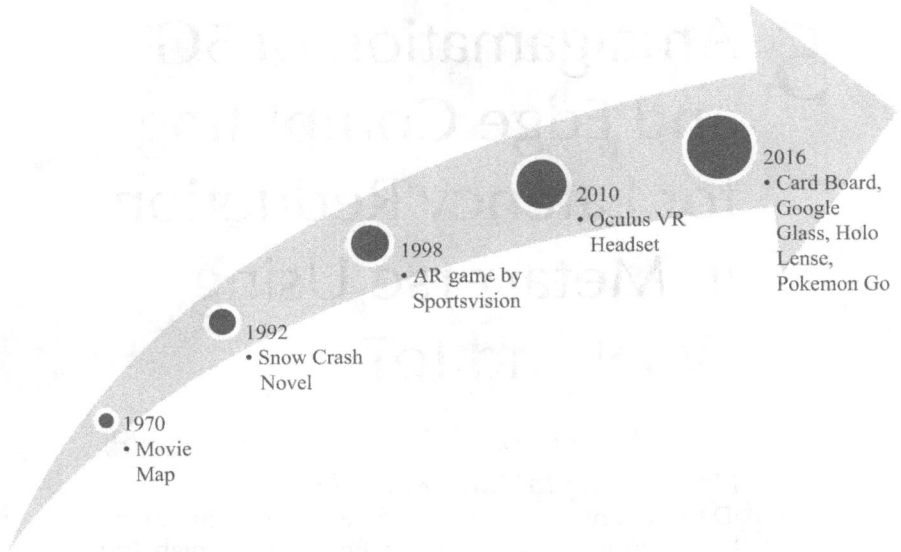

FIGURE 5.1   History of metaverse.

VR. The characters in some comic books used gadgets like "headsets" and "goggles" to enter a virtual world. Games based on virtual and augmented reality (AR) are being created today by further extending these notions [6]. Additionally, Sportsvision developed the first live game employing AR in 1998 that involved superimposing visuals over actual planes, which helped to shape the modern day metaverse concept. Palmer Luckey created the Oculus VR headset in 2010, enabling the use of 90° vision and powerful processing powers, sparking a significant advancement in VR technology. In the year 2018, Facebook purchased this Oculus VR in a $2 billion agreement. The novel Player One, which introduced the ground-breaking idea of the metaverse, was published around the same time and provides in-depth insights into the immersive environment. With the release of gadgets like Google Glass, Cardboard devices, and other XR-capable technologies and games like HoloLens and PokemonGo-style in the years 2014–2016, tech titans like Samsung, Microsoft, and Google further extended AR and VR to XR. Apple added Lidar (Light Detection and Ranging) to iPhones in 2020, preparing its products for XR technology in the future. The most notable development came in October 2021, when a tech giant company Facebook renamed it as Meta, ushering in the metaverse as a cutting-edge technological advancement [7–9].

ARCHITECTURAL LAYOUT AND CHALLENGES IN METAVERSE

Five layers can be identified in the metaverse's architectural design [6, 10]. In Figure 5.2, five layers make up the user-centric layout. The architecture entails the coordination of numerous existing technologies that cooperate to deliver a real-time experience. In the metaverse universe, the layers establish the conceptual portions of the experiences individuals will have and the huge metaverse applications [11–15]. The layer at the top is infrastructure and connectivity layer. Different technologies necessary to connect to the extensive network, such as 5G, cloud computing, IoT, and Wi-Fi, are a part of this layer.

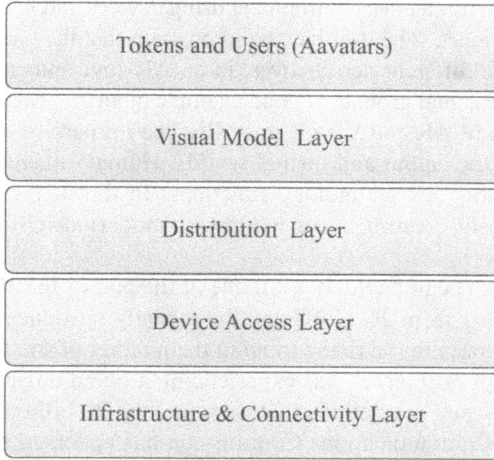

FIGURE 5.2 Different layers of metaverse framework.

Networking protocols such as direct 3D and OpenGL are also included in this layer. This layer also includes content delivery. The device access layer, which comprises the devices and access once used to be a part of the immersive reality, is the second layer. Smart wearables, kinaesthetic devices, VR headsets, smart glasses, and so on are included in this layer. This layer is crucial because it enables the application of forces and movements to provide a feeling of touch, motion, or vibration. The distribution layer is the third layer of the metaverse architecture. By allotting IP addresses to specific users by using DNS (domain name server), this layer is in charge of enabling decentralised access to the metaverse. Blockchain technology, AI, and edge computing are some of the other methods of achieving decentralisation. These technologies allow for data decentralisation, which enhances system security and offers data offloading. The fourth layer of the framework is visual model layer, which consists of 3D visualisation models created using augmented and VR. The usage of spatial computing allows for human connection with technology while also enabling robots and devices to store and control real environments and things, hence reducing barriers between the real and virtual worlds. The user interface layer comprises technology enabling the user to distinguish themselves from other immersive environments and is the fifth layer. It entails the production and acquisition of brand-new assets known as non-fungible tokens (NFTs) by users with avatars. These features are already available on some platforms, such as Roblox and Rec Room, with limited scalability. This layer will be a direct reflection of metaverse's upcoming advances. To this layer, new characteristics are constantly being added, including real-time presence, real-time experience, and community-driven content. It is possible to dissolve distance, things, and space thanks to this metaverse layer. This layer allows the application domain and future of metaverse to be expanded.

The duplication of physical items and processes by computers is known as a digital twin. It operates in conjunction and synchronisation with the actual world, reflecting any changes that occur in real-time. VR is a technology that creates a simulated environment with virtual items that are sometimes based on reality and sometimes constructed in the user's imagination.

The user enters an immersive environment using devices and headgear, giving them a close-to-real experience. The user has the impression that they are inside the virtual setting. In one of the different perspectives, in an AR environment, virtual items are superimposed on the actual ground. A good example of an AR game is PokemonGo.

The combination of AR and VR creates MR. The immersive environment is created by combining the actual and virtual worlds. Human-machine interactions are the foundation of how XR technology functions. In the metaverse, various ideas coexist simultaneously, creating a new world with modern dwellings, offices, traffic, and other amenities.

After cell phones became widely available at the end of the 1980s, communication standards ranging from 2G to 4G and the recently introduced long-term evolution changed in response to the rising trend in the number of smartphones and smart gadgets. The concept of Internet was experiencing a boom during this time, driven by ideas like the IoT and predictions that there will be 20 billion linked gadgets by 2020. The Federal Communications Commission has endorsed the latest new generation of communication standards, 5G, which is making its way into the market of metaverse to make options for its implementation on a massive scale.

The metaverse is formed by encompassing different vast networks and virtual immersive 3D environments. Avatars are the names given to the users of this virtual environment. As depicted in Figure 5.3 [16–18], various AR, VR, and XR platforms are combined to create an immersive experience. In an immersive setting, avatars are 3D icons that represent user accounts. Advanced computer games frequently employ them. Avatars will eventually be able to mimic users' movements and actions after metaverse is widely used.

Technologies used in metaverse demand advanced high-speed internet with minimised latency. With extremely high competence and very little latency, 5G enables speeds of up to Gbps [19, 20]. It won't be long before 5G totally replaces 4G, as it has already been embraced by several wealthy countries. By focusing on the extremely high repetition range of millimetre waves, 5G resolves all the problems of 4G. Despite having significant negative effects on the human body, millimetre waves' speed and scalability appear to overcome all of the metaverse's latency problems. Software-defined networks (SDN), user-plane access of the core network closer to the base station, and network functions virtualisation are some of the goals of 5G [18, 21, 22]. 5G can be divided into three groups based on the frequency bands: high-speed band, which has millimetre waves in the frequency range of 24–39 GHz, with a high speed of 20 Gbps and is best for applications involving AR and VR necessitating extremely high internet speeds. Low-speed band, which has frequency below 1 GHz, offers extended coverage and is widely used in rural areas. Mid-speed band, which includes millimetre waves having a frequency of 1–6 GHz, has high speed and is used for metropolitan networks. Various AR games like PokemonGo are anticipated to have improved quality of experience thanks to 5G. Currently, a project called HoloVerse is testing the 5G network in order to deploy several metaverse-related services [20, 23, 24].

In the metaverse, new technologies like edge computing can be used for data offloading and work distribution. Although data offloading refers to the switching of calculations and other jobs to nearby devices capable of doing them in real-time, tasks allocation is important in pushing the tasks to the edge of the network

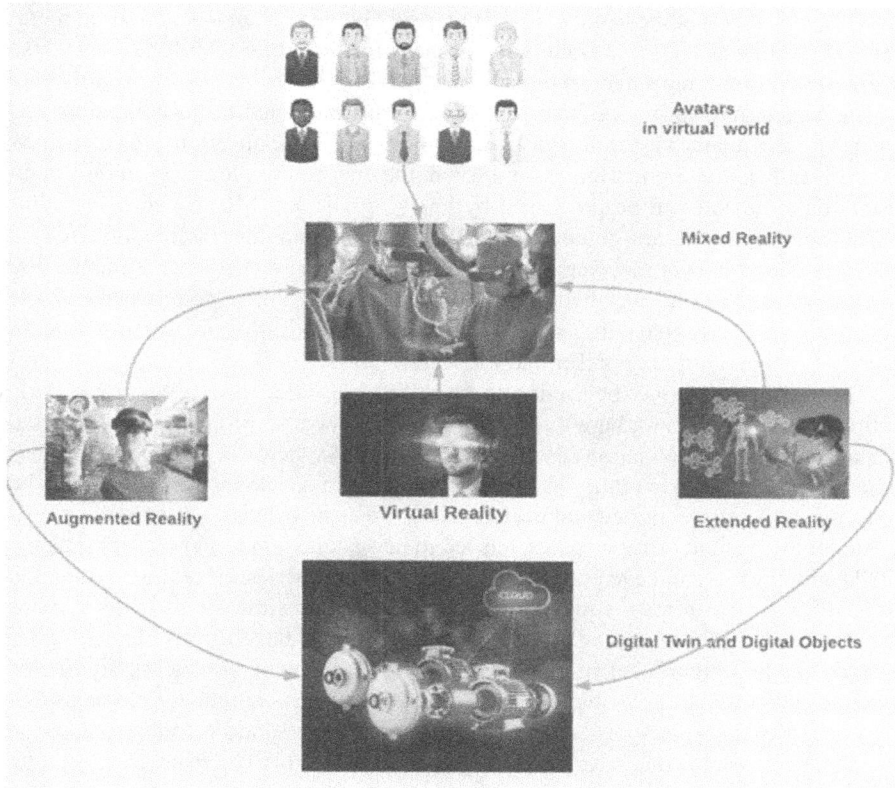

FIGURE 5.3   Enabling technologies for metaverse space.

and keeping resources in close vicinity of edge nodes. It offers high throughput, resource optimisation, and streaming and visualisation with minimal latency. Edge computing supports high-speed real-time computing activities by decentralising resources [25, 26].

The concepts of AR, VR, MR, and XR are further supporting technologies for the metaverse notion. Digital twins and completely immersive virtual worlds, which are essential components of the metaverse, are created with the aid of these technologies. Avatars, virtual space, and real space are all connected to one another through the use of XR and MR. Edge computing also serves the managing of processes and data analysis by leveraging data offloading. This helps to access the huge real-time multi-dimensional data created by various devices. This makes it a crucial part of metaverse because it helps mitigate latency to a larger level. Keeping data secure in the metaverse is another difficult task. By using blockchain, this problem can be solved. Blockchain can be used to overcome this problem. Blockchain employs decentralisation to guarantee data security by employing ledgers and avoiding the usage of a third party. In the metaverse, security and privacy can be attained by using digital assets and user-generated content. Some of the blockchain's components, such as NFT, which is specific to each user, are already being used in the actualised meta-verse. The user of NFT can trade and purchase real estate in the digital metaverse utilising smart contracts and cryptocurrency [27–29].

Metaverse may gain a lot from using deep learning which trains computers and smart devices using vast volumes of data. For various application areas, such as Smart Health Care, Smart Homes, and so on, the enormous volumes of multidimensional data created can be processed by applying DL and ML. By utilising computer vision and the brain-computer interface, human-computer interaction (HCI) is also an integral part of metaverse which allows hand-free movements without using input devices. The interface between robots and users will also be relatively simple because it will be simple to reflect body motions. For use in the metaverse, this technology is now undergoing preliminary research [30–32].

The metaverse is only beginning to be utilised in video games. Sandbox, Axie Infinity, and Decentraland are a few well-known metaverse projects with restricted scalability. Future applications of the metaverse in all spheres of life have enormous potential. (1) Immersive reality platforms, which are in developmental stages, can be used for shopping, giving virtual performances by using avatars, taking recreational virtual tours, virtual office cultures, has led to new social bonds and content sharing. (2) The metaverse can have potential application in industries for taking automation to next levels. It can also reduce the time needed to run simulations. Digital twins can be used in the metaverse to lower the quality-related hazards. (3) Metaverse has already made a significant impact on the gaming business, producing significant revenue, and it will do so going forward by incorporating more realistic elements that rely on multisensory responses. Roblox, Horizon Worlds, and Fortnite are a few of the metaverse games that have previously been released. (4) The metaverse has a lot of potential applications in the fields of education, health, and smart cities [4, 33–35].

The deployment of the metaverse is fraught with difficulties. Before the full deployment of metaverse, it is imperative that these difficulties be resolved. Several of the significant ones are [18, 25, 26, 29]:

- One of the biggest challenges is storing the multi-dimensional data that is produced by the numerous metaverse apps. This may have an impact on the system's capacity and use. Effective methods should be developed to make use of the huge amounts of data collected since it is crucial to use historical data to predict how systems will behave. In order to maintain sustainability, data pertaining to the insertion and deletion of new avatars must be handled carefully.
- Headaches, physical exhaustion, slight motion sickness, and postural problems can all be brought on by various metaverse devices, such as headphones and other sensing equipment. These gadgets can also lead to psychological problems in people like addiction and cyber syndrome.
- With 4G speed, it's difficult to render quickly or stream large volumes of data quickly in real-time. Users are prevented from having a fully immersive experience as a result of latency and delay.
- Due to the extreme intricacy of human activity, it is difficult to feel certain sensations in the metaverse, such as smell, wind, and mood fluctuations.

The real-like experience is limited in the metaverse due to complexity restrictions, the fact that humans can take on multiple identities, and other things.

- To make the metaverse a notion that is understood everywhere, it is crucial to design a cooperative open-source system that is compatible regardless of the vendor.
- Since the metaverse deals with the Avatar's specific data, security and confidentiality are crucial. Data sent across the network must be concretely authenticated and encrypted using blockchain technology and other tools. To stop crime and theft in the metaverse, a suitable framework is needed that will enforce rules and regulations and offer an ethical conscience.
- For the metaverse to be accessible to everyone, it is also necessary to optimise the price and size of equipment like headsets and eye-worn lenses. For people who have problems with their sense of sight or other birth-related impairments, there is an appetite for alternate methods to connect.

ORGANIZATION

The remaining of chapter is organised as follows: The work that has been done so far to address the latency and performance challenges in the metaverse is illustrated in Section 2. The architectural framework for metaverse is also discussed in this section. Section 3 discusses the literature review of the papers related to metaverse and edge computing. This section also highlights how literature on the metaverse has been impacted by 5G and edge computing. The unique approach used to identify research gaps and goals for more study in this area is outlined in detail in Section 4. This part encompasses the problem statement and the desired outcomes. The findings of the literature survey are described in Section 5. In Section 6, the chapter's conclusion and future focus are addressed.

LITERATURE REVIEW

Metaverse has a significant impact on different domains like gaming, social interaction, medical consultation, digital currency, and revenue generation, hence producing significant revenue. IoT by the use of wireless sensor networks (WSN) can help to accelerate the impact of metaverse by incorporating more realistic elements that rely on multisensory responses [33–35]. The metaverse has a lot of potential for use in different future applications which has been studied by different researchers given below:

In [36], Yuxuan et al. explored how 5G will affect the metaverse's implementation and suggested the Mobile Edge Assisted Metaverse Light-field-video System (MEAMLS). By combining edge computing with 5G, the author contends that technical issues with data delivery over current mobile networks would be resolved. In addition to real-time multidimensional data collection, the suggested solution for 5G also supports immersive world viewing and content production for the metaverse. This is accomplished by combining an integrated edge server with a light-filed video (LFV) array that uses ML algorithms to encode LFV. By combining edge computing with a sparse reconstruction technique for content production in the metaverse, the latency is significantly reduced, and extended reality is created. For the implementation of scalable, reliable edge service solutions, more research is necessary. Additionally, the suggested method has not been evaluated for huge networks incorporating real-time multidimensional data.

In Ref. [37], Fengxiao et al. explored the effects of 5G on metaverse and demonstrated the utilisation of 6G for metaverse's upcoming applications. The author claims that in order to enable a completely immersive user experience and a large number of users simultaneously, metaverse needs seamless connectivity, which 5G can provide. Although the 5G network standard offers low latency, dependability, and security, it will eventually need to be upgraded as user numbers rise. The integration of blockchain, multi-access edge computing, intelligent sensing, and digital twins is covered in the study. Low latency and huge connection are two of the biggest obstacles to using 5G in the metaverse that are listed in the research. The study focuses on developing 6G to optimise the metaverse's capabilities. In Ref. [38], Peng et al. discussed how to manage the metaverse's multi-dimensional data volume in a 5G setting. Future metaverse apps must be highly reliable and have very little latency; hence, 5G must develop into 6G in order to fully deploy metaverse.

The safety of the user, fending off hostile data attacks, handling substantial amounts of historical data, and improving storage system performance are some of the other important issues covered in this study. Obi et al. investigated the effects of electromagnetic energy and 5G waves (both high wave and low wave) on human skin in Ref. [39]. The impact of radiation due to prolonged skin exposure is disastrous, despite the many benefits of 5G including better communication. According to the study, prolonged exposure to electromagnetic radiation weakens the anti-cancer defence in living tissues. The analysis focuses on the importance for more comprehensive studies on 5G before it becomes fully operational all over the world. In Ref. [40], Zhaohui et al. put forward a mobility-aware metaverse optimisation with 5G. Because the metaverse demands high energy, quick caching, and consumer-rich background material, this chapter supports edge-caching and optimisation using 5G.

The balance of service latency, physical mobility, power consumption, and service decomposition are among the several elements taken into account. Making decisions using the suggested solution is ideal. The suggested approach can reduce delays by anywhere between 11.8% and 35.6% on average without consuming more resources or lowering user experience standards. Technology, virtual environments, and other metaverse research difficulties were examined by Lik-Hang et al. [41]. This chapter portrays the metaverse as a vast, persistent technology with enormous potential that is significantly influenced by advancements in 5G, ML, and XR. Eight key technologies—AR, MR, VR, XR, HCI, edge computing, ML, and 5G—are included. The main obstacles to the implementation of metaverse, according to the report, are latency and network scalability. In Africa, 5G bases for metaverse were investigated by Vuyelwa et al. [42]. The report outlines the difficulties faced by Africans in the implementation and formulation of concepts for the future workforce. The demand for immersive technology and a metaverse in Africa is therefore urgent, necessitating 5G's rapid, dependable transmission. The idea was put into practice at a university and given the label of metaversity to give people in remote locations access to educational resources. With financial support from organisations like Microsoft, Facebook, and

Nokia, a bottom-up strategy was used to put this idea into practice. To fully utilise metaversity, other technologies and ideas will need to be added in the future.

Sahraoui et al. in Ref. [43] studied response of edge computing in latency reduction in metaverse. The research suggests decentralised visualisation solutions that have low latency. The use of edge devices in distributed computing to reduce processing costs supported by a fog and edge-based hybrid model is proposed. The performance and cost optimisation are greatly enhanced by carrying out the physics simulation at the end devices. According to the simulation results, latency is reduced by 50% compared to other methods. Other open challenges in the metaverse have been identified, including user safety, privacy, and addiction. The function of 5G, edge computing, AI, and blockchain in the metaverse was researched by Pham et al. in Ref. [44]. Since edge computing and the cloud are combined, edge computing has been seen as the foundational technology for 5G. Edge computing places a strong emphasis on user proximity, which reduces network operations, lowers latency, and improves user experience. New Play to Earn games (P2E) governed by security from blockchain, web 3.0 which support token-based economic currencies like Bitcoin, Ethereum, and so on are just a few examples of the various applications given by edge computing and 5G in metaverse.

Yang et al. in Ref. [45] examine the metaverse's high streaming and data-intensive network needs. On the basis of mathematical interpretations, a cloud network flow model was suggested. However, the paper did not go into great detail about job distribution, latency control, or offloading.

Wang et al. [46] discussed the function of MEC (mobile edge computing) and fog computing in the metaverse implementation. Task distribution is done using mobile augmented reality (MAR) blockchain and edge computing in the metaverse.

Distributed computing was suggested as a remedy for metaverse issues by Yuna et al. in Ref. [47]. Through rendering, which calls for a variety of resources and their decentralisation, the immersive and engaging user experience is achieved. In order to manage real-time computations, the study suggests a framework for metaverse applications. A blockchain database is also used by the framework.

METHODOLOGY

A comprehensive systematic literature review (SLR) was carried out to investigate how 5G will help overcome the problems of scalability and high latency in metaverse. The research articles used for the study were sourced from a variety of sources, including Google Scholar, IEE Explorer, Elsevier, Hindawi, Taylor & Francis, and others. The search terms used were: "Meta AND "5G" AND "Edge computing", "Metaverse AND Challenges", "Issues in Metaverse Implementation", "Metaverse and Latency", and "Metaverse AND "decentralisation".

Research papers from 2018 and later were taken into consideration because the idea of the metaverse has advanced more in recent years.

A total of 120 articles, which we rated based on their high calibre, focus on the metaverse, and 5G. The study failed to include any articles with access limitations. Papers with restricted access and non-English literature were also excluded from consideration for study. Figure 5.4 illustrates the selection process for the

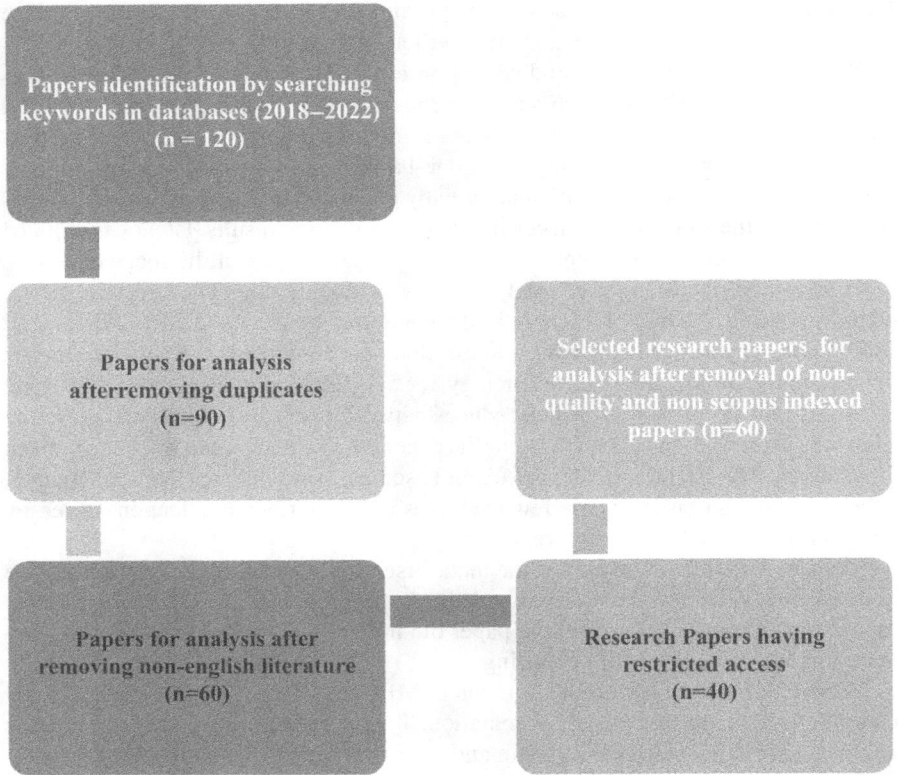

FIGURE 5.4 Procedure used for selecting research papers for SLR.

30 high-quality conference papers that were chosen for review after duplicates and low-quality conference papers with fewer than four pages were eliminated. The majority of the studies failed to address crucial problems as well as challenges associated with the implementation of the metaverse.

It was also noted in SLR that the dispersion of research publications across years on metaverse has increased. The number of research publications devoted to metaverse and related research issues has significantly increased till 2022, as shown in Figure 5.5, from a relatively low number in 2018. Despite the fact that the number of publications has significantly increased, there is a need for in-depth investigation to address the challenges faced.

Various research gaps were discovered after carefully examining the literature, and they are detailed below [36–47]:

1. **Decentralisation**: The metaverse requires real-time, high-speed data processing. A secure decentralised system built on blockchain and distributed computing is required to make immersive world applications available to everyone in the world. Task distribution is how the peer-to-peer (P2P) paradigm operates [48, 49].

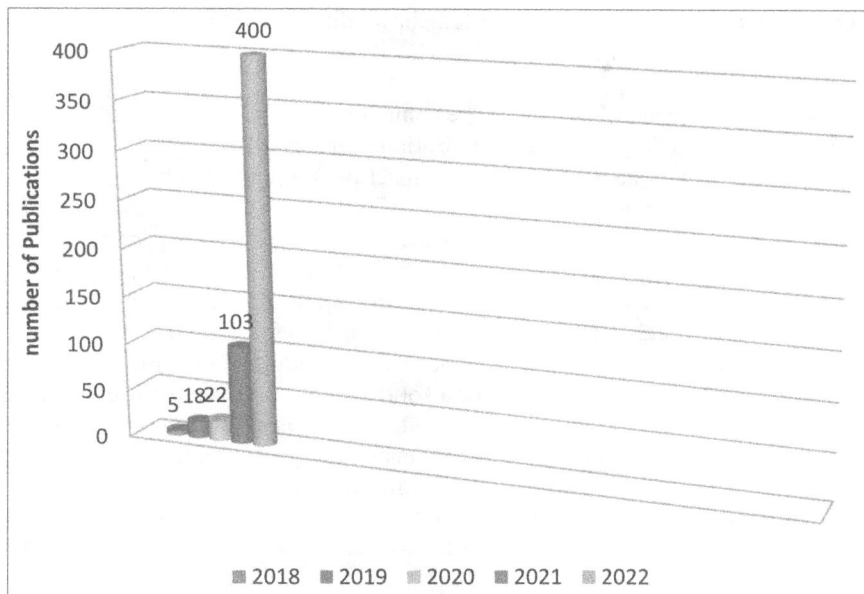

FIGURE 5.5   Research paper publications on metaverse and edge computing over the last five years.

2. **Reducing latency**: In the metaverse application, latency is a key factor in the immersive experience. Utilising integrated communication models and edge computing strategies, ultra-low latency can be attained. To design a system for no latency in the metaverse, offloading methods can be used with caching choices [50, 51].

3. **Rules and regulations setting**: By opening up new channels for socialisation and communication, metaverse will enhance human experiences. However, identities in the metaverse can be compromised and exploited for illegal purposes. There is sensitive data everywhere in the metaverse, and any intrusion could cause sensitive data to leak. The right definitions of rules and regulations are required. Users' safety is a challenge in these situations. To avoid any risk, strong norms and regulations must be put in place [52, 53].

4. **Connectivity and scalability**: In the field of virtual world technology, scalability is a significant problem. Scalability problems must be solved if the vast metaverse is to be created. Implementing edge and fog computing in conjunction with 5G technologies can address this.

5. Redesigning the protocols for data exchange between peer networks is also necessary [52–54] to enable multi-dimensional data. Cost, management of currencies and payments, device size, and data storage are further areas where there is a lack of research in the literature connected to the implementation of the metaverse [49–54].

On the basis of the challenges enlisted above, different (RQ) research questions were formulated which can be enlisted as:

RQ1: What is the significance of edge computing and 5G in metaverse?
RQ2: What is the impact of decentralisation in metaverse?
RQ3: How is meta keyword going to impact the research trends in different domains of computing?
RQ4: What are the possible applications of metaverse in the near future?

By overcoming network latency and centralisation difficulties, the metaverse may be completely realised and has enormous potential for the future of society. In the case of scalable networks, the current research and methodologies for latency reduction in the metaverse do not offer a solid answer. Because metaverse cannot exist in reality, a sturdy solution is required, as well as extremely fast speed and very little latency for an immersive experience. Due to the significant economic potential, there is a lot of room for research in this field, although there have been very few publications in recent years. Researchers have offered various ways; however, they are unreliable and don't handle additional issues like security and privacy.

LATENCY REDUCTION IN METAVERSE BY USING EDGE COMPUTING

In order to realise the full potential of metaverse, there is a need for ultralow latency, interoperability, security, and privacy. Currently, different applications based on metaverse use centralised clouds for data storage. However, they are unreliable and don't handle additional issues like security and privacy. Moreover, this also requires high-speed internet for running 3D animations, emulations, and collision detection. Higher number of users also a need to be incorporated which is limited in cloud-based server computations. The aforementioned issues restrict the implementation of a complete universally virtual world. In such scenarios, edge computing and fog computing can prove their mettle. The heavy computational load faced by a single server can be distributed to edge nodes situated near the end user, hence reducing the latency and improving the quality user experience [55–57].

The end devices perform tasks such as collision detection or physics emulation. By using edge computing in the metaverse, the edge devices are responsible to perform computation tasks related to the avatar's physical movement such as its momentum, mass, and physical forces of surrounding entities. As users traverse and teleport through virtual realms, embodied by their avatars, there arises a need for seamless correlation between these users in the physical realm and their virtual counterparts within the vast virtual universe. This notion of mapping the physical to the digital is not exclusive to the metaverse; it finds application in various other technologies like cyber-physical systems, digital twin, human-in-the-loop, hybrid human-artificial intelligence, and human-robot interaction, among others. These technologies will inevitably intertwine with the metaverse. However, the metaverse extends that to larger scale by mapping and simulating all our daily life activities to the cyberspace, and enriching such mapping with an immerse and interactive user experience. To achieve this purpose, the metaverse must operate in multi-layer architecture using edge computing [58–60].

Edge computing is a technology that brings computation and data storage closer to the location where it's needed, reducing latency and improving the overall user experience. While the metaverse and edge computing were not explicitly linked concepts at that time, it's possible that the two technologies could have a role in shaping the future of digital experiences and virtual worlds. There are different challenges which can be reduced considerably by using edge computing in metaverse which include the following [61–64]:

1. Reducing latency: In the metaverse, real-time interactions and responsiveness are critical for creating a seamless experience. Edge computing can help reduce latency by processing data closer to the end users, thus minimising delays and improving the overall performance of virtual environments.
2. Scalability: As the metaverse grows and attracts more users, the demand for computing resources will increase significantly. Edge computing can distribute the computational load across a network of edge nodes, enabling the metaverse platform to scale more efficiently.
3. Data processing and analytics: Edge computing can handle data processing and analytics tasks within the metaverse. As users interact with digital objects and environments, vast amounts of data are generated. Edge computing nodes can process this data in real-time, extracting insights and making dynamic adjustments to the virtual world.
4. Enhancing immersion: The metaverse aims to provide a highly immersive experience. Edge computing can facilitate this by enabling more sophisticated rendering and graphical capabilities at the edge nodes, resulting in higher quality visuals and more realistic virtual environments.
5. Privacy and security: Edge computing can enhance privacy and security in the metaverse. By processing sensitive data locally at the edge, personal information can be kept closer to the user and be subject to stricter control and encryption measures, reducing the risk of data breaches.
6. Decentralisation: The metaverse's vision often involves decentralisation to empower users and prevent single points of control. Edge computing aligns with this principle, as it distributes computation across a network of nodes, avoiding a central authority for data processing and storage.
7. Real-world integration: Edge computing can facilitate the integration of real-world data into the metaverse. For example, data from IoT devices and sensors could be processed at the edge and incorporated into the virtual environment, creating more dynamic and interactive experiences.

In the context of the metaverse, edge computing can support various applications and services. By bringing computational resources closer to the edge of the network, edge computing can reduce the reliance on centralised cloud servers and enable localised processing and data storage. This can result in faster response times and a richer, more interactive metaverse experience for users [65].

Edge computing can also support the distribution of computing resources across different geographical locations, helping to decentralise the metaverse infrastructure and reduce congestion on the network. This can ensure smoother and more reliable interactions within the metaverse environment [61–66].

Edge nodes are essential for the successful implementation of a metaverse. These nodes serve as critical points of the distributed computer network, enabling low latency, high-speed data transmission, and storage [1]. Here are some technologies required for using edge nodes in the metaverse [67–71]:

Edge computing: Edge computing is a distributed computing paradigm that brings computation and data storage closer to the edge of the network, reducing latency and enabling real-time processing. It is a critical technology required to support the processing and storage requirements of the metaverse.

5G with multi-access edge computing: The 5G network infrastructure with multi-access edge computing support enables edge devices such as smartphones, servers, and other smart devices to communicate with the centralised servers. This architecture ensures faster data processing, low latency transmission, and improves the overall performance of the metaverse.

Decentralised storage systems: Decentralised storage systems, such as P2P networks, are required for storing data and digital assets generated within the metaverse. These systems ensure that data remain persistent and highly available to users, even if nodes go offline.

Blockchain technology: Blockchain technology can ensure the security and integrity of data and digital assets within the metaverse. It facilitates secure and transparent exchanges, storage, and management of digital assets through smart contracts. Blockchain technology also enables users to verify ownership and transfer of digital assets, ensuring a trustless and secure ecosystem.

Additionally, edge computing can help address the immense data processing requirements of the metaverse. The metaverse generates massive amounts of data from user interactions, VR experiences, and artificial intelligence (AI) components. Edge computing allows for efficient data processing at the edge devices, reducing the burden on the central infrastructure and enabling faster decision-making.

Overall, edge computing is a key technology for supporting the metaverse by enabling low-latency interactions, distributed computing capabilities, and efficient data processing. It helps overcome the limitations of traditional cloud-based infrastructure and enhances the overall user experience in the metaverse.

The metaverse represents the next evolutionary stage of the 3D Internet, seamlessly integrating immersive and interoperable ecosystems accessible through user-controlled avatars. To achieve this, a large-scale mapping of entities from physical and social spaces to the virtual universe is required. Various cutting-edge technologies serve as the foundation for the metaverse, enabling its functionalities and experiences.

1. **Blockchain:** The metaverse's economy involves trading virtual assets, and in some cases, interactions with the real economy occur. To secure and manage the significant amount of big data resulting from virtual goods and services trading, persistent and transparent trading mechanisms are needed. Blockchain technology plays a crucial role in this regard, providing security and preserving the value of digital assets and services. By decentralising trading, blockchain facilitates P2P digital asset exchange, with one of the most promising applications being NFTs used to mark ownership of digital goods.

2. **AI technologies:** AI empowers the metaverse in various aspects. Deep learning and machine learning algorithms excel when learning from vast amounts of data, which metaverse activities can generate. AI plays a crucial role in creating realistic virtual 3D worlds and automating content generation. For example, AI-based frameworks like GANverse3D enable content creators to generate virtual replicas of real objects automatically, complete with lights, physics models, and PBR materials. AI research superclusters, such as Facebook's RSC, focus on leveraging AI to create metaverse scenes and content, enhancing the overall experience.

3. **Beyond 5G/6G networks:** Navigating the metaverse smoothly through VR technology heavily relies on low network latency. As VR technologies demand very short delays, communication with metaverse servers becomes crucial. Metaverse activities generate massive amounts of big data, including social communications and high-resolution interactive 3D animations, requiring high network bandwidth. Beyond 5G and 6G networks can potentially meet these communication requirements, enabling real-time, ubiquitous, and ultra-reliable communications for the vast number of metaverse devices.

4. **XR technologies:** XR technologies encompass VR, mixed reality (MR), and AR. XR head-mounted helmets and displays serve as the primary access points to the metaverse. These devices offer users real-time immersive and interactive experiences through multi-sensory large-scale 3D modelling. Haptic sensors and IoT smart sensors contribute to environment digitisation by sensing surrounding objects, enhancing the users' engagement with the virtual environment. XR devices provide a more comprehensive and realistic access point to the metaverse compared to traditional screen-based devices like laptops or smartphones.

The integration of VR, AR, XR, and DT into the metaverse in conjunction with existing networks gives rise to a range of challenges encompassing interconnectivity, communication, quality of service (QoS), and security. The current applications built on 3G and 4G frameworks might not be adequate to meet the demands for high-quality service in the future metaverse. To address this concern, 6G communication is anticipated to introduce multiple frequency bands within the electromagnetic spectrum, incorporating diverse radios like millimetre wave (mmWave) and Terahertz (THz) communication. Additionally, 6G is expected to capitalise on non-orthogonal multiple access to enhance bandwidth capacity and spectral efficiency by employing additional power levels. In conclusion, the metaverse relies on a convergence of advanced technologies to create a seamless and immersive digital space, offering users unprecedented experiences and interactions within a vast interconnected virtual universe.

ROLE OF 5G AND 6G IN METAVERSE APPLICATIONS

5G technology is a fundamental enabler of the metaverse, providing the necessary infrastructure for real-time, immersive, and interactive experiences and fostering the development of innovative applications and services within this evolving digital landscape. 5G technology plays a crucial role in enabling and enhancing various aspects of metaverse applications. 5G could provide the speed and power necessary to support the ongoing development of the metaverse, allowing digital worlds to function and enabling exciting uses like metaverse VR, metaverse XR, and Web 3.0 [72]. While 4G networks cannot handle the large file sizes associated with metaverse applications, 5G networks provide peak upload and download speeds that are significantly faster than 4G, making it possible to handle these bigger files and support the uninterrupted function of digital communities within the metaverse. 5G networks use a wide range of protocols, including the radio access network protocols such as 5G New Radio, as well as network protocols like 5G System. Meanwhile, the protocols used by the metaverse vary depending on the specific applications, hardware, and software used. Some of the protocols seen in this context may include VR-specific protocols like OpenVR [1], XR-specific protocols such as WebXR [2], as well as real-time communication protocols like WebRTC [73].

5G and metaverse technology share a symbiotic relationship, where 5G can offer the high data rates, low latency, and network-slicing technology required to enable fully immersive experiences in the metaverse [69–71]. 5G networks operate using several technological specifications, including enhanced mobile broadband (eMBB), ultra-reliable low-latency communication (URLLC), and massive machine-type communication (mMTC). These technologies are key to enabling faster data transfer, lower latency, and higher bandwidth requirements that the metaverse requires. eMBB provides high-speed, low-latency connectivity to enhance digital experiences and enable users to interact in virtual environments. URLLC improves response times to guarantee near-instant connectivity required for mission-critical applications, such as remote surgery and real-time gaming. On the other hand, mMTC provides users with the ability to connect a vast number of devices simultaneously with lower power consumption. However, 5G technology is still in its early stages and relies on emerging technologies and standards for its implementation, including the use of mmWave and sub-6 GHz radio frequencies for communication. The future metaverse, enabled by 5G and 6G technology, will require communication with minimal delays [72]. However, a challenging issue is the limited computational capabilities of current hardware and software. To address this, more powerful computation entities will be needed to handle the data transmission, reception, and synchronisation of sub-metaverses, thereby fulfilling the network and user needs. Building large-scale complex digital twin and digital human spaces also presents fundamental problems such as system power, heat, and fluid management, requiring collaborative efforts from various stakeholders. To ensure a computation-friendly paradigm in the metaverse, numerical calculations, simulations, and visualisations' use cases should be employed. Developing computation-friendly hardware, along with edge and cloud computing infrastructure, can significantly improve intra-system and inter-system latency while achieving better power consumption statistics. For instance, enhancing GPU computational capability can elevate the visual impact of the metaverse and cloud games, allowing stakeholders to model more realistic settings and objects. Furthermore, leveraging edge and cloud computing can enhance the performance of terminal equipment like networked devices and data centres, as data won't need to be constantly transferred back to the central server. Consequently, the overall processing speed increases, resulting in reduced response times for users. In the near future, it is predicted that the metaverse technology paradigm will witness a substantial increase in various types of connectivity, such as machine-to-sensor, sensor-to-sensor, machine-to-machine, and sensor-to-cyber. By the year 2030, this connectivity is projected to be 670 times greater than that of 2020. This rapid development in metaverse technology has spurred the academic community to consider the technical requirements for addressing a range of network problems, including spectrum allocation, communication efficiency, and energy efficiency. Current 5G technology already enables mMTC and eMBB connections of up to 20 Gbps. However, due to the expected exponential growth of metaverse technologies, it is anticipated that 5G will reach its limitations by 2030. Therefore, the development of 6G technology becomes essential for the advancement of the metaverse [69–73].

The advent of 6G technology is anticipated to bring significant improvements in data transmission rates, potentially achieving speeds of up to 1 Tbps (and possibly even up to 10 Tbps). However, both 5G and 6G technologies currently employ millimetre-wave transmission within the 20–100 GHz range, which poses various communication challenges like modulation, phase noise, non-linear power amplifiers, and hardware and software issues with analogue-to-digital converters. These challenges are primarily due to the high transmission and communication speeds of these technologies. To address these hurdles, experts believe that in 5G/6G-enabled communication, the frequency band would need to exceed 100 GHz and possibly extend to a few THz.

Managing communication processes in the future metaverse technology will require defining high data rates while considering the diverse range of services that clients are expected to demand. These services include AR, MR, human nano-chip implants, VR, autonomous systems, connected robotics, smart markets, online gaming, online ordering, and more.

With the induction of 6G technology, data transmission rates are expected to see significant improvement, potentially reaching up to 1 Tbps (and possibly up to 10 Tbps). However, both 5G and 6G technologies use millimetre-wave transmission within the 20–100-GHz range, which introduces various communication challenges such as modulation, phase noise, non-linear power amplifiers, and analogue-to-digital converter hardware and software issues, mainly due to their high transmission and communication speeds. To overcome these challenges, it is believed that in 5G/6G-enabled communication, the frequency band would surpass 100 GHz and could even extend to a few THz.

Managing the communication processes in the future metaverse technology would involve defining high data rates, considering the diverse services that clients are expected to demand. These services encompass AR, MR, human nano-chip implants, VR, autonomous systems, connected robotics, smart markets, online gaming, online ordering, and more.

Despite some limitations of 5G and 6G technologies, they are considered the most promising candidates to enable ultra-fast and reliable communication among the interconnected millions and billions of devices in the metaverse technology paradigm. As the metaverse continues to evolve, these advanced communication technologies will play a crucial role in shaping its future.

Despite some limitations of 5G and 6G technologies, they are considered the most promising options to enable ultra-fast and reliable communication among the interconnected millions and billions of devices in the metaverse technology paradigm. As the metaverse continues to evolve, these advanced communication technologies will play a crucial role in shaping its future.

In metaverse technology, the first step in understanding the technical details of the metaverse involves the definition of the concept. The metaverse is an immersive environment where users can simulate different aspects of real-world experiences, communicate with others, and interact with virtual objects [4]. Metaverse technology is based on cutting-edge hardware and software solutions such as VR, AR, and MR devices. The technology also relies on AI to enable realistic virtual avatars, natural language processing, and other interactive systems [5]. The nature of the metaverse requires ultra-low latency and high-speed connectivity to enable immersive, seamless, and interactive experiences. Furthermore, protocols such as OpenVR, WebXR, and WebRTC are used to enable specific functionalities in the development of the metaverse [71–74].

The key components interlinking 5G and 6G-enabled technologies with metaverse include small cells to manage network densification problems and enhance the coverage area with a high transmission rate. Small cells play a crucial role in managing network densification issues and expanding coverage with high transmission rates. For ultra-fast communication in these technology-driven environments, carrier aggregation is employed to offer clients and end-users higher bandwidth services by combining multiple carriers. However, carrier aggregation has implications as it operates across several frequency bands simultaneously, potentially impacting connectivity and services for end-users. To address this concern, a radio access network (C-RAN) is suggested as an alternative technology for 5G and 6G-enabled metaverse networks. C-RAN can resolve hardware-related issues associated with end-user connectivity and services. Nevertheless, it raises scalability concerns when large-scale network expansion is anticipated, leading to computation and calculation challenges on the client side.

SDNs and SDN controllers play a crucial role in client-side computations and calculations for 5G and 6G-enabled metaverse technologies due to their computation-friendly nature. However, SDNs are less effective in managing security concerns in heterogeneous networks like the metaverse as they struggle to establish a trustworthy relationship among user access applications and SDN controllers. Security issues in 5G and 6G-enabled metaverse technology must be addressed since these networks are vulnerable to various attacks, such as distributed denial of service, eavesdropping, message forgery, and device tampering. Improving security standards, policies, and protocols becomes essential to safeguard the vast number of interconnected actuators, cybers, sensors, IoT devices, users, and other relevant entities in the network.

The metaverse is a highly interconnected virtual world where users can interact, communicate, and engage with each other and digital objects in real-time. Here are some key roles of 5G in metaverse applications [71–75]:

1. **Low latency**: 5G networks offer significantly reduced latency compared to previous generations of cellular networks. In the metaverse, low latency is essential for creating a seamless and immersive user experience. Real-time interactions and responsiveness are critical, especially in VR and AR applications, where even the slightest delay can disrupt the sense of presence and immersion.

2. **High bandwidth**: Metaverse applications generate and transmit vast amounts of data, including high-definition 3D graphics, audio, and video content. 5G's high bandwidth capabilities support the smooth transmission of these data-intensive elements, ensuring users can enjoy high-quality, detailed, and interactive virtual experiences without buffering or lagging.

3. **Massive device connectivity**: The metaverse involves a multitude of connected devices, from VR headsets and smartphones to IoT sensors and smart objects. 5G's ability to support a massive number of connected devices allows for seamless communication and coordination between users and their avatars and enables IoT devices to contribute to the immersive virtual environment.

4. **Edge computing integration**: 5G networks can be combined with edge computing infrastructure, bringing computation and data storage closer to the users. This reduces the reliance on distant data centres, leading to lower latency and faster processing of data, which is crucial for real-time interactions in the metaverse.

5. **Enhanced VR/AR experiences**: 5G's high bandwidth and low latency enable more sophisticated and realistic VR and AR experiences in the metaverse. Users can interact with high-resolution 3D content, engage in real-time multiplayer VR gaming, and participate in immersive virtual events, all with minimal delay and high fidelity.

6. **Real-time collaboration and communication**: 5G facilitates real-time collaboration and communication between users in the metaverse. Whether it's participating in virtual meetings, attending virtual concerts, or collaborating on virtual projects, 5G's low latency and high bandwidth contribute to seamless and fluid interactions.

7. **Spatial computing**: 5G can enhance spatial computing in the metaverse. Spatial computing involves the understanding and processing of the physical environment and the users' movements and actions within it. 5G's low latency and real-time communication enable accurate and rapid updates to

the spatial data, ensuring that users' virtual representations (avatars) and objects in the virtual environment respond accurately to their real-world movements.

8. **Multi-user experiences**: 5G supports multi-user experiences in the metaverse, where large numbers of users can interact simultaneously in shared virtual spaces. The high bandwidth and low latency of 5G networks enable smooth communication and synchronisation between users, enabling complex and interactive social experiences.

The establishment and maintenance of connections between interconnected devices will pose numerous challenges in the realm of ultra-fast communication. Therefore, it is essential to reevaluate the current communication equipment, as well as the physical, network, and application layer protocols that will be employed in this technology. QoS plays a crucial role in 5G/6G-enabled systems. In the metaverse, lightweight devices like head-mounted displays, goggles, eyeglasses, and mobile devices enable the representation of virtual scenes. However, dealing with the ultra-fast communication of 5G and 6G technology introduces complexities in managing tasks on the client's devices. This is due to the integration of old and new hardware, leading to several issues, with QoS being one of them. Additionally, achieving high QoS standards and ensuring proper security protocols while streaming data become serious challenges, especially since current technologies rely on 3G and 4G communication paradigms.

Regarding system resources in 5G/6G-enabled metaverse technologies, careful optimisation becomes crucial due to the involvement of various components, including actuators, sensors, base stations, servers, and humans. Effectively managing resources among interconnected devices necessitates giving significant attention to the functionalities of these devices to ensure seamless interoperability. The support for fast data is a critical aspect of 5G and 6G communication systems, making resource allocation in metaverse technology, with regards to time and frequency during transmission, essential for reliable operation among interconnected devices. Moreover, efficient resource allocation is necessary to determine the burst frequency for network-friendly operation, considering the time intervals between transmissions from various periodic sensing devices [73–76].

DISCUSSION AND RESULTS

The following shortcomings of the current solutions for latency reduction and optimisation in the metaverse have been identified after careful reading and analysis of the literature: The many solutions offered by papers are based on LFV, which calls for large storage areas. The storage capacity needed for LFV is challenging to provide because there has already been an enormous amount of multidimensional data collected. LFV needs a tremendous amount of accurate, high-quality media-related information in order to function in metaverse. The expense of implementing LFV is likewise very significant. Several researchers advocate using 6G technologies to stream high-definition content in real-time in the metaverse.

As 6G technology is not yet present in the marker and its compatibility is not well understood, this also presents difficulties for the implementation of the metaverse. The expense of employing 6G in metaverse will likewise be significant. The health risks associated with 5G have also been noted by several researchers, yet these claims are debatable because they have been refuted by various international organisations following investigations. Different mobility-based optimisation strategies have also been put forth by researchers in order to decrease latency and boost throughput. But as the scale of the network grows, the designed approaches are unable to deliver excellent performance.

In short, the many optimisation approaches for eliminating latency utilising 5G and decentralisation in edge computing that have been developed so far are unreliable since they have not been deployed on larger user spaces and scalable networks. There is a need for more thorough research in this field to make metaverse a global reality in the near future because the research articles related to a latency reduction in metaverse are similarly scarce.

Based on the literature survey and bibliometric analysis performed, the solution to the different RQ's can be formulated as

RQ1: What is the significance of edge computing and 5G in metaverse?

Based on the search string, a total of 186 articles were found. On analysing the articles clearly, it was found that most of the research papers were aligned towards survey and review studies in computer science and engineering as shown in Figure 5.6. It can also be clearly seen that the leading countries focussing on overcoming metaverse challenges and its research are China and the United States, followed by other developed nations (Figure 5.7).

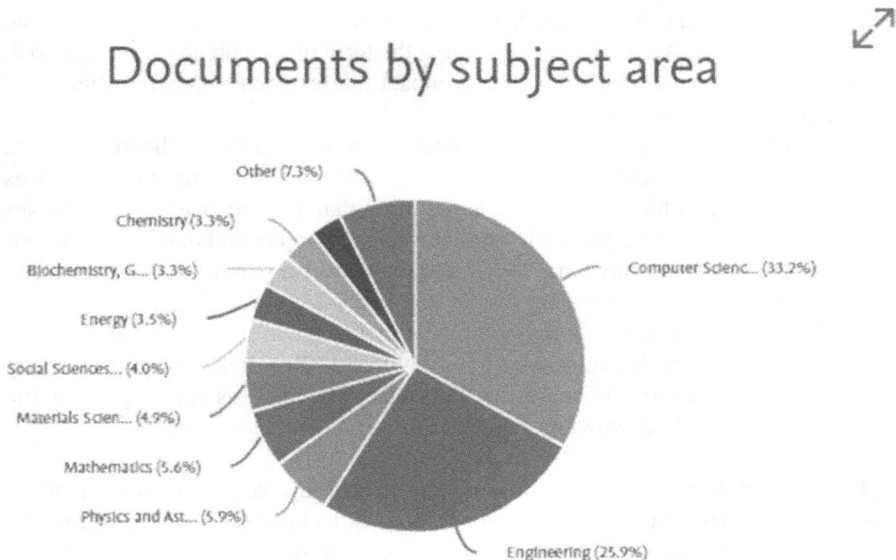

Documents by subject area

Other (7.3%)

Chemistry (3.3%)

Biochemistry, G... (3.3%)

Energy (3.5%)

Social Sciences... (4.0%)

Materials Scien... (4.9%)

Mathematics (5.6%)

Physics and Ast... (5.9%)

Computer Scienc... (33.2%)

Engineering (25.9%)

FIGURE 5.6   Research paper publications on metaverse, 5G, and edge computing.

Documents by country/territory

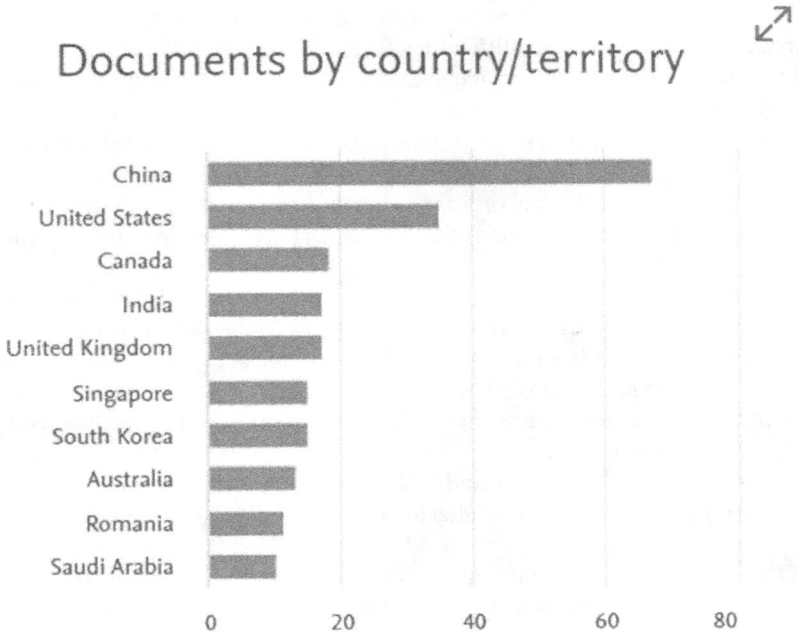

FIGURE 5.7   Leading countries focusing on research in metaverse and overcoming challenges.

RQ2: What is the impact of decentralisation in metaverse?

For analysing this research problem, a total of 91 articles were analysed. Based on the analysis, it was found that the implementation-based research papers were quite less as compared to the survey and review-based articles in the area of metaverse. This can be shown in the form of a graph given in Figure 5.8.

RQ3: How is meta keyword going to impact the research trends in different domains of computing?

After analysing the literature critically, it was found that in the initial years, the published articles in Scopus-indexed journals having meta as keyword was 6700 articles which has risen to about more than 65,000 articles. This makes the significance and popularity of metaverse even more elaborative. The trending increase in research articles related to metaverse over the last 25 years can be clearly shown in a graph given in Figure 5.9.

RQ4: What are the possible applications of metaverse in the near future?

Based on the literature survey, a total of 1900 articles are published on Scopus, related to applications of metaverse. The analysis can be given in the form of a graph as shown in Figure 5.10.

Based on analysis of literature, it is evident that metaverse has applications in all domains of life. However, most of the applications of metaverse are in alignment with computer science and engineering and require multiple technologies for implementation.

Documents by type

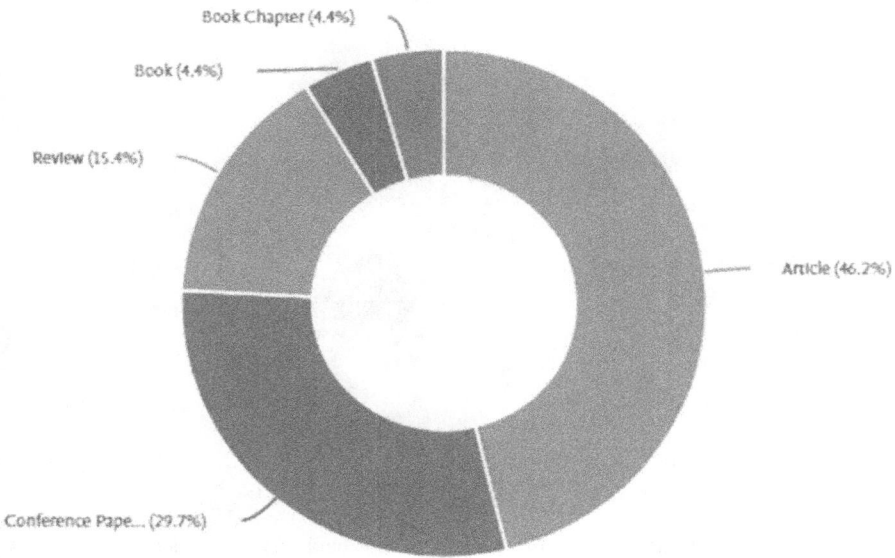

Book Chapter (4.4%)

Book (4.4%)

Review (15.4%)

Article (46.2%)

Conference Pape.... (29.7%)

FIGURE 5.8 Different types of articles related to decentralisation in metaverse.

Documents by year

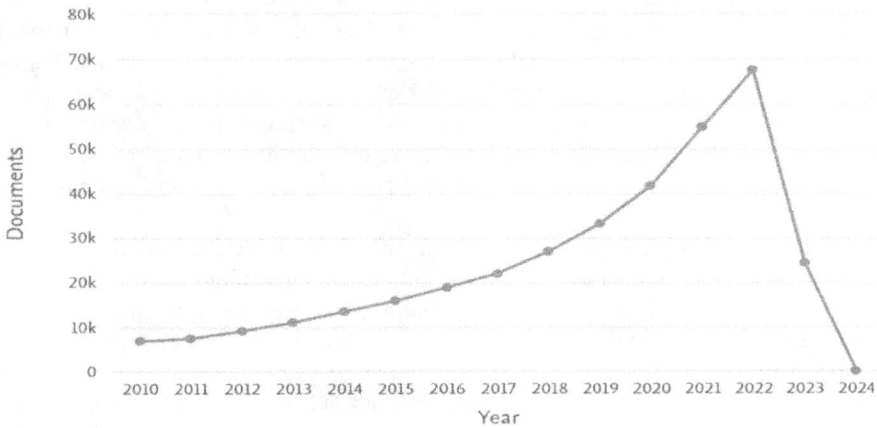

FIGURE 5.9 Different research article publications related to metaverse from 2010 onwards.

Documents by subject area

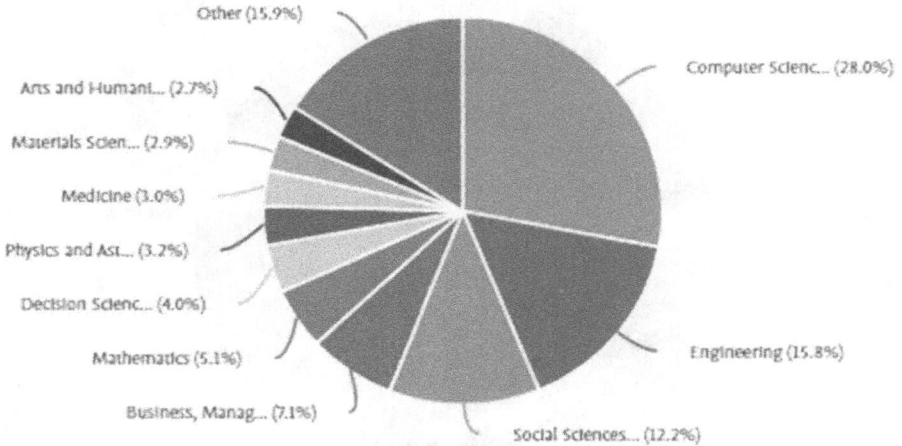

Other (15.9%)

Arts and Humani... (2.7%)

Materials Scien... (2.9%)

Medicine (3.0%)

Physics and Ast... (3.2%)

Decision Scienc... (4.0%)

Mathematics (5.1%)

Business, Manag... (7.1%)

Social Sciences... (12.2%)

Computer Scienc... (28.0%)

Engineering (15.8%)

FIGURE 5.10   different types of articles in different fields related to decentralisation in metaverse.

CONCLUSION

The metaverse, which will rule all facets of human life, is the internet's future. Numerous obstacles must be overcome in order to completely integrate metaverse. A few of these include the processing of growing data volumes and dimensions, seamless computation, security, and privacy, as well as full-coverage sensing, network scalability, cost, and high latency. In addition, metaverse requires a dependable, fast wireless network to deliver a top-notch user experience. We explored metaverse in this suggestion as a technology that will govern the future. It was discussed how the metaverse has changed over the past ten years. The five-layered architecture of the metaverse, together with its constituent parts and prospective applications, were examined.

In the metaverse, the goals for 5G and 6G communication paradigms are to achieve data transmission rates exceeding 1 Tb/s. It is important to acknowledge that the existing communication equipment, routing, and security protocols will likely be inadequate to handle the future metaverse technologies with such high data rates ranging from 1 Tb/s to 10 Tb/s, especially when integrating cutting-edge technologies like AR, VR, XR, and DT with the 5G and 6G communication frameworks. Current commercial wireless networks use the 3.4–3.8 GHz spectrum for broad signal coverage in various applications. However, efficiently utilising the spectrum in the future metaverse could prove to be a challenging task, particularly when dealing with the interoperability of new technologies and traditional communication platforms. Emerging technologies like AR, VR, XR, and DT will be capable of adapting to ultra-fast communication with Tb/s data rates, while existing technologies relying on 3G and 4G communication frameworks will struggle to keep up. Consequently, this creates numerous challenges for the research community, with security being one of the crucial issues that we will explore comprehensively in the upcoming sections, along with raising several questions in the open challenges section for further examination. To ensure the reliable operation of this emerging technology, multiple challenges need to be addressed collectively, ranging from general to complex issues. Existing IoT and WSN applications may not suffice for the complete execution of this technology, as it demands real-time intelligence, low latency, and high bandwidth. Although 5G and 6G technologies are considered high-speed communication paradigms for the metaverse, they are still in the emerging phase. To overcome the challenges and ensure reliable operation, new AI-enabled techniques and algorithms must be developed. By leveraging 5G and 6G technologies with these advancements, the communication of metaverse technology can effectively distribute services spontaneously and accurately. However, considering the future importance of this technology, it is also evident that attackers may target different vulnerability flaws to compromise client information and hijack application operations. Therefore, the metaverse technology will inevitably face new security and privacy challenges that could impact its applicability and scalability in the future [77].

The importance of IoT, WSN, 5G, and edge computing in resolving the metaverse's scalability and latency problems was examined. To examine every 5G and edge computing implementation on the metaverse, a thorough SLR was carried out. Based on this investigation, the results were analysed, and research gaps were identified. Different metaverse uses that might be possible were discussed. A five-year analysis of research trends on metaverse latency problems revealed an upsurge in publications in this field following Facebook's rebranding as Meta. Finally, goals were set and a suitable technique was proposed to realise the stated goals. The next stage will be to provide an algorithm and put the goals for eliminating latency into practice. The next step will be to propose an algorithm and put the goals into practice in order to get decentralised P2P networking in the metaverse to solve the speed and data handling problems.

REFERENCES

1. Deng, Y., Weng, Z., & Zhang, T. (2022). Metaverse-driven remote management solution for scene-based energy storage power stations. *Evolutionary Intelligence*, 1–12.
2. Mystakidis, S. (2022). Metaverse. *Encyclopedia*, 2(1), 486–497.
3. Aharon, D. Y., Demir, E., & Siev, S. (2022). Real returns from unreal world? Market reaction to metaverse disclosures. *Research in International Business and Finance*, 63, 101778.
4. Alvarez-Risco, A., Del-Aguila-Arcentales, S., Rosen, M. A., & Yáñez, J. A. (2022). Social cognitive theory to assess the intention to participate in the Facebook metaverse by citizens in Peru during the COVID-19 pandemic. *Journal of Open Innovation: Technology, Market, and Complexity*, 8(3), 142.
5. Njoku, J. N., Nwakanma, C. I., Amaizu, G. C., & Kim, D. S. (2022). Prospects and challenges of metaverse application in data-driven intelligent transportation systems. *IET Intelligent Transport Systems*, 17, 1–21.

6. Allam, Z., Sharifi, A., Bibri, S. E., Jones, D. S., & Krogstie, J. (2022). The metaverse as a virtual form of smart cities: Opportunities and challenges for environmental, economic, and social sustainability in urban futures. *Smart Cities*, *5*(3), 771–801.
7. Nath, K., Dhar, S., & Basishtha, S. (2014). Web 1.0 to Web 3.0-Evolution of the Web and its various challenges. In *2014 International Conference on Reliability Optimization and Information Technology (ICROIT)* IEEE, 86–89.
8. Tlili, A., Huang, R., Shehata, B., Liu, D., Zhao, J., Metwally, A. H. S., … Burgos, D. (2022). Is metaverse in education a blessing or a curse: A combined content and bibliometric analysis. *Smart Learning Environments*, *9*(1), 1–31.
9. Kerdvibulvech, C. (2022). Exploring the impacts of COVID-19 on digital and metaverse games. In *International Conference on Human-Computer Interaction* (pp. 561–565). Springer, Cham.
10. Cheng, R., Wu, N., Chen, S., & Han, B. (2022). Will metaverse be next G internet? vision, hype, and reality. *arXiv Preprint*, arXiv:2201.12894, 197–204.
11. Chen, T., Zhou, H., Yang, H., & Liu, S. (2022). A review of research on metaverse defining taxonomy and adaptive architecture. In *2022 5th International Conference on Pattern Recognition and Artificial Intelligence (PRAI)* (pp. 960–965). IEEE.
12. Far, S. B., & Rad, A. I. (2022). Applying digital twins in metaverse: User interface, security and privacy challenges. *Journal of Metaverse*, *2*(1), 8–16.
13. Huang, Y., Qiao, X., Wang, H., Su, X., Dustdar, S., & Zhang, P. (2022). Multi-player immersive communications and interactions in metaverse: Challenges, architecture, and future directions. *arXiv Preprint*, arXiv:2210.06802. https://ieeexplore.ieee.org/document/10368073/authors#authors
14. Zhou, B. (2022). Building a smart education ecosystem from a metaverse perspective. *Mobile Information Systems*, 1–10.
15. Jovanović, A., & Milosavljević, A. (2022). VoRtex metaverse platform for gamified collaborative learning. *Electronics*, *11*(3), 317.
16. Hendaoui, A., Limayem, M., & Thompson, C. W. (2008). 3D social virtual worlds: Research issues and challenges. *IEEE Internet Computing*, *12*(1), 88–92.
17. Pellas, N., Mystakidis, S., & Christopoulos, A. (2021). A systematic literature review on the user experience design for game-based interventions via 3D virtual worlds in K-12 education. *Multimodal Technologies and Interaction*, *5*(6), 28.
18. Park, S. M., & Kim, Y. G. (2022). A metaverse: Taxonomy, components, applications, and open challenges. *IEEE Access*, *10*, 4209–4251.
19. Kostoff, R. N., Heroux, P., Aschner, M., & Tsatsakis, A. (2020). Adverse health effects of 5G mobile networking technology under real-life conditions. *Toxicology Letters*, *323*, 35–40.
20. Karipidis, K., Mate, R., Urban, D., Tinker, R., & Wood, A. (2021). 5G mobile networks and health—A state-of-the-science review of the research into low-level RF fields above 6 GHz. *Journal of Exposure Science & Environmental Epidemiology*, *31*(4), 585–605.
21. Kaur, K., Mangat, V., & Kumar, K. (2022). A review on virtualized infrastructure managers with management and orchestration features in NFV architecture. *Computer Networks*, 109281.
22. Anderson, J., & Rainie, L. (2022). *The Metaverse in 2040*. Pew Research Center.
23. Clement, J. C., Indira, N., Vijayakumar, P., & Nandakumar, R. (2021). Deep learning based modulation classification for 5G and beyond wireless systems. *Peer-to-Peer Networking and Applications*, *14*(1), 319–332.
24. Eldowek, B. M., Bauomy, N. A. E. S., El-Rabaie, E. S. M., Abd El-Atty, S. M., & Abd El-Samie, F. E. (2021). A survey of 5G millimeter wave, massive multiple-input multiple-output, and vehicle-to-vehicle channel measurements and models. *International Journal of Communication Systems*, *34*(16), e4830.

25. Kim, W. S. (2021). Edge computing server deployment technique for cloud VR-based multi-user metaverse content. *Journal of Korea Multimedia Society, 24*(8), 1090–1100.
26. Ha, D. B., Truong, V. T., & Lee, Y. (2022). Intelligent reflecting surface assisted RF energy harvesting mobile edge computing NOMA networks: Performance analysis and optimization. *EAI Endorsed Transactions on Industrial Networks and Intelligent Systems, 9*(32).
27. Khan, L. U., Han, Z., Niyato, D., Hossain, E., & Hong, C. S. (2022). Metaverse for wireless systems: Vision, enablers, architecture, and future directions. *arXiv Preprint*, arXiv:2207.00413.
28. Gadekallu, T. R., Huynh-The, T., Wang, W., Yenduri, G., Ranaweera, P., Pham, Q. V., ... Liyanage, M. (2022). Blockchain for the metaverse: A review. *arXiv Preprint*, arXiv:2203.09738.
29. Rashid, M. M., Choi, P., Kwon, K. R., Kwon, S. G., & Lee, S. H. Emergence of the metaverse: How blockchain, AI, AR/VR, and digital transformation technologies will change the future world. *arXiv preprint*, arXiv:2207.00413. (pp. 266–268).
30. Aloqaily, M., Bouachir, O., Karray, F., Al Ridhawi, I., & El Saddik, A. (2022). Integrating digital twin and advanced intelligent technologies to realize the metaverse. *IEEE Consumer Electronics Magazine, 12*(6), 47–55.
31. Chengoden, R., Victor, N., Huynh-The, T., Yenduri, G., Jhaveri, R. H., Alazab, M., ... Gadekallu, T. R. (2022). Metaverse for healthcare: A survey on potential applications, challenges and future directions. *arXiv Preprint*, arXiv:2209.04160, vol 11, (pp. 12765–12795).
32. Wang, Y., Su, Z., Zhang, N., Xing, R., Liu, D., Luan, T. H., & Shen, X. (2022). A survey on metaverse: Fundamentals, security, and privacy. IEEE Commun. Surv. Tutor.
33. Duan, H., Li, J., Fan, S., Lin, Z., Wu, X., & Cai, W. (2021, October). Metaverse for social good: A university campus prototype. In *Proceedings of the 29th ACM International Conference on Multimedia* (pp. 153–161).
34. Suzuki, S. N., Kanematsu, H., Barry, D. M., Ogawa, N., Yajima, K., Nakahira, K. T., ... Yoshitake, M. (2020). Virtual experiments in metaverse and their applications to collaborative projects: The framework and its significance. *Procedia Comput. Sci., 176*, 2125–2132.
35. Garavand, A., & Aslani, N. (2022). Metaverse phenomenon and its impact on health: A scoping review. *Inform. Med. Unlocked*, vol. 32, 101029.
36. Chauhan, H. S., Babbar, H., & Rani, S. (2023). D2PG: Deep Deterministic Policy Gradient-Based Vehicular Edge Caching Scheme for Digital Twin-Based Vehicular Networks. *International Journal of Performability Engineering, 19*(5).
37. Tang, F., Chen, X., Zhao, M., & Kato, N. (2022). The roadmap of communication and networking in 6G for the metaverse. *IEEE Wirel. Commun, 30*(4), pp. 7281.
38. Peng, H., Chen, P. C., Chen, P. H., Yang, Y. S., Hsia, C. C., & Wang, L. C. (2022, August). 6G toward metaverse: Technologies, applications, and challenges. In *2022 IEEE VTS Asia Pacific Wireless Communications Symposium (APWCS)* (pp. 6–10). IEEE.
39. Agostino Di Ciaula., (2018). Towards 5G communication systems: Are there health implications?, *International Journal of Hygiene and Environmental Health, 221*(3), pp. 367–375, https://doi.org/10.1016/j.ijheh.2018.01.011.
40. Huang, Z., & Friderikos, V. (2022). Mobility aware optimization in the Metaverse. *arXiv Preprint*, arXiv:2209.08899, DOI: 10.1109/GCWkshps56602.2022.10008679.
41. Lee, L. H., Braud, T., Zhou, P., Wang, L., Xu, D., Lin, Z., ... Hui, P. (2021). All one needs to know about metaverse: A complete survey on technological singularity, virtual ecosystem, and research agenda. *arXiv Preprint*, arXiv:2110.05352.
42. Ruwodo, V., Pinomaa, A., Lannie, U. W. U., & Sutinen, E. (2022). 5G bases for a physical metaversity in Africa. In *Conference on M4D Mobile Communication Technology for Development* (p. 94).
43. Dhelim, S., Kechadi, T., Chen, L., Aung, N., Ning, H., & Atzori, L. (2022). Edge-enabled metaverse: The convergence of metaverse and mobile edge computing. *arXiv Preprint*, arXiv:2205.02764.

44. Huynh-The, T., Pham, Q. V., Pham, X. Q., Nguyen, T. T., Han, Z., & Kim, D. S. (2022). Artificial intelligence for the metaverse: A Survey. *arXiv Preprint*, arXiv:2202.10336 (vol 117), https://doi.org/10.1016/j.engappai.2022.105581.
45. Cai, Y., Llorca, J., Tulino, A. M., & Molisch, A. F. (2022). Compute- and data-intensive networks: The key to the metaverse. *arXiv Preprint*, arXiv:2204.02001. DOI: 10.1109/6GNet54646.2022.9830429.
46. Wang, Y., & Zhao, J. (2022). Mobile edge computing, metaverse, 6G wireless communications, artificial intelligence, and blockchain: Survey and their convergence. *arXiv Preprint*, arXiv:2209.14147. DOI: 10.1109/WF-IoT54382.2022.10152245.
47. Jiang, Y., Kang, J., Niyato, D., Ge, X., Xiong, Z., Miao, C., & Shen, X. (2022). Reliable distributed computing for metaverse: A hierarchical game-theoretic approach. *IEEE Trans. Veh. Technol*, 72(1), 1084–1100.
48. Ryskeldiev, B., Ochiai, Y., Cohen, M., & Herder, J. (2018, February). Distributed metaverse: Creating decentralized blockchain-based model for peer-to-peer sharing of virtual spaces for mixed reality applications. In *Proceedings of the 9th augmented human international conference* (pp. 1–3).
49. Frey, D., Royan, J., Piegay, R., Kermarrec, A. M., Anceaume, E., & Le Fessant, F. (2008, March). Solipsis: A decentralized architecture for virtual environments. In *1st International Workshop on Massively Multiuser Virtual Environments*.
50. Zhang, X., Huang, X., Yin, H., Huang, J., Chai, S., Xing, B., & Zhao, L. (2022). LLAKEP: A low-latency authentication and key exchange protocol for energy internet of things in the metaverse era. *Mathematics*, 10(14), 2545.
51. Van Huynh, D., Khosravirad, S. R., Masaracchia, A., Dobre, O. A., & Duong, T. Q. (2022). Edge intelligence-based ultra-reliable and low-latency communications for digital twin-enabled metaverse. *IEEE Wirel. Commun. Lett*, 11(8), 1733–1737.
52. Giannini, T., Bowen, J. P., Michaels, C., & Smith, C. H. (2022). Digital Art and Identity Merging Human and Artificial Intelligence: Enter the Metaverse. *Proceedings of EVA London 2022*, 1–7.
53. Bali, M. S., Gupta, K., Koundal, D., Zaguia, A., Mahajan, S., & Pandit, A. K. (2021). Smart architectural framework for symmetrical data offloading in IoT. *Symmetry*, 13(10), 1889.
54. Lamba, S. S., & Malik, R. (2022). Into the metaverse: Marketing to Gen Z consumers. In Applying Metalytics to Measure Customer Experience in the Metaverse (pp. 92–98). IGI Global.
55. Garg, D., Sidhu, J., & Rani, S. (2019). Improved TOPSIS: A multi-criteria decision making for research productivity in cloud security. *Computer Standards & Interfaces*, 65, 61–78.
56. Rezaei, A., & Noori, L. (2018). Novel low-loss microstrip triplexer using coupled lines and step impedance cells for 4G and WiMAX applications. *Turk. J. Electr. Eng. Comput. Sci.*, 26, 1871–1880.
57. Percaz, J. M., Chudzik, M., Arnedo, I., Arregui, I., Teberio, F., Laso, M. A., & Lopetegi, T. (2015). Producing and exploiting simultaneously the forward and backward coupling in EBG-assisted microstrip coupled lines. *IEEE Antennas Wirel. Propag. Lett.*, 15, 873–876.
58. Chinig, A., Errkik, A., Abdellaoui, L. E., Tajmouati, A., Zbitou, J., & Latrach, M. (2016). Design of a microstrip diplexer and triplexer using open loop resonators. *J. Microw. Optoelectron. Electromagn. Appl.*, 15, 65–80.
59. Keshavarz, S., Abdipour, A., Mohammadi, A., & Keshavarz, R. (2019). Design and implementation of low loss and compact microstrip triplexer using CSRR loaded coupled lines. *AEU-Int. J. Electron. Commun.*, 111, 152913.
60. Yang, T., Chi, P.-L., & Itoh, T. (2010). Compact quarter-wave resonator and its applications to miniaturized diplexer and triplexer. *IEEE Trans. Microw. Theory Tech.*, 59, 260–269.

61. Yang, T., & Rebeiz, G. M. A. (2016). 1.26–3.3 GHz tunable triplexer with compact size and constant bandwidth. *IEEE Microw. Wirel. Compon. Lett.*, *26*, 786–788.
62. Qian, J.-F., & Chen, F.-C. (2017). Wide stopband microstrip triplexer using common crossed resonator and uniform impedance resonator. *Prog. Electromagn. Res. Lett.*, *69*, 79–86.
63. Sugchai, T., Nattapong, I., & Apirun, C. (2015). Design of microstrip triplexer using common dual-mode resonator with multi-spurious mode suppression for multiband applications. *Appl. Mech. Mater.*, *763*, 182–188.
64. Shafiei, A., Jamshidi, M., Khani, F., Talla, J., Peroutka, Z., Gantassi, R., Baz, M., Cheikhrouhou, O., & Hamam, H. (2021). A hybrid technique based on a genetic algorithm for fuzzy multiobjective problems in 5G, internet of things, and Mobile edge computing. *Math. Probl. Eng.*, *2021*, 9194578.
65. Minh, Q. N., Nguyen, V.-H., Quy, V. K., Ngoc, L. A., Chehri, A., & Jeon, G. (2022). Edge computing for IoT-enabled smart grid: The future of energy. *Energies*, *15*, 6140.
66. Rezaei, A., Yahya, S. I., Noori, L., & Jamaluddin, M. H. (2022). Designing high-performance microstrip quad-band bandpass filters (for multi-service communication systems): A novel method based on artificial neural networks. *Neural Comput. Appl.*, *34*, 7507–7521.
67. Zhu, H. (2022). Metaaid: A flexible framework for developing metaverse applications via AI technology and human editing. *arXiv Preprint*, arXiv:2204.01614.
68. Zhang, T., Shen, J., Lai, C.-F., Ji, S., & Ren, Y. (2023). Multi-server assisted data sharing supporting secure deduplication for metaverse healthcare systems. *Future Gener. Comput. Syst.*, *140*, 299–310.
69. Zhang, Z., Wen, F., Sun, Z., Guo, X., He, T., & Lee, C. (2022). Artificial intelligence-enabled sensing technologies in the 5G/internet of things era: From virtual reality/augmented reality to the digital twin. *Adv. Intell. Syst.*, *4*(7), 2100228.
70. Zhang, C., Xu, C., Sharif, K., & Zhu, L. (2021). Privacy-preserving contact tracing in 5g-integrated and blockchain-based medical applications. *Comput. Stand. Interfaces*, *77*, 103520.
71. Zvarikova, K., Michalikova, K. F., & Rowland, M. (2022). Retail data measurement tools, cognitive artificial intelligence algorithms, and metaverse live shopping analytics in immersive hyper-connected virtual spaces. *Linguist. Philos. Investig.*, *21*, 9–24.
72. Zhernova, K., & Chechulin, A. (2022). Overview of vulnerabilities of decision support interfaces based on virtual and augmented reality technologies. In Proceedings of the Fifth International Scientific Conference "Intelligent Information Technologies for Industry" (IITI'21), pages 400–409. Springer International Publishing.
73. Zhang, L. J. (2022). MRA: Metaverse reference architecture. In Internet of Things–ICIOT 2021: 6th International Conference, Held as Part of the Services Conference Federation, SCF 2021, Virtual Event, December 10–14, 2021, Proceedings, pages 102–120, Cham. Springer International Publishing.
74. Zhang, J., Zhong, H., Cui, J., Tian, M., Xu, Y., & Liu, L. (2020). Edge computing-based privacy-preserving authentication framework and protocol for 5G-enabled vehicular networks. *IEEE Trans. Veh. Technol.*, *69*(7), 7940–7954.
75. Zhou, Y., Xiao, X., Chen, G., Zhao, X., & Chen, J. (2022). Self-powered sensing technologies for human metaverse interfacing. *Joule*, *6*(7), 1381–1389.
76. Zola, F., Segurola-Gil, L., Bruse, J. L., Galar, M., & Orduna-Urrutia, R. (2022). Attacking Bitcoin anonymity: Generative adversarial networks for improving Bitcoin entity classification. *Appl. Intell.*, *52*(15), 17289–17314.

6 Blockchain Revolution
Using Distributed Ledger Technology to Transform Healthcare in WSN-IoT Environment

Shivani Wadhwa[1], Divya Gupta[2], and Nagma[3]
[1]Chitkara University Institute of Engineering and Technology, Chitkara University, Punjab, India
[2]Department of Computer Science and Engineering, Chandigarh University, Mohali, Punjab, India
[3]Algoma University, Brampton Campus, Brampton, Ontario, Canada

INTRODUCTION

Due to rapid adoption of automation in healthcare, massive electronic patient records have been created. Such expansion necessitates extraordinary protection for healthcare data while in use and trade [23]. The emergence of blockchain technology as a transparent and secure platform for data storage and exchange opens up interesting possibilities for addressing critical confidentiality, security, and integrity challenges in healthcare. Blockchain has received significant attention in recent years from both industry and academia due to its potential to revolutionise the healthcare sector [39]. Continuously, novel blockchain applications and research initiatives are being unveiled, with healthcare standing out as a prominent domain.

Blockchain technology holds immense potential in addressing crucial aspects of healthcare, including privacy, security, data sharing, and data preservation [19]. Interoperability is a critical demand in the healthcare industry. Interoperability is the ability of two entities, whether human or machine, to precisely, efficiently, and reliably transfer information or data. The goal of healthcare interoperability is to enable the seamless interchange of health-related information, such as electronic health records (EHR), between medical providers and patients. This enables the distribution of information across different healthcare organisations and fosters collaboration and continuity of service [4].

Furthermore, the rising costs of healthcare infrastructure and software have put a significant burden on global economies. However, the introduction of blockchain technology is causing beneficial changes in the healthcare sector, benefiting both organisations and stakeholders. Blockchain is revolutionising healthcare outcomes by optimising

DOI: 10.1201/9781003438205-6

business processes, increasing patient outcomes, streamlining patient data administration, strengthening compliance, lowering costs, and enabling effective utilisation of medical information [1]. Another critical element is its potential to transform the transportation of pharmaceuticals and healthcare devices within the complex medical supply chain. The use of blockchain in healthcare supply chains has the potential to eliminate the risk of counterfeit drugs, which pose a substantial risk to patients all over the world. Furthermore, blockchain-based technologies are being investigated for a variety of healthcare applications, such as data management, storage solutions, and device management.

The usage of blockchain technology across the aforementioned application areas provides a plethora of benefits that significantly improve the quality of experience for various stakeholders and end users. Patients, caregivers, healthcare professionals, pharmaceutical companies, and insurance providers are among the stakeholders. The ability to safely communicate healthcare data while protecting consumers' privacy is a critical innovation that moves the medical sector towards higher intelligence [41]. Blockchain contributes to the enhancement of healthcare services and improves the entire user experience by providing data privacy and security, promoting a smarter and more efficient healthcare ecosystem.

BLOCKCHAIN TECHNOLOGY

Blockchain technology is a distributed ledger system that allows for the safe recording, storage, and verification of transactions and data among several network members or nodes [32]. It was first launched as the underpinning technology for the cryptocurrency Bitcoin, but it has subsequently found applications in fields other than finance, such as healthcare.

Although Bitcoin is the most well-known application of blockchain technologies, there are several additional applications that go well beyond digital currencies, such as banking services and products, property tracking, electronic identity verification, the electoral process, and so on. How to use blockchain technologies to construct smart contracts is a prominent topic [30]. In short, smart contracts are commitments among mutually sceptical participants that are autonomously implemented by the blockchain's consensus mechanism—without the need for an authority to be trusted.

It employs consensus techniques and encryption to validate the authenticity of a transaction in an untrustworthy and insecure environment. The receiver node in a decentralised blockchain network carefully examines incoming transactions before placing them in a block. Following that, a consensus technique, popularly known as "proof-of-work," is used to validate the information within each block. When the consensus process is finished, the block is added to the existing chain, and all nodes in the network accept and propagate the revised chain [24].

IMPORTANCE OF BLOCKCHAIN IN HEALTHCARE

The need for advancement is increasing at an unprecedented rate in the continually shifting healthcare scene. The contemporary healthcare landscape necessitates high-quality healthcare services supported by ground-breaking and innovative technologies. In this sense, blockchain technology is poised to play a major role in altering the

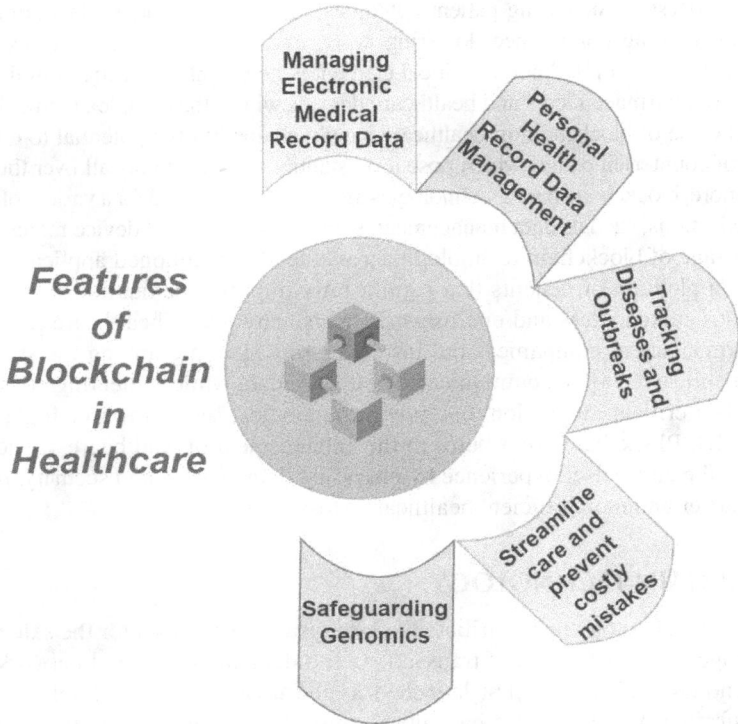

Features
of
Blockchain
in
Healthcare

Managing
Electronic
Medical
Record Data

Personal Health Record Data Management

Tracking Diseases and Outbreaks

Streamline care and prevent costly mistakes

Safeguarding
Genomics

FIGURE 6.1   Features of blockchain in healthcare.

healthcare sector, ushering in a revolution in the delivery and experience of healthcare, in this case (refer to Figure 6.1). In addition, the healthcare sector is going through a shift towards an approach that prioritises patients, with a focus on two critical elements: always-accessible services and suitable healthcare resources. Healthcare organisations can improve their capacity to provide efficient care and high-quality medical services to patients by leveraging the potential of blockchain. Furthermore, blockchain technology has the ability to address the time-consuming and recurrent processes that lead to rising healthcare expenses by providing quick and effective solutions [17]. Blockchain also allows citizens to actively participate in health research programs, improving research findings and facilitating sharing of information on public health. This, in turn, has the potential to improve treatment outcomes for a large number of people. While traditional healthcare systems rely on centralised databases, blockchain offers a decentralised and distributed way to manage healthcare systems and organisations [9, 11].

WSN IN HEALTHCARE

Wireless sensor networks (WSNs) have found applications in a wide range of disciplines by exploiting distributed wireless sensors with limited energy and processing power. While low-power hardware and communication protocols were once thought to be only suited for low-data-intensive jobs, recent advances in low-power hardware and communication protocols have extended their applicability to high-data-intensive applications such as visual sensing and picture communication. These developments in wireless sensing and networking technologies are critical for connecting the physical and cyber worlds, which is a critical step towards realising the Internet of Things (IoT) [36]. WSNs have the potential to significantly improve healthcare service provisioning and management in the medical domain.

Wireless medical sensors are critical in healthcare because they continuously monitor patients' physiological indicators such as heart rate, respiration, blood pressure (BP), and pulse, oxygen saturation (SpO_2). These compact, resource-constrained gadgets are usually battery-powered and attached to the patient's body, allowing for real-time data collecting in hospital or home settings. Furthermore, non-intrusive sensors can analyse physiological states and detect falls, providing doctors, nurses, and caregivers with critical information about the patient's medical condition.

VARIOUS CAPABILITIES OF BLOCKCHAIN TECHNOLOGY TO SUPPORT GLOBAL HEALTHCARE

Blockchain technology has numerous uses and functions in the field of healthcare. It allows for the secure communication of medical information, control over the drug supply chain, and the safe transfer of patient medical records. Furthermore, blockchain helps healthcare professionals investigate genetic code and gain fresh insights. Blockchain's technical aspects, such as healthcare data protection, genome management, electronic data management, medical records, interoperability, digitalised tracking, and epidemic control, contribute to its extraordinary capabilities in healthcare. The total digitalisation of blockchain technology, as well as its incorporation into healthcare applications, provides a solid platform for its widespread adoption and implementation [25].

Blockchain technology provides a transparent and verifiable mechanism for monitoring the entire prescription process, from drug production to medication availability in pharmacies. Congestion, products tracking, and performance may all be successfully monitored by merging the IoT and blockchain. This allows for more efficient acquisition management, which reduces disruptions and bottlenecks in hospitals, pharmacies, and other healthcare facilities that rely on specific prescriptions. Implementing blockchain-based technical frameworks aids in preventing unauthorised changes to logistical data, promoting confidence, and discouraging the illegal handling of documents, payments, and prescriptions by persons attempting to obtain pharmaceuticals illegally. These technological advances have the potential to improve patient care while remaining cost-effective [37].

Blockchain technology removes all barriers and constraints to establishing multi-level authenticity. Because it keeps an incorruptible, decentralised, and transparent record of all medical data, it is well-suited for security applications. While blockchain is based on a public ledger, it also protects privacy by utilising complicated and secure algorithms that protect the identity and confidentiality of individuals' medical information. This enables patients, healthcare providers, and medical practitioners to share vital knowledge and data in a timely and secure manner [34].

Patients may play a key role in pushing the move to interoperability by taking control of their healthcare records and assuring easy access by embracing blockchain technology. This gives patients more control over their sensitive data while also improving privacy and confidentiality. The use of blockchain technologies in healthcare addresses the issues of quantifying, implementing, and enforcing quality management and compliance requirements. Furthermore, blockchain-based solutions can help regulatory bodies identify between authentic and counterfeit drugs, protecting public health and safety [6]. This guarantees that all authorised parties interchange online transactions carrying the patient's information.

Blockchain technology is becoming more widely recognised and accepted in the healthcare industry. This technology is particularly appealing to early adopters in the medical industry. The main objective of blockchain in the near future revolutionising the healthcare sector is to address the difficulties that currently affect its structure [8]. By utilising blockchain, healthcare professionals, patients, and chemists receive instant access to full information, allowing for improved collaboration and informed decision-making (refer to Figure 6.2). Healthcare organisations are constantly investigating, experimenting, and delving into the potential of blockchain technology in the medical profession [15]. It has developed as a significant tool in healthcare, with advantages such as medication traceability, extended payment alternatives, and decentralised patient health information storage. Blockchain, along with significant technical breakthroughs such as machine learning and artificial intelligence, plays a critical role in the medical sector [21]. These developments, when combined, are defining the future of healthcare by spurring innovation and improving patient outcomes.

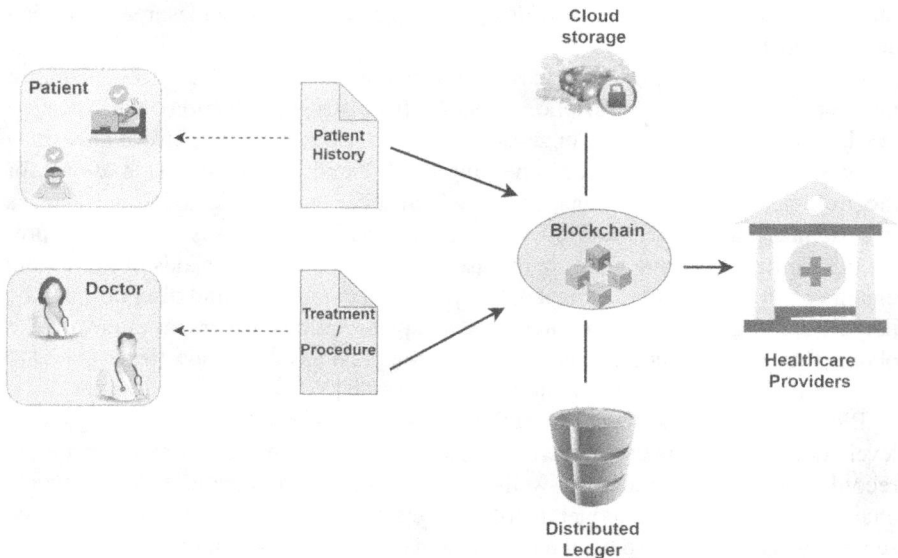

FIGURE 6.2 Data management in blockchain healthcare.

BENEFITS OF BLOCKCHAIN IN HEALTHCARE

In the healthcare industry, the adoption of blockchain technology brings forth a multitude of significant benefits:

1. Data security: The decentralised and unchangeable nature of blockchain makes sure patient data are secure and tamper-proof. Each transaction is cryptographically connected to the one before it, resulting in a transparent and verifiable data chain. This increased protection lowers the likelihood of incidents involving data breaches and illegal access to critical medical data.
2. Interoperability: Blockchain enables easy data interchange and compatibility among different healthcare systems, such as EHRs and health information exchanges. It enables heterogeneous systems to safely communicate and share data, resulting in greater management of care and better outcomes for patients.
3. Supply chain management: Blockchain can be used to track the whole supply chain of pharmaceuticals and medical devices in the pharmaceutical business. This increases transparency, lowering the risk of counterfeit items and ensuring pharmaceutical validity from manufacturer to patient.
4. Fraud reduction: The transparency and accountability of blockchain make it more difficult for fraudulent actions to occur within the healthcare system. This has the potential to reduce insurance fraud, duplicate billing, and other financial malpractices.
5. Decentralisation: The decentralised architecture of blockchain eliminates the need for a centralised governing body, lowering the possibility of single points of failure or data breaches. This improves system dependability and robustness.
6. Cost savings: By eliminating administrative overhead and optimising operations, blockchain has the potential to save healthcare organisations' money. By eliminating intermediaries and laborious processes, operations can become more efficient.
7. Clinical trials and research: By offering a secure and transparent platform for sharing research data, blockchain has the potential to completely transform the clinical research process. Researchers can gain access to anonymised data while preserving the confidentiality of patients, resulting in more rapid and collaborative medical innovations.

ROLE OF WSN IN BLOCKCHAIN AND HEALTHCARE

WSN constitute networks of restricted, self-contained sensor nodes outfitted with a variety of sensors, networking skills, and processing units. These sensors are distributed geographically to gather and send data from surroundings or particular locations of interest. WSNs have received a lot of attention and are being used in a variety of areas, notably healthcare, agriculture, environmental monitoring, industrial automation, and smart cities. WSN in Blockchain Healthcare is a fusion of two innovative technologies with the possibility to transform healthcare systems [28].

WSN comprises a network of small, self-contained sensors capable of collecting data from the human body and blockchain is a decentralised safe ledger that assures confidentiality and anonymity.

Sensor data gathering in healthcare settings is critical for effective patient monitoring and prompt treatments. Several variables, however, can contribute to data inaccuracy, resulting in false alarms and significant hazards to the healthcare system. Sensor malfunctions, sensor node resource limits (such as restricted power and transmission capability), sensor displacement, transmission interference, and even malicious data injection are some of the primary causes contributing to data errors. These mistakes can lead to false alarms, which have a negative impact on the healthcare system by increasing caregiver fatigue, decreasing service quality, and wasting precious resources. Figure 6.3 illustrates the functioning of WSN sensors with the blockchain technology and healthcare data. Integrating WSN with blockchain in healthcare can result in a number of advantages and unique applications:

1. Remote patient monitoring: WSN-enabled devices, including trackers and biosensors for example, can collect BP, biometric details, as well as additional health-related details from patients in real time. This information can be anonymously sent to the blockchain and accessed by authorised healthcare practitioners. It enables continuous tracking of patients' medical states, allowing quick medications and better outcomes for patients.
2. Exchanging data and interoperability: WSN gadgets can create huge amounts of personal information that must be exchanged among many healthcare bodies and systems. Exchange of information and interoperability across diverse healthcare systems become more efficient and simplified by exploiting blockchain's decentralised architecture. It encourages a uniform approach to healthcare data management, which improves care coordination and reduces redundancy of effort.

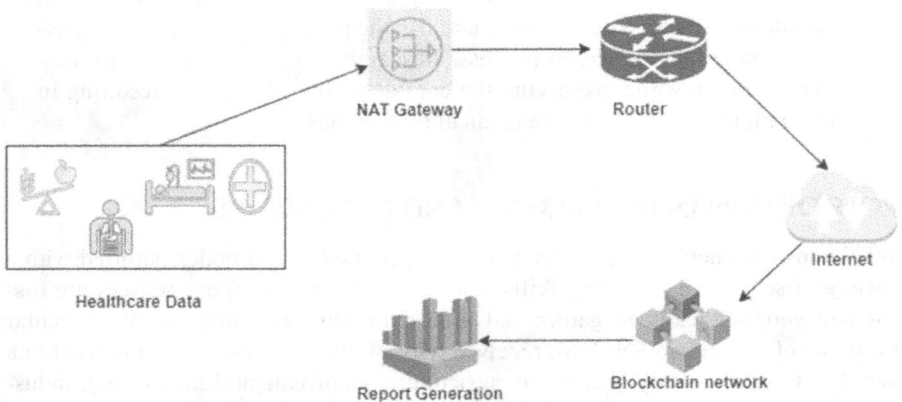

FIGURE 6.3   WSN in blockchain and healthcare data management.

3. Clinical trials and medical research: WSN in blockchain as well as healthcare can help improve the conduct of clinical trials and medical research. Data acquired from trial patients can be maintained on the blockchain, guaranteeing visibility and traceability. With patient approval, researchers can gain access to anonymised data, speeding the pace of research and medication development.

4. Supply chain management: Ensuring the validity and monitoring of medicinal products and devices for medical use is critical in the healthcare business. By combining WSN and blockchain, a transparent and immutable supply chain may be created, decreasing the risk of counterfeit drugs and assuring the authenticity of products from the producer to the consumer.

5. Smart contracts for healthcare services: The smart contract capability of the blockchain is capable of being used for WSN-based services in the healthcare sector. Smart contracts, for example, may facilitate and uphold commitments between patients and medical professionals, initiating payments depending on specific conditions like completing residence-based medications or achieving particular medical benchmarks.

6. Data ownership and consent management: Patients can own and control their own healthcare data created by WSNs. Patients can use blockchain to authorise or deny accessibility to their medical information to particular medical professionals or specialists, guaranteeing that their authorisation is respected and data privacy is enhanced.

7. Data integrity and security: The data created by WSNs are extremely sensitive, particularly in healthcare contexts. The intrinsic qualities of cryptographic security and the confidentiality that blockchain technology offers assure that data acquired from WSN sensors are secure and reliable. It protects confidentiality of patient information and aids in the development of trust between patients and medical professionals.

PROJECTS BASED ON BLOCKCHAIN AND HEALTHCARE

This section examines various notable blockchain and AI-powered telemedicine and remote monitoring of patient projects.

SHIVOM

Shivom is an innovative genomics and wellness platform that aspires to improve the lives of individuals by allowing them to safely store as well as manage their genomic data. It employs blockchain technology to protect the confidentiality of information, protection, and integrity while promoting partnerships among researchers, patients, and healthcare professionals. Shivom's mission is to build a worldwide genomics and personalised healthcare ecosystem in which individuals have sovereignty over their genomic data and can help advance scientific discoveries while profiting from personalised medical insights. Shivom intends to address privacy problems, improve data security, and stimulate genetic research by integrating blockchain technology [26].

CORAL HEALTH

Coral Health is developing a platform based on blockchain technology that will allow individuals to securely keep and distribute their health data. It uses artificial intelligence to analyse health records, deliver personalised insights, and contribute to medical research while keeping data privacy in mind. Coral Health distinguishes itself in the healthcare industry by focusing on blockchain technology and powerful data analytics. The company's goal is to improve healthcare data management, enable personalised medicine, and foster collaboration and innovation in the healthcare business by addressing data interoperability and security concerns [38].

ENCRYPGEN

EncrypGen is a genomic information platform where users are able to safely store and sell their genetic data. It makes use of blockchain technology to secure data contributors' transparency, security, and fair reward, while also easing access to information for researchers. EncrypGen's goal is to strengthen individuals by giving them authority over their genomic data, while also encouraging collaborations in research and advances in personalised treatment. The platform solves issues about data privacy, security, and equitable pay by integrating blockchain technology, hence accelerating genomics innovation and discovery [33].

HEALTH WIZZ

Health Wizz is an application for smartphones that allows people to organise and keep track of their health data from numerous sources. It employs blockchain technology to maintain security and confidentiality of information while allowing consumers to share information on their own terms with healthcare professionals and researchers. It promises to provide people more control over their own medical records while also allowing for smooth data interchange with medical professionals and other stakeholders [9].

MEDICALCHAIN

Medicalchain is powered by blockchain telemedicine network that uses AI to analyse data and track patients. Patients can authorise healthcare practitioners to have access to their health data through their blockchain-based safe and transparent health record management service, and their AI models analyse the information in order to generate individualised treatment strategies. The focus of Medicalchain on using blockchain technology for private medical record administration and improved interoperable leads to more effective and compassionate healthcare delivery. Medicalchain promises to alter the way patient data are kept and transmitted by improving privacy, security, and accessibility, resulting in better healthcare results [22].

HEALTHEREUM

Healthereum is a platform built on blockchain that uses artificial intelligence to engage patients as well as monitor them remotely. Patients can securely save their wellness records on their platform, get prizes for completing health-related tasks, and participate in telemedicine discussions with medical experts. The focus of Healthereum on patient participation, suggestions, and incentivisation, along with secure communication and consent management, aims to improve the patient experience and overall healthcare quality. The platform emphasises openness, data security, and participation by patients in healthcare interactions by integrating blockchain technology [18].

APPLICATIONS OF BLOCKCHAIN IN HEALTHCARE

Blockchain technology is rapidly expanding and gaining popularity, opening up new avenues for healthcare delivery. Its capacity to promote seamless and speedy information sharing among all key stakeholders and healthcare providers aids in the creation of cost-effective drugs and innovative therapies for a variety of diseases. As a result, the healthcare business will continue to grow and expand in the next years. The benefits of blockchain technology have recently been proved in the transportation industry, and the healthcare sector is no exception. Healthcare is one of the first industries to embrace digitisation and innovation due to its direct impact on quality of life. At the same time, blockchain technology is gaining traction, notably in the financial industry [40]. It provides a plethora of vital and useful services in the healthcare industry. It provides a plethora of substantial and amazing potential in the healthcare industry, covering scientific research, logistics, and increasing relationships between practitioners and patients [15]. Figure 6.4 illustrates the application of blockchain in the field of healthcare.

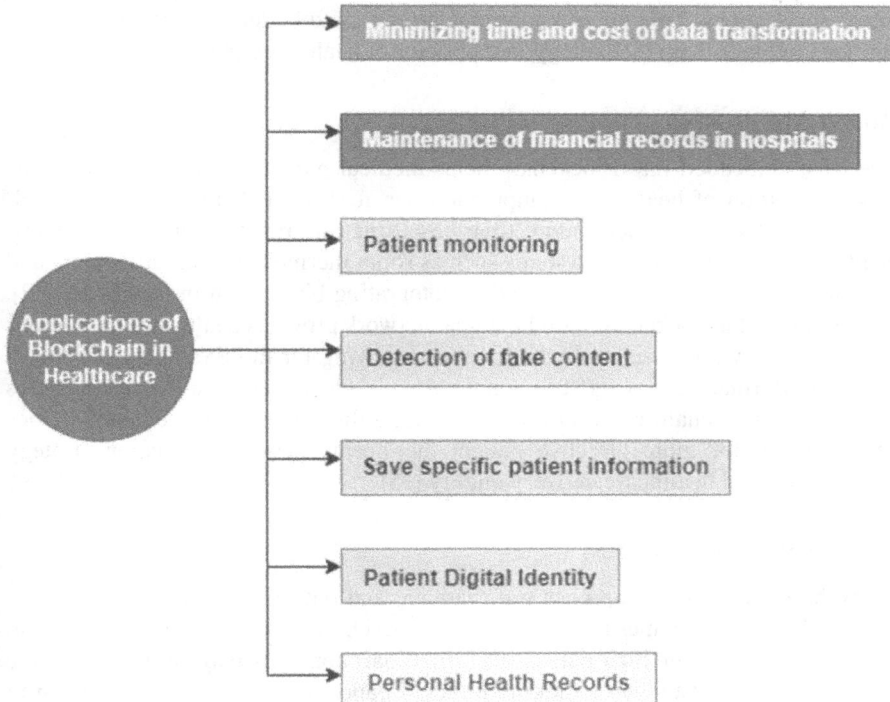

FIGURE 6.4 Applications of blockchain in healthcare.

MINIMISING TIME AND COST OF DATA TRANSFORMATION

Blockchain networks have considerable benefits in terms of lowering the time and cost of data transformation. They also have the ability to address the issue of swiftly and properly validating medical qualifications. Patient identification and data security can be secured by harnessing blockchain networks, clearing the path for disruptive ideas and discoveries that can revolutionise healthcare delivery globally. Implementing blockchain technology makes it easier to build efficient and private data exchange networks that can be monetised [5].

Healthcare systems benefit from blockchain because it is a worldwide distributed computer system that offers secure time stamping of transaction history and trustworthy record-keeping. Each node in this network is critical in authenticating and recording all information, assuring its integrity and transparency.

MAINTENANCE OF FINANCIAL RECORDS IN HOSPITALS

In the bookkeeping process, having correct financial records is critical. Clinical studies, too, necessitate efficient operations and evaluation. Blockchain firms have developed solutions to improve accounting and reporting processes in various fields. Individuals, for example, can use blockchain-based systems to arrange appointments with healthcare practitioners in advance and submit the relevant paperwork, reducing the need for patients to wait in lines and saving time. We may learn about the dangers, benefits, and challenges of blockchain in healthcare by examining real-world applications. This direct knowledge allows us to better comprehend the possibilities of blockchain technology in the sector of healthcare [42].

PATIENT MONITORING

The trust embedded into Blockchain helps medical personnel guarantee that they have possession of healthcare equipment when it is needed. Blockchain technology allows doctors to spend more time observing and remotely treating patients' health concerns. Monitoring systems such as room thermostats, bed utilisation, and resource availability can be improved by integrating blockchain in healthcare [12]. The creation of a blockchain-based medical network provides healthcare institutions and providers with a secure digital identity, allowing for more efficient collaboration and information sharing. The combination of blockchain and IoT technologies increases supply chain reaction time and traceability, improving medical logistics and allowing for more effective patient monitoring. This combination strategy improves overall healthcare delivery quality [35].

DETECTION OF FAKE CONTENT

Blockchain technology improves transparency and detects fake information more effectively. When it comes to clinical trials, blockchain ensures that individuals and customers can validate their data easily [16]. Smart contracts play an important role in obtaining approvals while making protocol paperwork and conclusions open to public scrutiny [13]. The general public gains the capacity to closely monitor the progress of clinical studies by employing blockchain, boosting trust and account-ability. This technology's primary goal is to deliver a user-friendly experience while providing patients with secure and immediate access to their medical and insurance information.

SAVE SPECIFIC PATIENT INFORMATION

A substantial amount of patient information and medical data is generated during the various phases of clinical investigations. Laboratory testing, quality assess-ments, calculations, and fitness surveys are all examples of this. These data may be conveniently stored and retrieved via blockchain technology, preserving its valid-ity and integrity. Healthcare practitioners can access the stored data and validate its authenticity by comparing it to the original data saved in the blockchain system. Blockchain technology's cryptographic techniques provide a safe platform for data sharing. In an EHR format, healthcare practitioners record key information such as the patient's name, date of birth, diagnosis, therapies, and outpatient history in the patient details [7].

PATIENT DIGITAL IDENTITY

Accurately locating patient records in healthcare databases utilising unique data sets is a critical part of health information sharing. Systems such as the Master Patient Index as well as Enterprise Patient Master Index have been created to help healthcare organisations or trusted networks handle patient identities more efficiently. Due to patient identity mismatches, duplicate records of patients may be developed, as well as the existence of missing or erroneous medical data.

PERSONAL HEALTH RECORDS

Personal Health Records (PHRs) are patient-focused programmes that empower indi-viduals as the genuine owners of their health information, as opposed to the current practice of provider-centric EHRs for managing patient data. The primary goal of PHRs is to give patients secure and convenient access to their comprehensive health data, which is collected from a variety of sources such as provider visit records, immunisation history, prescription details, and data from smartphone devices track-ing physical activity, among others. Patients can take ownership of their health infor-mation by employing PHRs, ensuring control, accuracy, and verification of their comprehensive health records.

APPLICATIONS OF WSN IN HEALTHCARE

The use of WSN in healthcare creates new opportunities for patient-centred care, remote healthcare services, and data-driven decision-making. The ability to securely and efficiently gather and transmit real-time health data has a tremendous impact on healthcare delivery, leading to improved patient outcomes and quality of life [29]. WSN technology's uses in healthcare are predicted to revolutionise the sector and pave the way for novel and personalised healthcare solutions as it evolves [20]. Few WSN applications in healthcare are:

1. Remote patient monitoring: WSN-enabled wearable devices and biosensors capture vital signs, biometric data, and other health-related information from patients in real time. These real-time data are wirelessly provided to healthcare providers, allowing for remote monitoring and timely interventions, which is especially beneficial for patients with chronic diseases or those in remote places [3].

2. Ambient-assisted living: WSN-based smart home technology can help the elderly and disabled with their everyday tasks. These systems monitor activity, detect falls, and provide timely notifications or assistance, promoting independence and safety [27].

3. Telemedicine: WSN devices provide real-time data transmission from patients to healthcare providers, enabling telemedicine services [2]. This technology enables doctors to remotely monitor patients' health conditions, conduct consultations, and alter treatment programmes without the need for patients to visit the doctor.

4. Environmental monitoring: WSNs can be used in healthcare facilities to monitor environmental characteristics such as air quality, temperature, and humidity. These monitoring systems aid in maintaining a safe and healthy environment for patients and healthcare personnel.

5. Drug delivery systems: WSN-enabled drug delivery systems can monitor and adjust pharmaceutical dosages based on a patient's state, delivering optimal therapeutic outcomes while minimising negative effects.

6. Clinical trials and research: WSNs serve an important role in medical research by allowing continuous data collection from study participants. The ability to collect data in real time improves the efficiency and accuracy of clinical trials and medical research projects.

7. Clinical trials and research: WSNs serve an important role in medical research by allowing continuous data collection from study participants. The ability to collect data in real time improves the efficiency and accuracy of clinical trials and medical research projects [10, 14].

CHALLENGES OF BLOCKCHAIN IN HEALTHCARE

Blockchain technology has been employed in the healthcare business to address specific difficulties. The lack of knowledge and awareness is one of the most significant barriers to utilising this revolutionary technology in hospitals. Blockchain applications are still in their infancy, necessitating additional technological investigation and development. They do, however, have enormous potential for satisfying the needs of medical groups and regulatory organisations. The healthcare system is primed for change, and blockchain is likely to play an important part in its progress [31]. Blockchain technology's uses in healthcare will grow as it improves, offering vital information about treatment outcomes and development. Blockchain technology is critical for providing secure and trustworthy healthcare transactions and information exchanges.

Blockchain technology holds immense promise for the healthcare business, but it is not without its hurdles. Some of the significant challenges of blockchain in healthcare are as follows:

1. Interoperability: A big problem is achieving interoperability among various healthcare systems and blockchain networks. Healthcare organisations store and exchange data using a variety of systems and standards, and connecting these diverse systems with blockchain can be difficult.
2. Cost and infrastructure requirements: Implementing blockchain technology in healthcare may necessitate considerable investments in infrastructure, such as hardware, software, and network resources. Blockchain technologies' cost-effectiveness and interoperability with existing healthcare IT systems must be assessed.
3. Regulatory and legal concerns: Healthcare is a highly regulated industry, and incorporating blockchain technology necessitates adherence to numerous legal and regulatory frameworks. Concerns around data ownership, consent management, and liability must be addressed before blockchain may be used in healthcare.
4. Data privacy and security: While blockchain technology has intrinsic security characteristics such as encryption and immutability, protecting the privacy and security of sensitive healthcare data remains a difficulty. Patient confidentiality and adherence to data protection standards are crucial factors.
5. User adoption and education: Because blockchain is a new technology, both healthcare professionals and patients may be unaware of its potential benefits. To increase user adoption and facilitate the integration of blockchain solutions into existing healthcare workflows, education and training programmes are required.
6. Scalability: When it comes to processing a large volume of transactions fast, blockchain networks, particularly public ones, encounter scalability challenges. Healthcare creates massive volumes of data, and blockchain must be capable of handling the scalability needs of efficiently storing and managing this data.
7. Ethical considerations: The use of blockchain in healthcare poses ethical concerns about data ownership, permission, and the possibility of prejudice or bias in health data analysis. It is critical to provide ethical rules and frameworks for blockchain implementation.
8. Power consumption: Since blockchain requires significant computational resources, it is power-intensive. Nevertheless, the majority of WSN devices use restricted energy sources, such as batteries, which necessitate conservative consumption of electricity in order to extend their lifespan. Integrating blockchain into WSNs may place additional burden on already limited power budgets, which could result in diminished network lifespan and reliability of operation.
9. Heterogeneity: WSNs are frequently made up of dissimilar devices from multiple vendors, resulting in variability in software, hardware, and protocols for communication. Incorporating blockchain across these different devices necessitates the use of standardised interfaces and protocols, which can be difficult to install and manage in practice.

10. Processing time: Due to the decentralised feature of the blockchain, obtaining consensus among WSN nodes with differing processing rates and capacities can be difficult. As a result, the time required for confirmation and recording transactions on the blockchain can vary, compromising the adaptability and efficiency of time-sensitive WSN services.

CONCLUSION AND FUTURE SCOPE

Since its inception with Bitcoin, blockchain technology has matured into a versatile tool that can be applied in a variety of industries, including healthcare. According to our findings, blockchain can help the healthcare industry save time and money by enabling remote patient monitoring, improving content authenticity verification, maintaining medical data security, and more. Several initiatives, like Shivam, Coral Health, EncryptGen, and Health Wizz, have emerged to offer novel healthcare solutions using blockchain technology. However, further research is needed to completely understand, characterise, and evaluate the potential of blockchain in healthcare. Continued investigation will aid in the discovery of new insights and the realisation of the full potential of blockchain technology in the healthcare area.

Blockchain technology has the ability to authenticate and record transactions with the approval of network participants in the near future. This breakthrough has the potential to transform health information sharing by offering patients with numerical security via public and private key encryption. Blockchain solutions have numerous benefits, including secure patient record management, breach prevention, interoperability promotion, process efficiency, drug and prescription control, and monitoring of medical and supply chains. When integrating blockchain within organisational structures, it is critical to take a strategic approach, carefully examining the possible benefits and costs involved with adopting and managing blockchain-based technology. For example, organisations must examine the ongoing costs of blockchain, taking into account the high energy consumption associated with running blockchains. Nonetheless, the future of blockchain in healthcare is promising.

REFERENCES

1. Aggarwal, S., Chaudhary, R., Aujla, G. S., Kumar, N., Choo, K.-K. R., and Zomaya, A. Y. (2019). Blockchain for smart communities: Applications, challenges and opportunities. Journal of Network and Computer Applications, 144:13–48.
2. Al Ameen, M., Liu, J., and Kwak, K. (2012). Security and privacy issues in wireless sensor networks for healthcare applications. Journal of Medical Systems, 36:93–101.
3. Alemdar, H. and Ersoy, C. (2010). Wireless sensor networks for healthcare: A survey. Computer Networks, 54(15):2688–2710.
4. Alzahrani, A. G., Alhomoud, A., and Wills, G. (2022). A framework of the critical factors for healthcare providers to share data securely using blockchain. IEEE Access, 10:41064–41077.
5. Ashima, R., Haleem, A., Bahl, S., Javaid, M., Mahla, S. K., and Singh, S. (2021). Automation and manufacturing of smart materials in additive manufacturing technologies using Internet of Things towards the adoption of industry 4.0. Materials Today: Proceedings, 45:5081–5088.

6. Bhavin, M., Tanwar, S., Sharma, N., Tyagi, S., and Kumar, N. (2021). Blockchain and quantum blind signature-based hybrid scheme for healthcare 5.0 applications. Journal of Information Security and Applications, 56:102673.

7. De Aguiar, E. J., Faical, B. S., Krishnamachari, B., and Ueyama, J. (2020). A survey of blockchain-based strategies for healthcare. ACM Computing Surveys (CSUR), 53(2):1–27.

8. Dhiman, P., Kukreja, V., Manoharan, P., Kaur, A., Kamruzzaman, M., Dhaou, B., and Iwendi, C. (2022). A novel deep learning model for detection of severity level of the disease in citrus fruits. Electronics, 11(3):495.

9. Dimitrov, D. V. (2019). Blockchain applications for healthcare data management. Healthcare Informatics Research, 25(1):51–56.

10. Ejaz, M., Kumar, T., Kovacevic, I., Ylianttila, M., and Harjula, E. (2021). Health-blockedge: Blockchain-edge framework for reliable low-latency digital healthcare applications. Sensors, 21(7):2502.

11. Farouk, A., Alahmadi, A., Ghose, S., and Mashatan, A. (2020). Blockchain platform for industrial healthcare: Vision and future opportunities. Computer Communications, 154:223–235.

12. Galetsi, P., Katsaliaki, K., and Kumar, S. (2023). Exploring benefits and ethical challenges in the rise of mHealth (mobile healthcare) technology for the common good: An analysis of mobile applications for health specialists. Technovation, 121:102598.

13. Griggs, K. N., Ossipova, O., Kohlios, C. P., Baccarini, A. N., Howson, E. A., and Hayajneh, T. (2018). Healthcare blockchain system using smart contracts for secure automated remote patient monitoring. Journal of Medical Systems, 42:1–7.

14. Haque, S. A., Rahman, M., and Aziz, S. M. (2015). Sensor anomaly detection in wireless sensor networks for healthcare. Sensors, 15(4):8764–8786.

15. Houtan, B., Hafid, A. S., and Makrakis, D. (2020). A survey on blockchain-based self-sovereign patient identity in healthcare. IEEE Access, 8:90478–90494.

16. Hussien, H. M., Yasin, S. M., Udzir, N. I., Ninggal, M. I. H., and Salman, S. (2021). Blockchain technology in the healthcare industry: Trends and opportunities. Journal of Industrial Information Integration, 22:100217.

17. Ismail, L. and Materwala, H. (2020). Blockchain paradigm for healthcare: Performance evaluation. Symmetry, 12(8):1200.

18. Jaiswal, H., Misra, A., and Misra, P. K. Health record management system using block chain technologies.

19. Khezr, S., Moniruzzaman, M., Yassine, A., and Benlamri, R. (2019). Blockchain technology in healthcare: A comprehensive review and directions for future research. Applied Sciences, 9(9):1736.

20. Ko, J., Lu, C., Srivastava, M. B., Stankovic, J. A., Terzis, A., and Welsh, M. (2010). Wireless sensor networks for healthcare. Proceedings of the IEEE, 98(11):1947–1960.

21. Kukreja, V. and Dhiman, P. (2020). A deep neural network-based disease detection scheme for citrus fruits. In 2020 International Conference on Smart Electronics and Communication (ICOSEC), 97–101. IEEE.

22. Kumar, R. and Tripathi, R. (2019). Traceability of counterfeit medicine supply chain through blockchain. In 2019 11th International Conference on Communication Systems & Networks (COMSNETS), 568–570. IEEE.

23. Lee, J.-H. and Pilkington, M. (2017). How the blockchain revolution will reshape the consumer electronics industry [future directions]. IEEE Consumer Electronics Magazine, 6(3):19–23.

24. Lin, I.-C. and Liao, T.-C. (2017). A survey of blockchain security issues and challenges. International Journal of Network Security, 19(5):653–659.

25. Mettler, M. (2016). Blockchain technology in healthcare: The revolution starts here. In 2016 IEEE 18th International Conference on e-Health Networking, Applications and Services (Healthcom), 1–3. IEEE.

26. Motsi-Omoijiade, I. and Kharlamov, A. (2021). Blockchain for healthcare applications and use cases. In Blockchain and Public Law, 157–190. Edward Elgar Publishing.

27. Neves, P., Stachyra, M., and Rodrigues, J. (2008). Application of wireless sensor networks to healthcare promotion. Journal of Communications Software and Systems, 4(3):181–190.

28. Nguyen, C. V., Nguyen, M. T., Le, T. T., Tran, T. A., and Nguyen, D. T. (2021). Blockchain technology in wireless sensor network: Benefits and challenges. ICSES Transactions on Computer Networks and Communications:1–4.

29. Omar, I. A., Jayaraman, R., Debe, M. S., Salah, K., Yaqoob, I., and Omar, M. (2021). Automating procurement contracts in the healthcare supply chain using blockchain smart contracts. IEEE Access, 9:37397–37409.

30. Opoku, D.-G. J., Perera, S., Osei-Kyei, R., and Rashidi, M. (2021). Digital twin application in the construction industry: A literature review. Journal of Building Engineering, 40:102726.

31. Pham, H. L., Tran, T. H., and Nakashima, Y. (2018). A secure remote healthcare system for hospital using blockchain smart contract. In 2018 IEEE Globecom Workshops (GC Wkshps), 1–6. IEEE.

32. Politou, E., Casino, F., Alepis, E., and Patsakis, C. (2019). Blockchain mutability: Challenges and proposed solutions. IEEE Transactions on Emerging Topics in Computing, 9(4):1972–1986.

33. Rey, C. M. (2018). Banking on blockchain. Clinical OMICs, 5(2):14–15.

34. Srivastava, G., Parizi, R. M., and Dehghantanha, A. (2020). The future of blockchain technology in healthcare Internet of Things security. Blockchain Cybersecurity, Trust and Privacy, 161–184. Springer.

35. Tanwar, S., Parekh, K., and Evans, R. (2020). Blockchain-based electronic health care record system for healthcare 4.0 applications. Journal of Information Security and Applications, 50:102407.

36. Tennina, S., Di Renzo, M., Kartsakli, E., Graziosi, F., Lalos, A. S., Antonopoulos, A., Mekikis, P. V., and Alonso, L. (2014). Wsn4qol: A WSN-oriented healthcare system architecture. International Journal of Distributed Sensor Networks, 10(5):503417.

37. Tripathi, G., Ahad, M. A., and Paiva, S. (2020). S2HS: A blockchain based approach for smart healthcare system. Healthcare, 8:100391.

38. Vora, J., Nayyar, A., Tanwar, S., Tyagi, S., Kumar, N., Obaidat, M. S., and Rodrigues, J. J. (2018). Bheem: A blockchain-based framework for securing electronic health records. In 2018 IEEE Globecom Workshops (GC Wkshps), 1–6. IEEE.

39. Yli-Huumo, J., Ko, D., Choi, S., Park, S., and Smolander, K. (2016). Where is current research on blockchain technology?—A systematic review. PLoS One, 11(10):e0163477.

40. Zhang, P., Schmidt, D. C., White, J., and Lenz, G. (2018). Blockchain technology use cases in healthcare. Advances in Computers, 111:1–41.

41. Zhou, L., DeAlmeida, D., and Parmanto, B. (2019). Applying a user-centered approach to building a mobile personal health record app: Development and usability study. JMIR mHealth and uHealth, 7(7):e13194.

42. Zulkifl, Z., Khan, F., Tahir, S., Afzal, M., Iqbal, W., Rehman, A., Saeed, S., and Almuhaideb, A. M. (2022). Fbashi: Fuzzy and blockchain-based adaptive security for healthcare IoTs. IEEE Access, 10:15644–15656.

7 Blockchain in Internet of Medical Things
Insights on Healthcare Systems

Sushmita Jain[1], Saurabh Manoj Kothari[2], and Satyam Kumar Agrawal[1]

[1]Centre for in vitro Studies and Translational Research, Chitkara School of Health Sciences, Chitkara University Institute of Engineering and Technology, Chitkara University, Punjab, India
[2]College of Engineering Design & Physical Sciences, Brunel University, Middlesex, London, UK

INTRODUCTION

Blockchain, introduced alongside Bitcoin in 2008 by Satoshi Nakamoto [1], is a data structure that establishes an immutable and irreversible transactional system through the creation of a digital ledger. Blockchain functions as a peer-to-peer (P2P) system for distributed data sharing and processing and makes it possible for anonymous people to participate in a variety of network transactions even in the absence of trust between them. It serves as data structures that can easily track and store information from a wide range of devices by removing the requirement for a centralized cloud. Operating as a tamper-proof digital ledger, blockchain diligently preserves an expanding collection of data records without relying on a centralized approach or a master computer. Transactions among nodes are conducted using public key cryptography, with the ensuing records securely stored on a shared ledger. These records form a chain of blocks, interconnected through cryptographic means, thereby rendering any alteration or removal of recorded data on the blockchain ledger impossible. Users must complete a cryptographic puzzle known as proof-of-work in order to contribute new data to the blockchain. The first block in the blockchain is called the "Genesis block," and following blocks keep a cryptographic hash of the previous block (Figure 7.1). While participants can view transactions, this does not imply that everyone can access the actual content. The private key serves as a protective measure for safeguarding the actual content [2].

Blockchain technology extends beyond Bitcoin due to its numerous advantageous features [3]. Its secure, decentralized, and autonomous characteristics make it an optimal solution for addressing security issues within the Internet of Medical Things (IoMT) [4]. Blockchain technology and artificial intelligence (AI) systems

DOI: 10.1201/9781003438205-7

129

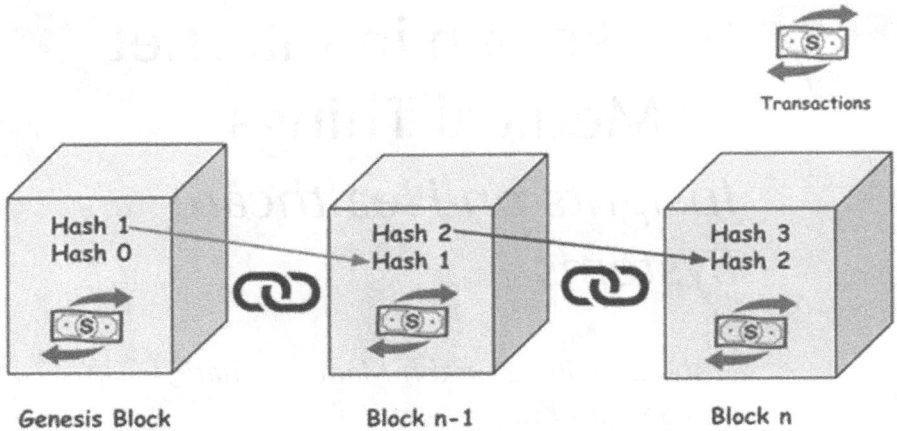

FIGURE 7.1   Schematic representation of data structure and blocks of blockchain.

have gained popularity in IoMT systems, mostly as a result of their successful application in other industries like finance [5]. Blockchain technology is frequently used in IoMT systems as a way to maintain security and share data between patients, physicians, and insurance providers. In addition, anomalous behavior that can be signs of impending attacks in network flows and patient data is crucially identified by AI systems. However, a number of obstacles prevent the widespread use of these methods in IoMT systems [6]. Given the data quantities and communication requirements of IoMT systems, one major issue is caused by the intrinsic properties of blockchain technology, which may lead to latency, storage restrictions, and communication overhead [7]. Due to its distributed architecture and decentralization, public blockchain technology frequently displays excessive latency. Real-time systems may think about implementing private blockchains, which perform better, to allay this worry [8]. On the other side, AI systems are less suitable for recognizing uncommon attacks that happen infrequently since they require significant amounts of data to detect abnormalities properly. Despite these obstacles, blockchain and AI are gradually making their way into IoMT systems, primarily at the cloud layer [9].

In recent years, blockchain technology has emerged as a highly secure and decentralized platform with numerous significant features. Its unique attributes, such as the elimination of third-party intermediaries and tamper-resistant properties, make it an ideal solution for ensuring data confidentiality and privacy. By leveraging blockchain, the problems related to security and management in the Internet of Things (IoT) platform can be significantly mitigated [10–13]. Blockchain technology offers a decentralized architecture that removes the need for a central authority or intermediary to validate transactions or manage data. This decentralization enhances security by eliminating single points of failure and reducing the vulnerability to malicious attacks or unauthorized access. Instead, the blockchain relies on a distributed network of nodes, each independently verifying and recording transactions, ensuring the integrity of the system [14]. One of the key advantages of blockchain technology is its ability to limit tampering and unauthorized modifications to the data stored on the platform. Once a transaction is recorded on the blockchain, it becomes virtually immutable, meaning that it cannot be altered or deleted without consensus from the network participants. This tamper-resistant nature ensures the integrity and authenticity of the data, reducing the risk of data breaches or manipulation [15]. Furthermore, the use of blockchain technology can address security and management challenges inherent in the IoT platform [16]. The IoT involves numerous interconnected devices that collect and transmit vast amounts of data. Ensuring the security and privacy of this data is crucial to prevent unauthorized access or misuse. By leveraging blockchain, the IoT platform can benefit from enhanced security measures, such as encryption, digital signatures, and access controls. These measures provide an added layer of protection to IoT data, reducing the risks associated with data breaches or privacy violations [17]. Additionally, blockchain technology offers transparency and auditability, enabling stakeholders to track and verify the provenance and integrity of IoT data. The decentralized nature of the blockchain ensures that data records are shared and replicated across multiple nodes, making it difficult for malicious actors to manipulate or corrupt the data without detection. This transparency and auditability can enhance trust and accountability in the IoT ecosystem, enabling stakeholders to verify the accuracy and reliability of the data. In conclusion, the emergence of blockchain technology as a highly secure and decentralized platform has brought numerous significant features to the forefront. By eliminating the need for intermediaries and providing tamper-resistant properties, blockchain ensures data confidentiality and privacy [18]. Leveraging blockchain in the IoT platform can effectively address security and management challenges, providing enhanced security measures, transparency, and auditability. As a result, the integration of blockchain technology into the IoT holds great promise for mitigating security risks and ensuring the integrity of data in this interconnected ecosystem [19, 20].

FACTORS RESPONSIBLE FOR LIMITATIONS OF BLOCKCHAIN TECHNOLOGY IN IoMT

IoT/IoMT devices that lack sufficient computational power, memory, energy, or bandwidth are referred to as resource-constrained IoT/IoMT devices. Running the blockchain directly on IoT/IoMT devices with low processing power is not suitable for a few reasons. The reasons that limit the blockchain to be introduced in IoMT are discussed in this section.

LACK OF COMPUTATIONAL RESOURCES

It can be a challenge for IoT/IoMT devices as they often have limited processing power and memory. This can make implementing complex technologies such as blockchain difficult, which require significant computational resources [21], for example, a little IoT device with a low processing power, like a smart thermostat. This device primarily focuses on gathering sensor data and managing temperature. The computational demands of performing consensus algorithms, cryptographic procedures, and data verification could overwhelm this device's limited resources, resulting in sluggish reaction times, delayed temperature adjustments, or even system crashes, if the blockchain were to be hosted directly on it.

INSUFFICIENT BANDWIDTH

It can also be a challenge for IoT/IoMT devices, particularly those that require real-time data transmission. Blockchain technology can be data-intensive, which may not be feasible for devices with limited bandwidth. For example, consider placing an IoT device to track soil moisture levels in a rural agricultural area. The device only occasionally connects to the network because of its position. Maintaining a stable copy of the blockchain and remaining in sync with other nodes would be difficult if the blockchain were hosted directly on this device. The device might find it difficult to establish and maintain connections, which could lead to data transmission delays, potential gaps in the blockchain's data, and issues taking part in the consensus process.

POWER PRESERVATION

This is another important consideration for IoT/IoMT devices, as they often operate on battery power or other limited energy sources. Implementing blockchain technology may require additional energy consumption, which could impact the device's battery life and overall performance. Overall, while blockchain technology has the potential to enhance security and privacy in IoT/IoMT applications, it is important to consider these practical challenges when implementing the technology in real-world scenarios [22]. For instance, a wearable health monitor that monitors vital signs needs to conserve electricity to ensure longer battery life. Running resource-intensive blockchain processes on such a gadget would soon deplete the battery, jeopardizing its primary role of monitoring and reporting health data.

ROLE OF TOPOLOGIES IN BLOCKCHAIN NETWORK

Topologies play an important role in the design and operation of a blockchain network. The topology of a blockchain network refers to the way in which nodes are connected to each other and how they communicate [23]. There are several different types of topologies that can be used in a blockchain network, including P2P, client-server, and hybrid topologies. In a P2P topology, all nodes in the network are equal and communicate with each other directly. This type of topology is often used in public blockchains like Bitcoin and Ethereum, where anyone can join the network as a node (Figure 7.2) [24, 25].

In a client-server topology, there is a central server that manages the network and communicates with all nodes. This type of topology is often used in private or permissioned blockchains where access to the network is restricted. Hybrid topologies combine elements of both P2P and client-server topologies [26]. For example, some nodes may communicate directly with each other while others communicate through a central server. The choice of topology depends on several factors such as security requirements, scalability needs, and the type of blockchain being used (Figure 7.3). Each topology has its own advantages and disadvantages, so it is important to carefully consider which one is best suited for your specific use case [27].

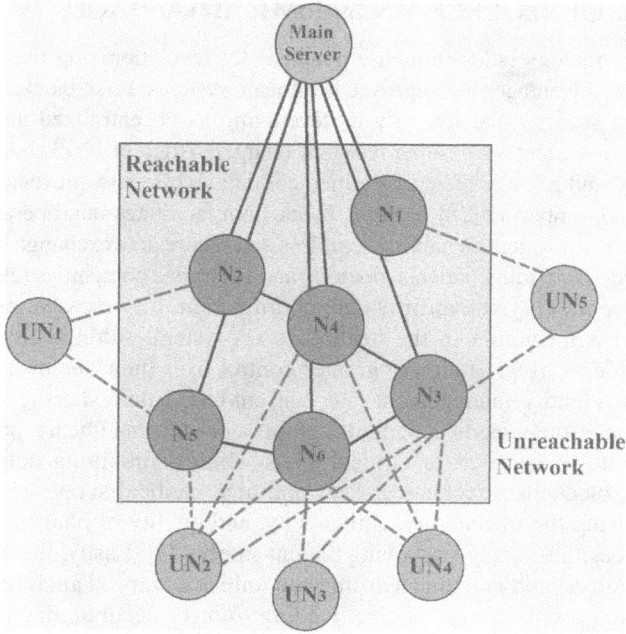

FIGURE 7.2 P2P (peer-to-peer) topology of Bitcoin.

A powerful backend server hosts the database management system

Client computers access the database on the server via a front end web interface or an API in a client installed application

Client/server Architecture

A database is attached to a server and stored on a SAN or RAID system

FIGURE 7.3 Client-server topology.

POTENTIAL OF BLOCKCHAIN IN IoMT: ADVANTAGES

Blockchain technology holds immense potential for revolutionizing the IoMT sector, offering several advantages for improved healthcare systems. First, blockchain provides enhanced data security and integrity by leveraging its decentralized and immutable nature. The transparent yet tamper-resistant characteristics of blockchain ensure the confidentiality and privacy of sensitive medical data, while also preventing unauthorized access and tampering [28]. Second, blockchain facilitates interoperability among different healthcare systems, enabling seamless and secure data exchange between various stakeholders, including patients, doctors, and insurance companies [29].

This interoperability streamlines care coordination, improves patient outcomes, and reduces fragmentation in the healthcare ecosystem. Additionally, blockchain empowers patients by giving them greater control over their health data, enabling self-sovereign identity management and consent-based data sharing. Patients can securely manage their medical records, grant access to healthcare providers, and even participate in research or clinical trials while maintaining data ownership. Furthermore, blockchain technology can optimize medical supply chain management by ensuring the traceability, authenticity, and quality of pharmaceuticals and medical devices, ultimately improving patient safety [30]. Lastly, the integration of blockchain with AI and machine learning can unlock advanced analytics, predictive modeling, and personalized medicine, leading to more accurate diagnostics, treatment optimization, and proactive healthcare interventions (Figure 7.4).

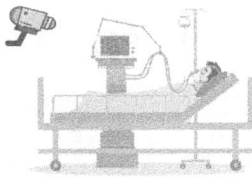

Real Time Observation

Managing Patient Reports

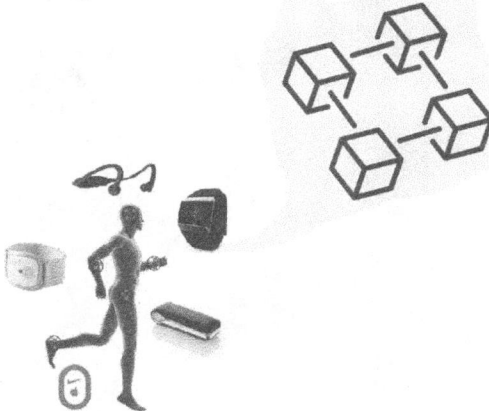

Remote Sensing by Body Sensors

Remote Consultation by Physician

FIGURE 7.4   Applications of blockchain in IoMT.

Overall, blockchain technology offers significant advantages in the IoMT sector, promoting data security, interoperability, patient empowerment, supply chain optimization, and advanced healthcare analytics [31].

LIMITATIONS OF BLOCKCHAIN TECHNOLOGY IN IoMT: DISADVANTAGES

There are several security challenges associated with using blockchain in IoMT. These should be considered and kept in mind while using blockchain technology. The same has been summarized below.

SCALABILITY

It represents a significant challenge in maintaining the security and integrity of the blockchain as the number of devices and transactions on the network grows. With an expanding network, ensuring the robustness of the blockchain becomes increasingly complex. The rise in device and transaction volume introduces difficulties in upholding the desired level of security and integrity for the system. Consequently, addressing scalability is a pressing concern for blockchain technology [32].

PRIVACY

It emerges as another challenge when considering the application of blockchain technology in healthcare. Although the transparency and immutability inherent in blockchain offer advantages, they can pose a significant hurdle to protect sensitive healthcare data. The transparent nature of the blockchain may conflict with the necessity to safeguard confidential medical information. Striking the right balance between transparency and privacy becomes crucial when incorporating blockchain into healthcare systems [15]. The medical device can use cryptographic methods to check the software/firmware's integrity and authenticity prior to installing an update or patch. To get the digital signature or hash of the trusted update, the device can access the blockchain. The device can verify that it hasn't been tampered with or compromised by comparing the update received with the one that was recorded on the blockchain.

INTEROPERABILITY

It stands as a formidable challenge due to the variations in protocols and standards across different blockchain networks. Each network may operate on distinct rules, creating difficulties when attempting to integrate them with existing healthcare systems. Achieving seamless interoperability between diverse blockchain networks and established healthcare infrastructures becomes a complex task, requiring careful consideration and harmonization of protocols and standards [33]. Interoperability in the healthcare industry may be improved by combining blockchain with IoT devices. Wearable fitness trackers and remote patient monitoring systems are two examples of IoT devices that can safely transmit data to the blockchain, where it can be accessed by healthcare professionals and other authorized parties. As a result, real-time monitoring, data analysis, and customized healthcare services are made possible, promoting the seamless integration of IoT data into the healthcare ecosystem.

THE CONSENSUS MECHANISMS

Employed in blockchain networks, consensus mechanisms can introduce security challenges. Consensus mechanisms are fundamental for ensuring agreement on the validity and ordering of transactions in a decentralized manner. However, certain consensus mechanisms, such as proof of work, are susceptible to attacks like 51% attacks or Sybil attacks [34]. These attacks exploit vulnerabilities in the consensus mechanisms, jeopardizing the security of the entire network. Therefore, designing robust consensus mechanisms that are resistant to such attacks is imperative for the secure implementation of blockchain technology. Smart contracts, which are self-executing contracts running on the blockchain, introduce their own set of vulnerabilities. While smart contracts offer automation and efficiency, they can also be susceptible to bugs or exploits that compromise the security of the network [35]. With the use of the blockchain network, a consensus process involving numerous parties, including producers of medical equipment, authorities in charge of regulation, and healthcare practitioners, can be implemented. With the help of this consensus, the devices are protected from malicious or unverified upgrades by making sure that only authorized and confirmed modifications are recorded on the blockchain.

FLAWS IN THE CODE OF A SMART CONTRACT

Administrative procedures like insurance claims and billing can be automated and streamlined using blockchain-based smart contracts. Self-executing contracts known as "smart contracts" have specified rules that are encoded on the blockchain. These contracts in the healthcare industry can automate payment settlements, eligibility checks, and claims processing, decreasing paperwork, eliminating errors, and speeding up transactions. Blockchain-based smart contracts have the potential to improve efficiency, reduce costs, and foster confidence in the healthcare payment ecosystem by getting rid of middlemen and enhancing transparency. But if there is a flaw in it, then it can result in unintended consequences that may lead to financial or operational damage. Thus, meticulous code review and comprehensive testing are essential to mitigate smart contract vulnerabilities and ensure the overall security of the blockchain network [36].

In conclusion, blockchain technology holds immense potential for enhancing security on the IoMT. However, it is crucial to address and overcome the associated challenges. Scalability must be tackled to maintain the integrity and security of the blockchain as the network expands. Privacy concerns must be carefully managed to protect sensitive healthcare data while leveraging the transparency of the blockchain. Achieving interoperability between various blockchain networks and existing healthcare systems necessitates the harmonization of protocols and standards. Robust consensus mechanisms are indispensable in guarding against attacks and ensuring the security of the network. Lastly, vulnerabilities must be mitigated through diligent code review and thorough testing. By considering and addressing these challenges, blockchain technology can be effectively employed to bolster security in the IoMT ecosystem [37].

FUTURE OF BLOCKCHAIN IN IoMT

The future of blockchain in the IoMT sector is poised to be transformative, with blockchain technology playing a pivotal role in revolutionizing healthcare systems [38]. Several key areas highlight the potential of blockchain in shaping the future of IoMT (Figure 7.5).

IMPROVED DATA SECURITY

Blockchain's decentralized and immutable nature ensures enhanced data security in IoMT. The tamper-resistant and transparent properties of blockchain provide a robust framework for protecting sensitive medical data, mitigating the risk of data

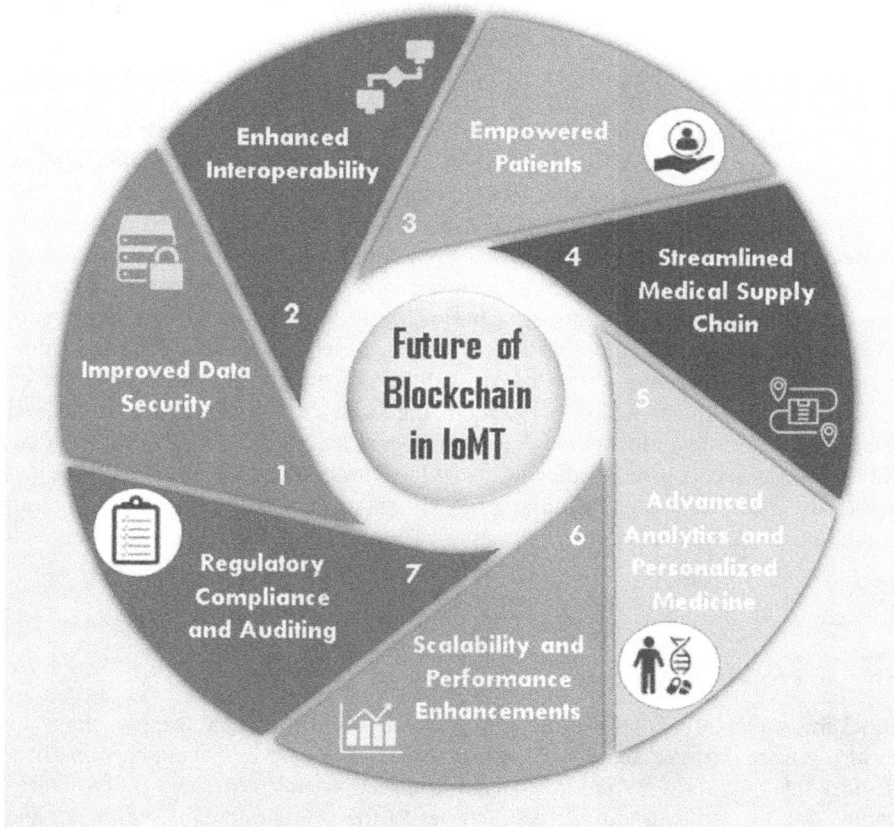

FIGURE 7.5   Future of blockchain technology in IoMT.

breaches and unauthorized access. By leveraging cryptographic techniques and smart contracts, blockchain enables the secure and auditable sharing of medical records, ensuring the confidentiality and privacy of patient information. Blockchain can also provide protection against fraud, security and transparency, access control, pseudonymity, and full decentralization. However, it is important to properly test IoMT/IoT devices before integrating them with blockchain technology to ensure that they will not cause damage to the system and to prevent physical damage from occurring [39].

ENHANCED INTEROPERABILITY

Blockchain has the potential to address the challenge of interoperability in healthcare by establishing a standardized platform for seamless data exchange among different healthcare providers and systems. The rapid growth in the number of IoMT/IoT devices has introduced scalability challenges, which refers to the system's ability to handle future growth without negatively impacting performance. Using common protocols and standards, blockchain enables secure and efficient sharing of healthcare data, promoting better care coordination and improved patient outcomes [40].

EMPOWERED PATIENTS

Blockchain technology empowers patients by giving them greater control over their health data. With self-sovereign identity solutions and smart contracts, patients can securely manage their medical records, grant access to healthcare providers, and even participate in research or clinical trials. This shift toward patient-centric data ownership fosters trust, transparency, and collaboration between patients and healthcare providers [41]. Huge amounts of patient-generated data can be gathered via IoT devices and wearables. By allowing them to own and control these data, blockchain can provide patients with more authority. Patients have the option to share this information with healthcare professionals, researchers, or insurance, thereby promoting personalized healthcare and possibly acquiring knowledge about their own medical conditions. Blockchain technology can help secure and open telemedicine services. Blockchain-based platforms allow for virtual consultations between patients and healthcare professionals while maintaining patient privacy, data integrity, and secure communication. The decentralized nature of blockchain technology also makes it possible for cross-border telemedicine, which eliminates distance restrictions and gives patients access to healthcare services from anywhere.

STREAMLINED MEDICAL SUPPLY CHAIN

Blockchain can revolutionize the management of the medical supply chain by ensuring traceability, authenticity, and quality control of pharmaceuticals, medical devices, and supplies. By recording each transaction and movement on the blockchain, stakeholders can easily track and verify the origin, storage conditions, and distribution of healthcare products, reducing the risks associated with counterfeit or substandard items. Since patient information is now dispersed among numerous healthcare organizations, it is challenging to obtain and share essential data when required. A patient-centric strategy that gives patients ownership over their medical information and the ability to provide healthcare providers access to that data can be implemented by using blockchain. The immutability of blockchain protects the accuracy of medical records, and its decentralized structure enables safe and frictionless data sharing, enhancing patient outcomes and care coordination [39].

ADVANCED ANALYTICS AND PERSONALIZED MEDICINE

The integration of blockchain with AI and machine learning has the potential to unlock advanced analytics, predictive modeling, and personalized medicine. By leveraging the secure and transparent infrastructure of blockchain to collect and share vast amounts of medical data, AI algorithms can gain valuable insights to improve diagnostics, treatment optimization, and proactive healthcare interventions. AI algorithms can find intricate patterns and indicators that might be signs of various diseases and ailments by analyzing massive amounts of medical data stored on the blockchain. By helping them diagnose patients more accurately, cut down on errors, and ensure that interventions are made on time, healthcare practitioners can benefit from advanced analytics. AI algorithms can identify risk factors and forecast the possibility of specific medical diseases or complications by routinely analyzing patient data and keeping track of health trends. By intervening sooner and perhaps halting the progression of diseases, this proactive strategy equips healthcare professionals to take preventive action. AI algorithms can develop individualized treatment regimens that are catered to the particular needs of each patient by reviewing a person's medical history, genetic information, lifestyle factors, and treatment outcomes. Sensitive patient data are stored securely and privately on the blockchain, and AI uses these data to deliver targeted and effective healthcare interventions [42].

SCALABILITY AND PERFORMANCE ENHANCEMENTS

As blockchain technology continues to evolve, efforts are underway to address its scalability and performance limitations. The development of layer 2 solutions, such as sidechains and off-chain transactions, aims to improve transaction throughput and reduce latency, making blockchain more suitable for real-time IoMT applications [43].

REGULATORY COMPLIANCE AND AUDITING

Blockchain's transparency and immutability simplify compliance with regulatory requirements in healthcare. By maintaining an auditable record of transactions and data access, blockchain facilitates regulatory audits, streamlines compliance processes, and ensures adherence to data protection and privacy regulations [44].

In summary, the future of blockchain in IoMT holds immense promise for transforming healthcare systems. Blockchain technology offers improved data security, enhanced interoperability, patient empowerment, streamlined supply chains, advanced analytics, and regulatory compliance. As the technology advances and matures, addressing scalability challenges and fostering collaboration among stakeholders will be crucial to fully realize the potential benefits of blockchain in IoMT. Healthcare industry research and development can be facilitated by the combination of blockchain and AI. Blockchain protects data privacy while still allowing AI algorithms to analyze huge datasets by securely distributing anonymized and aggregated data among institutions and researchers. This cooperative strategy can stimulate creativity, hasten medical advancements, and speed up research in fields like drug discovery, genomics, and precision medicine [45].

CONCLUSION

Revolutionizing healthcare systems by enhancing data security, promoting interoperability, empowering patients, streamlining the medical supply chain, enabling advanced analytics and personalized medicine, addressing scalability and performance challenges, and facilitating regulatory compliance and auditing—these advancements have the potential to significantly improve patient outcomes, enhance the efficiency of healthcare processes, and foster innovation in the IoMT ecosystem [46]. However, it is important to address the current limitations and challenges associated with blockchain technology, such as scalability, privacy concerns, interoperability, consensus mechanisms, and smart contract vulnerabilities. By addressing these challenges and leveraging the potential of blockchain technology, the future of IoMT can be shaped toward a more secure, interconnected, and patient-centric healthcare ecosystem [47]. In the coming years, as technology develops and matures, more IoMT applications of blockchain are anticipated to be adopted. To fully reap the rewards of blockchain in IoMT, it will be essential to resolve scalability concerns and promote stakeholder collaboration. Blockchain's safe and decentralized structure can make it easier to gather, share, and analyze enormous amounts of medical data, which will help AI algorithms produce insightful data for better diagnosis, treatment optimization, and proactive healthcare interventions [47].

Supply chain management, patient data protection, interoperability, and joint research projects are a few applications of blockchain technology that show potential which is also discussed in the above sections. Assuring the integrity and confidentiality of patient information, it offers a framework for safe and effective data interchange. Blockchain technology can improve patient outcomes, allow for personalized healthcare strategies, and spur innovation in the healthcare sector by fostering trust and openness. IoMT applications including smart contracts, decentralized clinical trials, telemedicine, and remote patient monitoring are anticipated to come into use in the future as blockchain technology continues to advance. These innovations have the power to fundamentally alter the way healthcare is provided, resulting in greater effectiveness, lower costs, and higher quality of patient care.

REFERENCES

1. (a) Nakamoto, S. Bitcoin: A Peer-to-peer Electronic Cash System. Technical Report, 2008.
2. Miraz, M. H.; Ali, M. Applications of blockchain technology beyond cryptocurrency. Annals of Emerging Technologies in Computing (AETiC) 2(1), 2018.

3. Halpin, H.; Piekarska, M. Introduction to security and privacy on the blockchain. In European Symposium on Security and Privacy Workshops (EuroS & PW), 2017, IEEE Computer Society.

4. Clayton, R., & Laurie, B. Proof of Work Proves not to Work. In 3rd Annual Workshop on Economics and Information Security, 2004, 1–9.

5. Soni, P.; Pal, A. K.; Islam, S. H. An improved three-factor authentication scheme for patient monitoring using WSN in remote health-care system. *Comput. Methods Programs Biomed.* 2019, *182*, 105054.

6. Moosavi, S. R.; Gia, T. N.; Rahmani, A.-M.; Nigussie, E.; Virtanen, S.; Isoaho, J.; Tenhunen, H. SEA: A secure and efficient authentication and authorization architecture for IoT-based healthcare using smart gateways. *Procedia Comput. Sci.* 2015, *52*, 452–459.

7. Alzubi, J. A. Blockchain-based Lamport Merkle digital signature: Authentication tool in IoT healthcare. *Comput. Commun.* 2021, *170*, 200–208.

8. Tahir, M.; Sardaraz, M.; Muhammad, S.; Khan, M. S. A lightweight authentication and authorization framework for blockchain enabled IoT network in health informatics. *Sustainability* 2020, *12*, 6960.

9. Nguyen, D. C.; Pathirana, P. N.; Ding, M.; Seneviratne, A. BEdgeHealth: A decentralized architecture for edge-based IoMT networks using blockchain. *IEEE Internet Things J.* 2021, *8*, 11743–11757.

10. Chiuchisan, I.; Costin, H.-N.; Geman, O. Adopting the Internet of Things technologies in health care systems. In Proceedings of the 2014 IEEE International Conference and Exposition on Electrical and Power Engineering (EPE), Iasi, Romania, 16–18 October 2014.

11. Asghar, M. H.; Negi, A.; Mohammadzadeh, N. Principle application and vision in Internet of Things (IoT). In Proceedings of the 2015 IEEE International Conference on Computing, Communication & Automation, Luxembourg, 8–11 September 2015.

12. Darshan, K.; Anandakumar, K. A comprehensive review on usage of Internet of Things (IoT) in healthcare system. In Proceedings of the 2015 IEEE International Conference on Emerging Research in Electronics, Computer Science and Technology (ICERECT), Mandya, India, 17–19 December 2015.

13. Janbi, N.; Katib, I.; Albeshri, A.; Mehmood, R. Distributed artificial intelligence-as-a-service (DAIaaS) for smarter IoE and 6G environments. *Sensors* 2020, *20*, 5796.

14. Joyia, G. J.; Rao, M.; Liaqat, A. F.; Rehman, S. Internet of medical things (IoMT): Applications, benefits and future challenges in healthcare domain. *J. Commun.* 2017, *12*, 240–247.

15. Zhou, L.; Wang, L.; Sun, Y.; Lv, P. Beekeeper: A blockchain-based IoT system with secure storage and homomorphic computation. *IEEE Access* 2018, *6*, 43472–43488.

16. Dorri, A.; Kanhere, S. S.; Jurdak, R. Towards an optimized blockchain for IoT. In Proceedings of the 2017 IEEE/ACM Second International Conference on Internet-of-Things Design and Implementation (IoTDI), Pittsburgh, PA, 18–21 April 2017.

17. Niranjanamurthy, M.; Nithya, B.; Jagannatha, S. Analysis of blockchain technology: Pros, cons and SWOT. *Cluster Comput.* 2019, *22*, 14743–14757.

18. Aitizaz, A.; Almaiah, M. A.; Hajjej, F.; Pasha, M. F.; Fang, O. H.; Khan, R.; Teo, J.; Zakarya, M. An industrial IoT-based blockchain enabled secure searchable encryption approach for healthcare systems using neural network. *Sensors* 2022, *22*, 572.

19. Epiphaniou, G.; Pillai, P.; Bottarelli, M.; Al-Khateeb, H.; Hammoudesh, M.; Maple, C. Electronic regulation of data sharing and processing using smart ledger technologies for supply-chain security. *IEEE Trans. Eng. Manage.* 2020, *67*, 1059–1073.

20. Kumar, A.; Sharma, S.; Goyal, N.; Singh, A.; Cheng, X.; Singh, P. Secure and energy-efficient smart building architecture with emerging technology IoT. *Comput. Commun.* 2021, *176*, 207–217.

21. Almalki, J.; Al Shehri, W.; Mehmood, R.; Alsaif, K.; Alshahrani, S. M.; Jannah, N.; Khan, N. A. Enabling blockchain with IoMT devices for healthcare. *Information* 2022, *13*(10), 448.
22. Jan, M. A.; Cai, J.; Gao, X.-C.; Khan, F.; Mastorakis, S.; Usman, M.; Alazab, M.; Watters, P. Security and blockchain convergence with internet of multimedia things: Current trends, research challenges and future directions. *J. Netw. Comput. Appl.* 2021, *175*, 102918.
23. Thomas, S.; Schwartz, E. A Protocol for Interledger Payments, 2015.
24. Kan, L.; Wei, Y.; Muhammad, A. H., et al. A multiple blockchains architecture on inter-blockchain communication. In 2018 IEEE International Conference on Software Quality, Reliability and Security Companion (QRS-C). IEEE, 2018, 139–145.
25. Wang, H.; Cen, Y.; Li, X. Blockchain router: A cross-chain communication protocol. In Proceedings of the 6th International Conference on Informatics, Environment, Energy and Applications, 94–97, 2017.
26. Wood, G. *Polkadot: Vision for a Heterogeneous Multi-chain Framework.* 2016, 21(2327), 4662.
27. Greenspan, G. *Multi-chain Private Blockchain, White Paper.* 2015.
28. Sun, W.; Cai, Z.; Li, Y.; Lui, F.; Fang, S.; Wang, G., Security and privacy in the medical Internet of Things: A review. Security and Communication Networks, 2018, *2018*, 9, Article ID 5978636.
29. Abou-Nassar, E. M.; Iliyasu, A. M.; El-Kafrawy, P. M.; Song, O. Y.; Bashir, A. K.; Abd El-Latif, A. A. DITrust chain: Towards blockchain-based trust models for sustainable healthcare IoT systems. *IEEE Access* 2020, *8*, 111223–111238.
30. Trnka, M.; Cerny, T.; Stickney, N. Survey of authentication and authorization for the Internet of Things. *Secur. Commun. Netw.* 2018, *2018*, 4351603.
31. Ellouze, F.; Fersi, G.; Jmaiel, M. Blockchain for internet of medical things: A technical review. The Impact of Digital Technologies on Public Health in Developed and Developing Countries, pp. 259–267, Springer, Cham, 2020.
32. Bormann, C.; Castellani, A. P.; Shelby, Z. Coap: An application protocol for billions of tiny internet nodes. *IEEE Internet Comput.* 2012, *16*, 62–67.
33. Bataineh, M. R.; Mardini, W.; Khamayseh, Y. M.; Yassein, M. M. B. Novel and secure blockchain framework for health applications in IoT. *IEEE Access* 2022, *10*, 14914–14926.
34. Yaga, D.; Mell, P.; Roby, N.; Scarfone, K. Blockchain Technology Overview. National Institute of Standards and Technology, Gaithersburg, MD, 2018.
35. Katib, A. F. M. R.; Albogami, I.; Albeshri, N. N. A. Data fusion and IoT for smart ubiquitous environments: A survey. *IEEE Access* 2017, *5*, 9533–9554.
36. Abdmeziem, M. R.; Tandjaoui, D. A cooperative end to end key management scheme for e-health applications in the context of Internet of Things. In Ad-Hoc Networks and Wireless, pp. 35–46, Springer, Berlin-Heidelberg, 2014.
37. Jurcut, A. D.; Ranaweera, P.; Xu, L. *Introduction to IoT Security*, 2020, 27–64.
38. Tagde, P.; Tagde, S.; Bhattacharya, T.; Tagde, P.; Chopra, H.; Akter, R.; Kaushik, D.; Rahman, M. H. Blockchain and artificial intelligence technology in e-health. *Environmental Science and Pollution Research International* 2021, *28*(38), 52810–52831.
39. Zubaydi, H. D.; Varga, P.; Molnár, S. Leveraging blockchain technology for ensuring security and privacy aspects in Internet of Things: A systematic literature review. *Sensors* 2023, *23*(2), 788.
40. Schmeelk, S.; Kanabar, M.; Peterson, K.; Pathak, J. Electronic health records and blockchain interoperability requirements: A scoping review. JAMIA Open 2022, *5*(3), ooac068.
41. Harrell, D. T.; Usman, M.; Hanson, L.; Abdul-Moheeth, M.; Desai, I.; Shriram, J.; Bautista, J. R.; Meyer, E. T.; Khurshid, A. Technical design and development of a self-sovereign identity management platform for patient-centric health care using blockchain technology. Blockchain in Healthcare Today 2022, *5*.

42. Fiore, M.; Capodici, A.; Rucci, P.; Bianconi, A.; Longo, G.; Ricci, M.; Sanmarchi, F.; Golinelli, D. Blockchain for the healthcare supply chain: A systematic literature review. *Applied Sciences* 2023, *13*(2), 686.

43. Islam, M. S. U.; Kumar, A.; Hu, Y. Context-aware scheduling in fog computing: A survey, taxonomy, challenges and future directions. *Journal of Network and Computer Applications* 2021, *180*, 103008.

44. Marbouh, D.; Simsekler, M. C.; Salah, K.; Jayaraman, R.; Ellahham, S. A blockchain-based regulatory framework for mHealth. *Data* 2022, *7*(12), 177.

45. Khezr, S.; Moniruzzaman, M.; Yassine, A.; Benlamri, R. Blockchain technology in healthcare: A comprehensive review and directions for future research. *Applied Sciences* 2019, *9*(9), 1736.

46. Ghosh, P. K.; Chakraborty, A.; Hasan, M.; Rashid, K.; Siddique, A. H. Blockchain application in healthcare systems: A review. *Systems* 2023, *11*(1), 38.

47. Adere, E. M. Blockchain in healthcare and IoT: A systematic literature review. *Array* 2022, *14*, 100139.

8 Wireless Sensor Network-Based Automated Computer-Assisted Diagnostic Approaches for Identification of Brain Tumor Disease in IoT-Based Healthcare

Kamini Lamba and Shalli Rani
Chitkara University Institute of Engineering and
Technology, Chitkara University, Punjab, India

BACKGROUND

The brain is one of the sensitive organs of the human body which is responsible for monitoring multiple functionalities inside the brain for interaction with outer world in distinct aspects according to every situation. However, appropriate functionality of brain can be disrupted in case of sudden injury caused to the brain due to any incident which may give the worst impact on growth of tissues lying within brain and may result into development of tumor inside brain. Although magnetic resonance imaging, computed tomography, positron emission tomography, and so on are used by radiologists for diagnosing presence of tumor inside brain. However, this process could be time-consuming and symptoms of development of tumor inside brain can remain unidentified due to very small size. So, there is a deep of developing an automated system which can be used by radiologists to make their decision regarding presence of tumor at an initial stage accurately for timely treatment of patients to save their lives. To do so, various systems based on deep learning techniques [1] are proposed by researchers due to the property of having multiple layers to extract significant features from the input data. On the contrary, traditional approaches have comparatively less capability for extraction of significant features due to the presence of single layer for transformation of input data to the desired output [2].

Thus, various factors such as multiple hidden layers, their inter-connectivity, and capacity for learning features from given input make great difference in conventional and deep learning-based networks. Other than this, traditional artificial neural

DOI: 10.1201/9781003438205-8

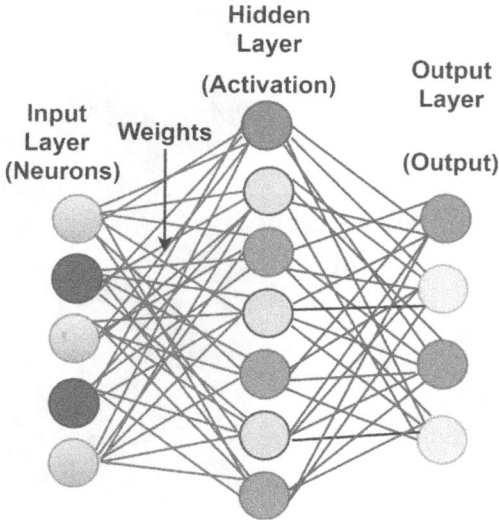

FIGURE 8.1   Basic architecture of traditional artificial neural network.

networks generally comprise three layers which are given training to achieve representations based on supervised learning for performing particular tasks [3]. On the contrary, deep neural networks have ability to represent patterns outcomes based on the given inputs via the following concept of unsupervised learning [4]. Figure 8.1 illustrates the pattern of traditional artificial neural network comprising three layers responsible for input, processing, and extraction of features where neurons are residing at the input to identify pixel values of given data which are further sent to hidden layer with respect to their weights for identifying edges based on which output layer extracts features for generating output at the final stage and Figure 8.2 represents deep neural networks having multiple layers for input, processing, and extraction of features which are further fed to fine-tune the network using algorithm of back-propagation to perform end-to-end tasks.

It is also clear that models based on deep learning help in processing complex patterns as well as objects to achieve accurate and efficient outcomes while predicting and classifying the given data due to which these models have demonstrated revolutionized potential in healthcare sector. Thus, healthcare sector has a high demand of emerging deep neural networks in its respective domain due to the dynamic as well as complicated nature of associated data such as medical imaging, electronic health records, data extracted from IoT (Internet of Things)–based sensors, wearables, and so on to improve patient outcomes when it comes to diagnose brain tumor disease in the healthcare. Thus, it is observed that deep learning networks have the ability to extract significant information based on distinct patterns that could be even challenging for existing approaches to analyze accurately.

Additionally, deep neural networks also support analysis of laboratory procedures which may further include chemicals, diagnosing glucose level, blood pressure, hemoglobin report, electronic health records, clinical notes, and specific tests

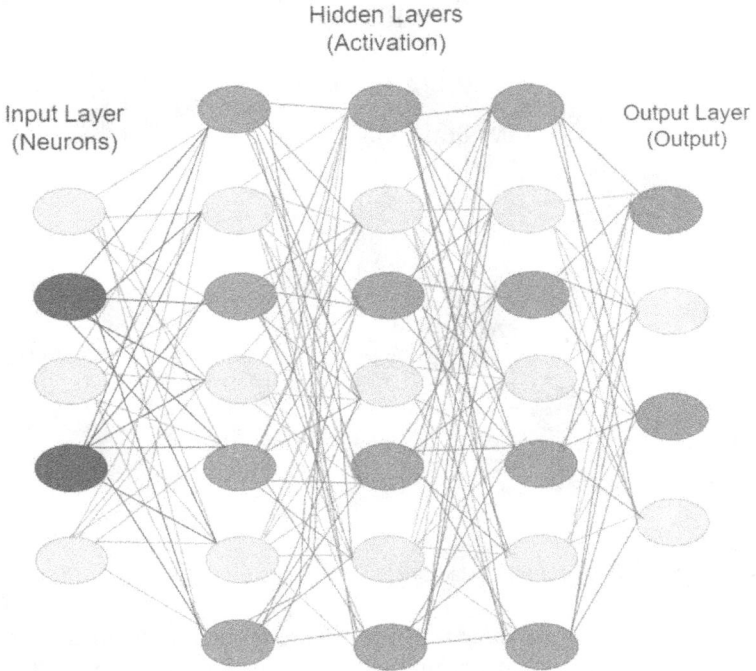

FIGURE 8.2   Basic architecture of deep neural networks.

for identifying presence of disease along with presence of drugs in an individual which could help medical professionals to provide recommendations to their patients for their speedy recovery. Thus, IoT-based sensors, wearables, and so on can be efficiently used for collection of data from multiple sources. Pre-processing is performed on the acquired data for keeping values of input data between "0" and "1" to perform normalization and resizing can also be done at this stage for removal of unwanted noise from the input data and to maintain uniformity. Moreover, augmentation of data can also be done if considered classes of data are not balanced to avoid overfitting and underfitting issues to achieve efficient and accurate outcomes in the healthcare sector.

Pre-processed data are generally forwarded to deep neural networks for extraction of hidden features from the input data which is the key parameter of utilizing deep learning models as these networks are capable enough for analyzing input data. Most of the researchers utilized transfer learning models at this point which are pre-trained so consume comparatively less time in providing training to the network and diagnosing presence of tumor. However, classifiers such as support vector machines, random forest, and so on can further be deployed at the output for differentiating infected images from healthy ones. After performing such operation, validation of model can also be done based on the test data to identify performance of the proposed model in terms of its performance metrics such as accuracy, sensitivity, specificity, and F1 score. Based on such parameters, deep learning models are generally

TABLE 8.1

Comparison of Approaches in the Healthcare

Technique	References	Features
Convolutional neural network	[1]	Excellent performance in analysis
Recurrent neural network	[2]	Ability to model temporal dependency
Generative adversarial network	[4]	Generates synthetic data
Transformer-based model	[5]	Efficient natural language processing tasks
Long short-term memory	[6]	Vanishes gradient issues
Deep reinforcement learning	[7]	Handles complex decision-making tasks

recommended by radiologists to have an alternate while making any decisions with respect to tumor presence.

LeCun et al. [1] have summarized a systematic overview comprising architectures as well as approaches of deep neural networks to represent the potential of its applications in healthcare. Rajkomar et al. [5] discussed the significance of deep learning in healthcare sector for analyzing images, decision-making processes for providing clinical support. Authors also presented in-depth analysis of techniques based on machine learning to provide emphases of deep learning in healthcare. Table 8.1 has shown comparison of existing approaches of deep neural networks in the healthcare along with its specific approaches, description, advantages, and drawbacks. Thus, deep learning has shown great potential to offer various advantages as well as probabilities to transform existing approaches in the healthcare sector.

For instance, models comprising convolutional neural networks showed amazing results during analysis of medical images. These automated systems can help in diagnosing numerous diseases such as tumor, lung cancer, skin cancer [6], Alzheimer's, Parkinson's, skin lesions, and so on an early stage due to which experts can consider these systems while making any decision regarding patients' health. Even algorithms based on deep neural networks [7] can also consider vast amount of clinical data for processing as well as analyzing purpose which further may include clinical notes, electronic health records, and so on. Due to this property, such systems are capable enough to extract features from the given data to give contribution in decision-making process which may ultimately result in efficient as well as accurate outcomes.

Table 8.2 represents difference between IoT and deep learning approaches in the healthcare based on key parameters.

CLASSIFICATION OF DEEP LEARNING MODELS FOR DIAGNOSING BRAIN TUMOR

Numerous deep learning techniques have been characterized into various algorithms which are required in smart healthcare industries to provide efficient and accurate outcomes while diagnosing brain tumor disease. Integration of such IoT and deep learning technology demonstrates great promise in healthcare via including wearable, medical sensors, health monitoring devices, predictive approaches, clinical

TABLE 8.2
Parameter-Based DL and IoT Approaches

Parameters	IoT Approaches	Deep Learning Approaches
Data collection	Sensors, wearables	Depends upon pre-existing data or requires labeled data
Data volume	Ability to handle real-time data at large scale	Can handle vast amount of data
Data variety	Distinct collection such as environmental factors, vital signs	Structured data
Real-time processing	Monitors real-time applications and performs analysis	Needs significant computational time
Complexity	Simple and inexpensive	Complex and require much infrastructure
Flexibility	Ability to adapt in distinct healthcare settings	Ability to perform distinct healthcare tasks
Interpretability	Depends upon additional analytics	Depends upon feature extraction
Training data	Limited or biased	Needs vast and diverse datasets
Model updation	May ask for updation in firmware	possible without hardware updations
Privacy and security	Concerned	Needs safeguards
Scalability	Limited to device capacity	Requires high-performance computing
Applications	Real-time monitoring, remote-patient monitoring, smart healthcare	Diagnosing disorders, medical imaging, drug discovery

records management, and so on. These classification approaches provide clear vision of utilizing them in the specific research applications of healthcare domain while discussing its benefits well as challenges.

Hybrid Models

To analyze images as well as sequence-based medical data, hybrid models come in effect via collaborating two or more deep learning models to have efficient and accurate outcomes in the smart healthcare sector for diagnosing brain tumor disease along with providing recommendations if any. For instance, hybrid models have the ability to perform analysis of clinical notes and electronic health records which may comprise structured form representing lab results, patients' health records, and so on or unstructured form representing patients' symptoms over time.

Transfer Learning

It utilizes the concept of pre-trained deep learning networks for extracting features in terms of pre-trained weights based on large dataset to overcome issues of less availability of dataset. It can also reduce time-consumption process in diagnosing diseases as these networks come with pre-training facility due to which they can learn from the past experiences and hence result in improved performance of networks in the smart healthcare industries. Thus, the patterns or features by one model during training process can be utilized by another model without any training to provide quick and efficient outcomes in diagnosing brain tumor. For instance, a model which is pre-trained on MRI images can be provided with fine-tuning feature to have efficient outcomes while diagnosing brain tumor or performing segmentation of brain stroke. Thus, obtained features can be transferred from one place to another for performing the desired operation at another level in healthcare domain. Figure 8.3 illustrates effective hybrid model for diagnosing brain tumor disease in

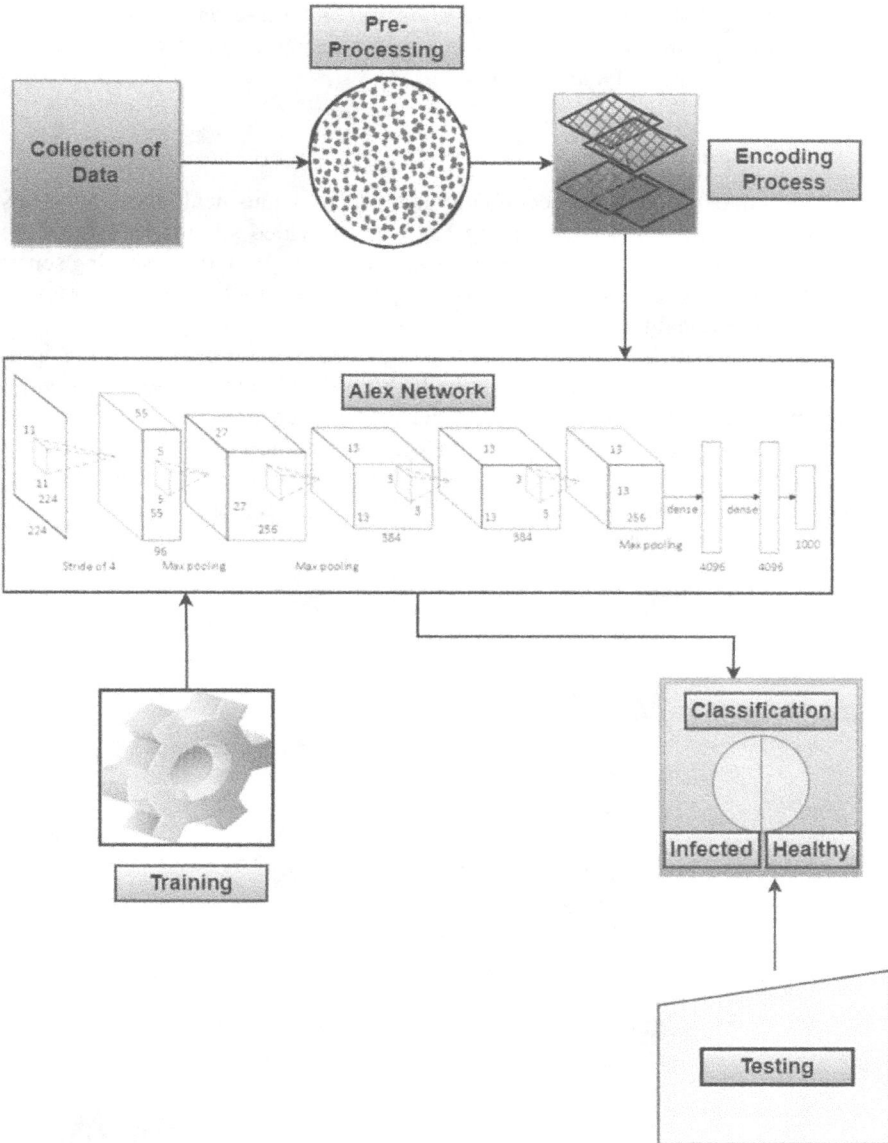

FIGURE 8.3   Basic architecture of effective hybrid model with style-transfer learning.

collaboration with style-transfer learning [8] where data have been gathered with the help of IoT sensors, wearables, and so on and undergone pre-processing for removal of unwanted distortions or balancing its respective classes with the help of data augmentation, resizing, and so on and conversion of images is done into respective pixels with the help of encoding operation.

Training has been provided to the model with the help of hybrid Alex network which further utilized concept of transfer learning for extraction of significant features from input data while diagnosing brain tumor disease and obtained output is evaluated based on test data to achieve efficient results in terms of its accuracy, specificity, time complexity, and sensitivity.

FEDERATED LEARNING

To maintain data privacy, this technology is used in various healthcare industries as shown in Figure 8.4 where training has been provided simultaneously to the multiple interconnected devices on decentralized server [9] without sharing sensitive information to central hub comprising aggregator which leads to privacy of data in the smart healthcare industries. Moreover, this technology also minimizes cost of communicating between multiple devices as procedure of training is performed at local level on edges of devices which ultimately requires less bandwidth consumption too. Thus, initial global model helps in achieving privacy of data in the smart healthcare at its most to provide efficient outcomes while diagnosing brain tumor disease due to its property of supporting updations in the local model whenever required and verification of model is performed to ensure effective outcomes in the smart healthcare.

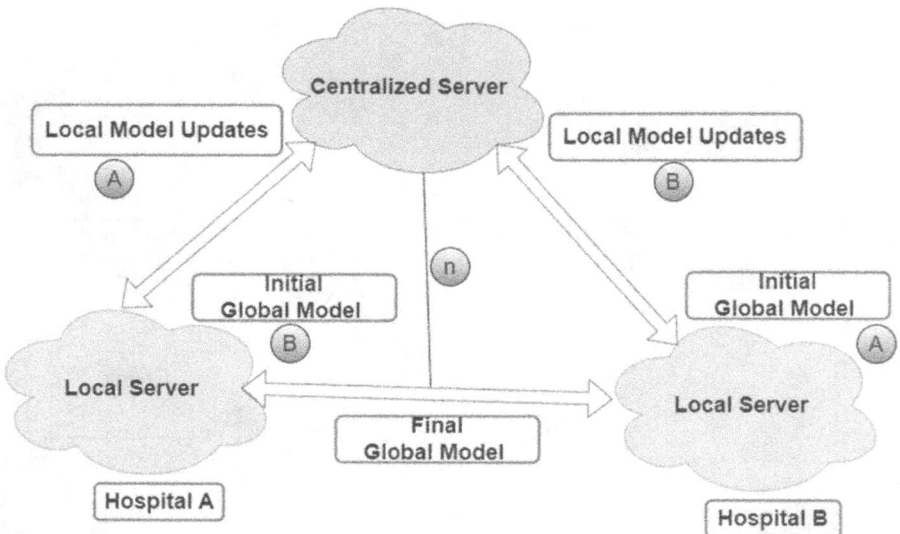

FIGURE 8.4 Basic architecture of federated learning.

REINFORCEMENT LEARNING

This technology helps in training models on the current experiences via sending the agent in the open environment. For each and every action, agents get reward for positive action which enables him to reach destination and penalty for each negative action that moves him away from reaching his goal. Thus, it helps in making intelligent decisions based on the feedback they received to improve prediction analysis and enhance performance of networks in the smart healthcare industries.

Based on these classification approaches, Bolhasani et al. [10] discussed benefits, applications, and challenges of multiple deep learning techniques to be used in the smart healthcare industries for diagnosing disorders at an early stage, health monitoring, risk management, decision-making process, and so on to have efficient and accurate results.

DEEP LEARNING ARCHITECTURES FOR DIAGNOSING BRAIN TUMOR DISEASE

There are various frameworks based on deep learning models which have been designed to be used in the special applications of smart healthcare which ultimately enables the task of analyzing, interpreting, and extracting meaningful information from the given input to enhance and improve patients' outcomes in the healthcare industries. Following is the description of a few deep learning architectures utilized for diagnosing brain tumor in smart healthcare sector.

CONVOLUTIONAL NEURAL NETWORK

Convolutional neural network architecture as shown in Figure 8.5 comprises convolutional, pooling, fully connected layers to perform analysis of given data in the smart healthcare systems for extraction of meaningful information and providing clear vision of understanding the brain tumor disease at an early stage to initiate timely treatment and risk management purposes. To achieve such tasks, convolutional layers utilize filters to remove noise from images and detection of patterns or features accurately.

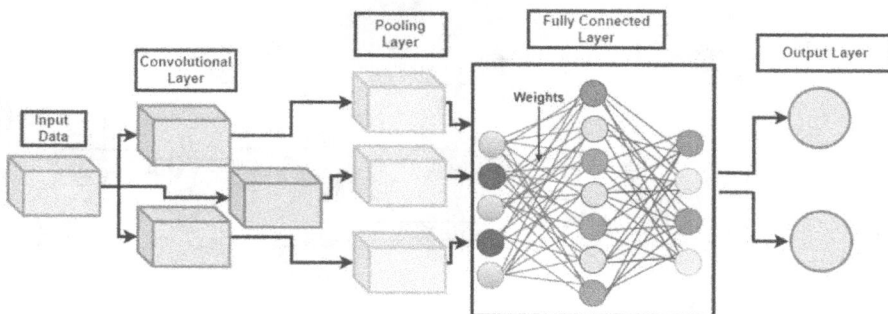

FIGURE 8.5   Basic architecture of convolutional neural networks.

Pooling layer is responsible for reduction of high-dimensionality features to minimize complexity of networks to have significant features. Fully connected layers are used for making predictions based on learned features, thus providing efficient outcomes in smart healthcare. Thus, these networks can be deployed in multiple applications of healthcare sector, such as identification of Alzheimer's, Parkinson's, tumor from MRI images, classification of skin cancer, glaucoma, diabetes, and so on, and diagnose any abnormality lying within scans.

Litjens et al. [11] also discussed various aspects of deep learning models comprising convolutional neural networks too for analyzing medical images which also convey its benefits and challenges and highlight the efficient and accurate outcomes in smart healthcare industries.

RECURRENT NEURAL NETWORK

Recurrent neural networks as shown in Figure 8.6 are used in analysis of sequence-based data for real-time applications in smart healthcare systems as they have the ability to capture dependencies as well as variations in lengthy variables collected from IoT devices.

These networks have features of feedback connectivity [12] which ultimately allows data to pass from one place to other based on distinct time zones. Hidden layer lying within these networks is responsible for maintaining experiences of past data which keeps itself updating and helps for future tasks to be performed. Thus, recurrent layers compute output at each step and send signal back to the network for accurate analysis based on time-series. Due to these properties, these networks are used to analyze clinical notes, electronic health records too for prediction of progress of brain tumor disorder and risk management and remote-based patient monitoring in the smart healthcare sectors. Choi et al. [13] also provided studies in which recurrent neural networks have been utilized for prediction of clinical events with the help of collecting electronic health records to achieve efficient results in making intelligent decisions while working in smart healthcare industries.

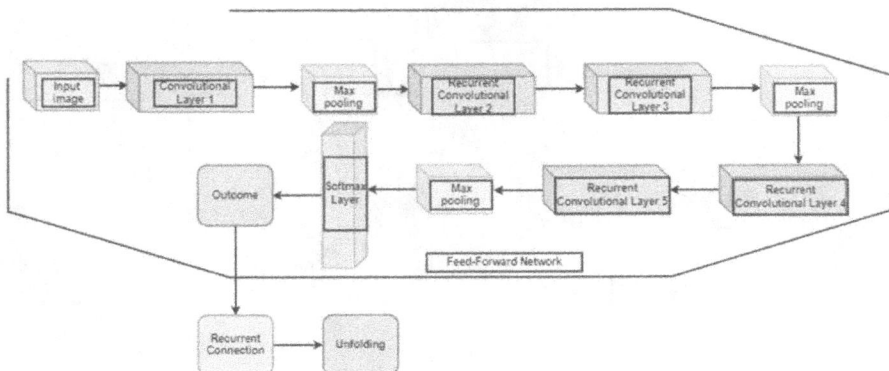

FIGURE 8.6   Basic architecture of recurrent neural networks.

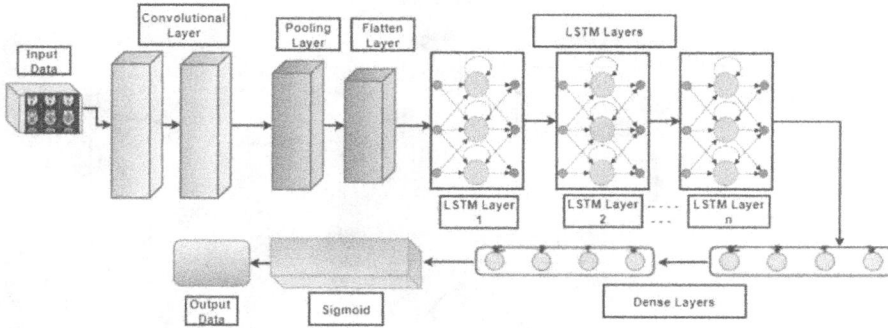

FIGURE 8.7 Basic architecture of long short-term memory model (LSTM).

LONG SHORT-TERM MEMORY MODELS

Long short-term memory models as shown in Figure 8.7 have the ability to tackle dependencies of long term while analyzing input data based on the time series for clinical notes, electronic health records, and so on and these networks are able to resolve issues raised due to vanishing gradient. Due to the availability of memory cells comprising three elements, namely, input gate responsible for regulating flow of data towards memory cell, forget gate which helps in monitoring information need to be removed from memory cell, and output gate used for determination of output value on the base of given input and state of memory cell, it has vast storage capacity for the information to be shared from one network to other to access portability for providing facilities to the patients for diagnosing and treatment of disorder if any at any time which ultimately minimize chances of losing medical data thus provide efficient outcomes in the smart healthcare industries. Lipton et al. [14] discussed various techniques of these networks used in collaboration with recurrent neural networks in diagnosing disease based on time-series while working on sequential medical data where long-term dependencies exist in actual and detection of abnormalities from input MRI images is required at the priority level to avoid risk of losing patients' life and improve outcomes of patients' health while diagnosing brain tumor in smart healthcare industry.

AUTOENCODERS

Autoencoder networks as shown in Figure 8.8 follow unsupervised learning technique which help in reconstruction of given data and identifying abnormalities lying within signals or images provided as an input in healthcare systems which have been generally collected via IoT sensors or wearable. These networks have ability to identify any deviation from the actual pattern as they comprise encoding-decoding parameter [15] where encoding is used for mapping input data to capture significant features where decoding is used for reconstructing the actual input data based on the representations stored in its network for achieving accurate results. Moreover, autoencoders have feature of data imputation that aids in filling missing or noisy medical data collected by various IoT sensors or variables on the basis of learned patterns representations. Generation of synthetic data is also possible to attain with the help of autoencoders which can be performed to achieve operations of augmenting data

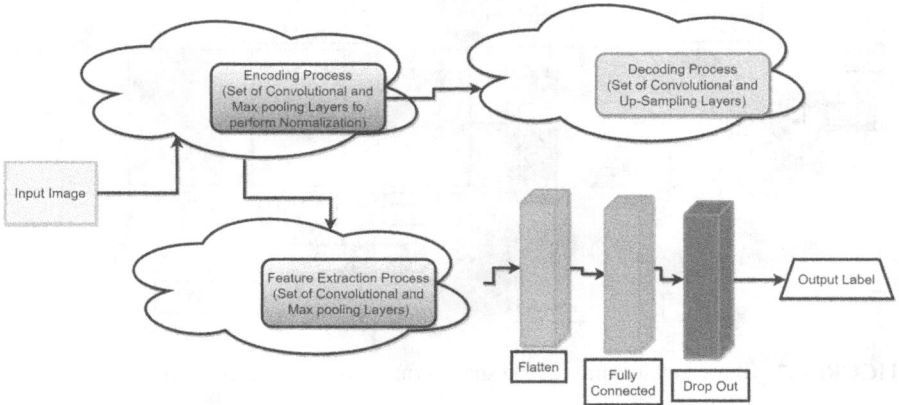

FIGURE 8.8 Basic architecture of convolutional based autoencoder.

in case of less availability of data or to balance the classes while diagnosing diseases in smart healthcare sector. Kim et al. [16] have also summarized the significance of using autoencoders in healthcare systems for capturing various patterns that can be complex too and identifying abnormalities if any to analyze and reconstruct medical data for improvement in the patients' outcomes while diagnosing brain tumor in smart healthcare applications.

GENERATIVE ADVERSARIAL NETWORK

Generative adversarial networks as shown in Figure 8.9 are utilized to generate synthetic medical data for improving pixel resolution of images due to which these networks can further perform operations based on data augmentation, filling missing or noisy values in input data while training model to achieve accurate and efficient results. These

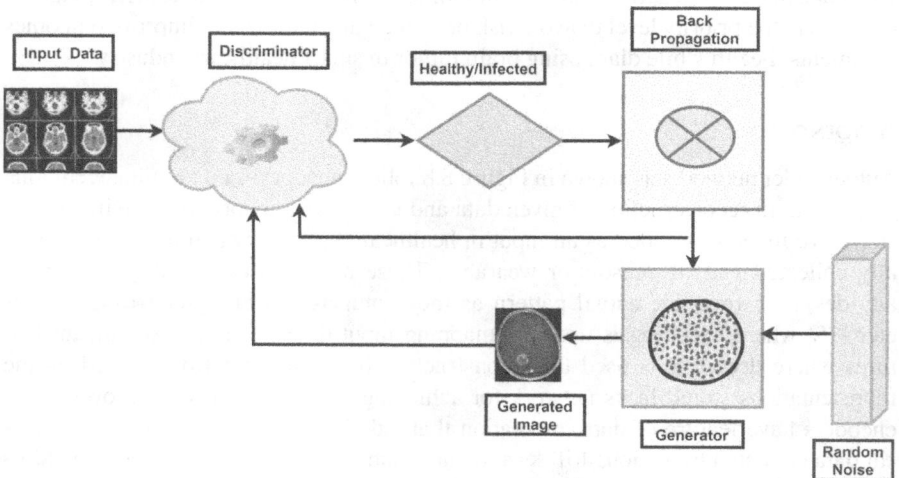

FIGURE 8.9 Basic architecture of generative adversarial network.

networks are generally comprised two elements, namely, a generator network responsible for obtaining synthetic data via mapping of vectors to achieve required distribution of data and discriminator network responsible for providing feedback to generator to help them improve their capability for generation of more accurate data. Thus, both networks give significant contributions in analyzing and predicting diseases where accessing data are restricted due to regulatory frameworks in healthcare while preserving data to be transferred from one place to other. Yi et al. [17] have also discussed various studies in which these networks have been used in specific applications of healthcare to obtain synthetic data, image transformation as well as diagnosis of abnormalities if any. These networks also demonstrate great promise in generating accurate and efficient outcomes while diagnosing brain tumor in smart healthcare industry.

Moreover, Table 8.3 represent techniques proposed by various researchers for diagnosing brain tumor disorder in the smart healthcare along with the respective objectives.

TABLE 8.3
State-of-Art Techniques for Diagnosing Brain Tumor Disease

Year	References	Techniques	Objective
2020	[20]	Deep neural network-transfer learning	Automated brain tumor classification
2019	[21]	Convolutional neural network, discrete wavelet transformation, long short-term memory network	Distinguishing liver and brain tumor
2022	[22]	Deep convolutional neural network	Diagnosing brain tumor disease
2019	[23]	Deep convolutional neural network with transfer learning	Classification of brain tumor
2022	[24]	Deep learning and VGG16	Analysis of brain tumor
2019	[25]	Convolutional neural network	Classification of brain waves
2019	[26]	Integrated noise-to-image and image-to-image	Data augmentation for diagnosing brain tumor
2018	[27]	Deep neural networks	Classification of brain tumor
2017	[28]	Deep neural network	Diagnosing brain cancer
2016	[29]	Contourlet transform and Zernike moments–based data fusion	Diagnosing brain tumor disease
2020	[30]	Integrated optimal wavelet statistical texture and recurrent neural network	Diagnosing and classifying brain tumor disease
2019	[31]	Convolutional neural networks	Identification and classification of abnormal brain images
2020	[32]	Deep neural network	Detection and classification of brain tumor disease
2022	[33]	Generative adversarial network-style transfer	To generate brain tumor image
2020	[34]	Enhanced Softmax loss function-deep neural network	Diagnosing brain tumor disease
2021	[35]	Deep neural network: machine learning classifier	Classifying brain tumor disorder
2021	[36]	Deep convolutional neural network: fully optimized framework	To perform multi-classification of brain tumor

(Continued)

TABLE 8.3 *(Continued)*

Year	References	Techniques	Objective
2020	[37]	Block-wise fine-tuning: transfer learning	To classify brain tumor
2019	[38]	DNN: machine learning algorithm	Brain tumor diagnosis
2019	[39]	Convolutional neural network: genetic algorithms	Classification of brain tumor grades
2019	[40]	Deep neural network	Brain tumor diagnosis
2019	[41]	Capsule: network	To identify the impact of pre-processing on brain tumor classification
2019	[42]	Convolutional neural network	To classify brain images
2021	[43]	Convolutional dictionary learning-local constraint	To classify brain tumor
2021	[44]	3D-CNN and feature selection	To diagnose microscopic brain tumor
2019	[45]	Deep CNN-extensive data augmentation	To classify multi-grade brain tumor
2021	[46]	CNN-SVM	To identify brain tumor disease
2021	[47]	Residual network-global average pooling	To classify multi-class brain tumor
2020	[48]	Deep neural network-robust feature selection	To diagnose and classify multimodal brain tumor
2020	[49]	Integrated VGG16-CNN	To classify brain tumor disease
2018	[50]	Hybrid learning	To diagnose and classify brain tumor
2018	[51]	3D convolutional neural network	To classify pancreatic tumor
2017	[52]	Deep neural network	To segment brain tumor
2019	[53]	Deep neural network	To diagnose and segment brain tumor
2017	[54]	Deep neural network	Theoretical framework to segment brain tumor
2016	[55]	CNN: transfer learning	Theoretical Framework to diagnose brain tumor
2019	[56]	Deep neural network	To assess glioma burden for automatic volumetric and bidirectional measurement
2016	[57]	Convolutional neural network	To segment MRI Brain images automatically
2019	[58]	Transfer learning: fine-tuning	To diagnose and classify brain tumor
2019	[59]	R-convolutional neural network	To diagnose and classify brain tumor
2020	[60]	Convolutional neural network	To diagnose brain tumor using MRI scans
2015	[61]	Watershed segmentation	To diagnose brain tumor

INNOVATIVE NEW TECHNIQUES FOR DIAGNOSING BRAIN TUMOR IN SMART HEALTHCARE

Most of the researchers discussed about traumatic injury [18] in brain along with its origin and causes and some authors provided pharmacological preconditioning of brain [19]. Based on such research, following are the innovative new techniques responsible for incorporating existing approaches with advanced technology to predict disorder at an early stage, enabling remote patient monitoring and improve patient outcomes.

INTERNET OF THINGS AND WEARABLE

To monitor patients' health remotely, collaboration of IoT [62] and wearable is required for collecting data based on real-time applications to make predictive analysis for preventing patients from deadly disease via providing timely treatment.

ARTIFICIAL INTELLIGENCE AND MACHINE LEARNING

To make intelligent decisions while diagnosing disorders and predicting progress [63] of disorder, techniques of artificial intelligence and machine learning play major role in analyzing vast amount of data for effective and accurate outcomes.

BLOCKCHAIN

To ensure integrity of data, interoperability, and management of medical records, this technology [64] enhances data privacy as well as security while sharing data from one place to another.

TELE-MEDICINE AND VIRTUAL REALITY

To consult experts remotely, this technology [65] provides medium to patients which ultimately minimize the need to visit in-person and promotes telemedicine experience.

PREDICTIVE ANALYTICS AND BIG DATA

For identification of patterns to make accurate and efficient predictions, techniques of predictive analytics and big data [66] come in effect.

NEW OPPORTUNITIES IN THE 21ST CENTURY

Following are the new opportunities which have been brought ahead due to the integration of IoT and deep learning techniques for diagnosing brain tumor in smart healthcare as shown in Figure 8.10.

DISEASE DIAGNOSIS

There are various algorithms based on deep learning have demonstrated great potential in diagnosing diseases based on medical images given as input data. Such techniques are capable enough to perform analysis of medical images and diagnose diseases if any. Convolutional neural networks have been considered in such techniques which give major contributions in healthcare while diagnosing and classification of diseases [6] to provide efficient and accurate results.

FIGURE 8.10   Opportunities in smart healthcare.

HEALTH MONITORING

Smart healthcare systems enable the collection of physiological data for monitoring patients' health with the help of IoT devices [5] which help to diagnose variations, perform prediction of health-related issues and suggest precautions to maintain good health.

PREDICTIVE ANALYSIS

Recurrent neural networks have the ability for analyzing data based on times-series from IoT-based devices to identify symptoms of any disorder experienced by an individual at an early stage, prediction of progress of disorder, and enhance life expectancy of patients [67].

REMOTE PATIENT MONITORING

Deep learning techniques are capable enough for monitoring patients remotely with the help of IoT wearable or sensors to make them aware of any abnormalities diagnosed within body via giving alert to consult with specialists without any delay [68].

DATA SECURITY

Integration of deep learning and IoT results in ensuring data securing feature while sharing it from one point to another. To do so, tensorflow federated framework has also been suggested to provide confidentiality during analysis of any medical data sensitive in nature [69].

CURRENT CHALLENGES IN IOT-BASED SMART HEALTHCARE

Although integration of deep learning techniques with IoT devices has demonstrated great potential in smart healthcare systems, there is a high requirement of addressing a few challenges experienced during its process as shown in Figure 8.11.

DATA PRIVACY

Collection of patients' data by IoT wearable or sensors has raised concern for privacy of data which may experience any unauthenticated accessing while transmitting leading to issues in preventing sensitive medical data [70].

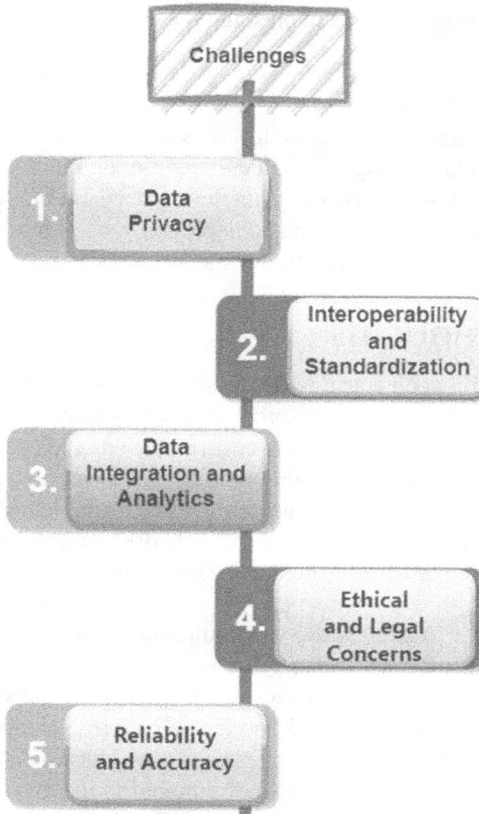

FIGURE 8.11 Current challenges in IoT-based smart healthcare.

INTEROPERABILITY AND STANDARDIZATION

Most of the systems comprise IoT sensors that may not support interoperability as well as standardization which are essential key components while sharing any information [71] from one place to another due to which it becomes challenge for such systems to ensure efficient outcomes, especially in healthcare sector [72].

DATA INTEGRATION AND ANALYTICS

Although IoT-based systems have the ability to analyze medical data but the possibility of such data in various forms such as images, signals, records, and so on can raise challenge for such systems during tasks of pre-processing and extracting features to have clear vision for achieving efficient outcomes [73].

ETHICAL AND LEGAL CONCERNS

IoT-based systems for smart healthcare may experience issues regarding ethical and legal concerns from patients whose data are supposed to be shared from one place to another as things [74] must be transparent from end-to-end tasks which is quite challenging job as it may also lead in biasing [75] which is against research ethics.

RELIABILITY AND ACCURACY

Considering patients' data collected via IoT sensors and wearable may result in low-quality or noisy signals or images which can deteriorate quality of actual data and lead to unreliable and less accuracy of data during decision-making process [76] as keeping noise-free, reliable input data and deploying data validation approaches are necessary to provide efficient results in healthcare sector [77].

CONCLUDING THOUGHTS

Collaborating IoT with deep learning demonstrates great potential in smart healthcare while diagnosing brain tumor disorder at an early stage, performing predictive analysis, health monitoring, and so on to ensure efficient and accurate outcomes. To achieve so, there are various IoT-based sensors and wearable responsible for collection of huge data which further relies on deep learning techniques such as convolutional neural networks, recurrent neural networks, transfer learning, federated learning, and so on for analyzing medical data to make decision-making process accurately. Moreover, these systems have the ability to monitor patients remotely to avoid any delay in their treatment too which ultimately enhance the life expectancy of patients. While working with such systems, there are also a few challenges that need to be taken care such as data privacy, inter-operability and standardized issues, data integration and analytics, ethical and legal concerns, reliability and accuracy, and so on. In spite of these challenges, integrated deep learning models with IoT provide efficient and accurate outcomes to aid radiologists in decision-making process while diagnosing brain tumor disease for providing timely treatment to patients in smart healthcare sector.

REFERENCES

1. Y. LeCun, Y. Bengio, and G. Hinton, "Deep learning," *Nature*, vol. 521, no. 7553, pp. 436–444, 2015.
2. Y. Bengio, A. Courville, and P. Vincent, "Representation learning: A review and new perspectives," *IEEE Transactions on Pattern Analysis and Machine Intelligence*, vol. 35, no. 8, pp. 1798–1828, 2013.
3. D. E. Rumelhart, G. E. Hinton, and R. J. Williams, "Learning representations by back-propagating errors," *Nature*, vol. 323, no. 6088, pp. 533–536, 1986.
4. Y. Bengio, "Learning deep architectures for AI," *Foundations and Trends® in Machine Learning*, vol. 2, no. 1, pp. 1–127, 2009.
5. A. Rajkomar, J. Dean, and I. Kohane, "Machine learning in medicine," *New England Journal of Medicine*, vol. 380, no. 14, pp. 1347–1358, 2019.
6. A. Esteva, B. Kuprel, R. A. Novoa, J. Ko, S. M. Swetter, H. M. Blau, and S. Thrun, "Dermatologist-level classification of skin cancer with deep neural networks," *Nature*, vol. 542, no. 7639, pp. 115–118, 2017.
7. R. Miotto, F. Wang, S. Wang, X. Jiang, and J. T. Dudley, "Deep learning for health-care: Review, opportunities and challenges," *Briefings in Bioinformatics*, vol. 19, no. 6, pp. 1236–1246, 2018.
8. N. A. Samee, N. F. Mahmoud, G. Atteia, H. A. Abdallah, M. Alabdulhafith, M. S. Al-Gaashani, S. Ahmad, and M. S. A. Muthanna, "Classification framework for medical diagnosis of brain tumor with an effective hybrid transfer learning model," *Diagnostics*, vol. 12, no. 10, p. 2541, 2022.
9. S. Nazir, and M. Kaleem, "Federated learning for medical image analysis with deep neural networks," *Diagnostics*, vol. 13, no. 9, p. 1532, 2023.
10. H. Bolhasani, M. Mohseni, and A. M. Rahmani, "Deep learning applications for IoT in health care: A systematic review," *Informatics in Medicine Unlocked*, vol. 23, p. 100550, 2021.
11. G. Litjens, T. Kooi, B. E. Bejnordi, A. A. A. Setio, F. Ciompi, M. Ghafoorian, J. A. Van Der Laak, B. Van Ginneken, and C. I. Sánchez, "A survey on deep learning in medical image analysis," *Medical Image Analysis*, vol. 42, pp. 60–88, 2017.
12. S. Srinivasan, P. S. M. Bai, S. K. Mathivanan, V. Muthukumaran, J. C. Babu, and L. Vilcekova, "Grade classification of tumors from brain magnetic resonance images using a deep learning technique," *Diagnostics*, vol. 13, no. 6, p. 1153, 2023.
13. E. Choi, M. T. Bahadori, A. Schuetz, W. F. Stewart, and J. Sun, "Doctor AI: Predicting clinical events via recurrent neural networks," in *Machine Learning for Healthcare Conference*. PMLR, 2016, pp. 301–318.
14. Z. C. Lipton, D. C. Kale, C. Elkan, and R. Wetzel, "Learning to diagnose with LSTM recurrent neural networks," *arXiv Preprint*, arXiv:1511.03677, 2015.
15. F. Bashir-Gonbadi, and H. Khotanlou, "Brain tumor classification using deep convolutional autoencoder-based neural network: Multi-task approach," *Multimedia Tools and Applications*, vol. 80, pp. 19909–19929, 2021.
16. J.-C. Kim, and K. Chung, "Multi-modal stacked denoising autoencoder for handling missing data in healthcare big data," *IEEE Access*, vol. 8, pp. 104933–104943, 2020.
17. X. Yi, E. Walia, and P. Babyn, "Generative adversarial network in medical imaging: A review," *Medical Image Analysis*, vol. 58, p. 101552, 2019.
18. K. Thapa, H. Khan, T. G. Singh, and A. Kaur, "Traumatic brain injury: Mechanistic insight on pathophysiology and potential therapeutic targets," *Journal of Molecular Neuroscience*, vol. 71, no. 9, pp. 1725–1742, 2021.
19. A. K. Rehni, T. G. Singh, A. S. Jaggi, and N. Singh, "Pharmacological preconditioning of the brain: A possible interplay between opioid and calcitonin gene related peptide transduction systems," *Pharmacological Reports*, vol. 60, no. 6, p. 904, 2008.

20. A. Rehman, S. Naz, M. I. Razzak, F. Akram, and M. Imran, "A deep learning-based framework for automatic brain tumors classification using transfer learning," *Circuits, Systems, and Signal Processing*, vol. 39, pp. 757–775, 2020.
21. H. Kutlu, and E. Avcı, "A novel method for classifying liver and brain tumors using convolutional neural networks, discrete wavelet transform and long short-term memory networks," *Sensors*, vol. 19, no. 9, p. 1992, 2019.
22. A. Chattopadhyay, and M. Maitra, "MRI-based brain tumour image detection using CNN based deep learning method," *Neuroscience Informatics*, vol. 2, no. 4, p. 100060, 2022.
23. S. Deepak, and P. Ameer, "Brain tumor classification using deep CNN features via transfer learning," *Computers in Biology and Medicine*, vol. 111, p. 103345, 2019.
24. A. Younis, L. Qiang, C. O. Nyatega, M. J. Adamu, and H. B. Kawuwa, "Brain tumor analysis using deep learning and VGG-16 ensembling learning approaches," *Applied Sciences*, vol. 12, no. 14, p. 7282, 2022.
25. S. R. Joshi, D. B. Headley, K. Ho, D. Paré, and S. S. Nair, "Classification of brain-waves using convolutional neural network," in *2019 27th European Signal Processing Conference (EUSIPCO)*. IEEE, 2019, pp. 1–5.
26. C. Han, L. Rundo, R. Araki, Y. Nagano, Y. Furukawa, G. Mauri, H. Nakayama, and H. Hayashi, "Combining noise-to-image and image-to-image GANS: Brain MR image augmentation for tumor detection," *IEEE Access*, vol. 7, pp. 156 966–156 977, 2019.
27. H. Mohsen, E.-S. A. El-Dahshan, E.-S. M. El-Horbaty, and A.-B. M. Salem, "Classification using deep learning neural networks for brain tumors," *Future Computing and Informatics Journal*, vol. 3, no. 1, pp. 68–71, 2018.
28. M. Kachwalla, M. Shinde, R. Katare, A. Agrawal, V. Wadhai, and M. Jadhav, "Classification of brain MRI images for cancer detection using deep learning," *International Journal of Advanced Research in Computer and Communication Engineering*, vol. 3, pp. 635–637, 2017.
29. A. A. Pandian, and R. Balasubramanian, "Fusion of contourlet transform and Zernike moments using content based image retrieval for MRI brain tumor images," *Indian Journal of Science and Technology*, vol. 9, no. 29, pp. 1–8, 2016.
30. S. S. Begum, and D. R. Lakshmi, "Combining optimal wavelet statistical texture and recurrent neural network for tumour detection and classification over MRI," *Multimedia Tools and Applications*, vol. 79, pp. 14009–14030, 2020.
31. P. M. Krishnammal, and S. S. Raja, "Convolutional neural network based image classification and detection of abnormalities in MRI brain images," in *2019 International Conference on Communication and Signal Processing (ICCSP)*. IEEE, 2019, pp. 0548–0553.
32. A. Raj, A. Anil, P. Deepa, H. A. Sarma, and R. N. Chandran, "Brainnet: A deep learning network for brain tumor detection and classification," in *Advances in Communication Systems and Networks: Select Proceedings of ComNet 2019*. Springer, 2020, pp. 577–589.
33. D. Mukherkjee, P. Saha, D. Kaplun, A. Sinitca, and R. Sarkar, "Brain tumor image generation using an aggregation of GAN models with style transfer," *Scientific Reports*, vol. 12, no. 1, p. 9141, 2022.
34. S. Maharjan, A. Alsadoon, P. Prasad, T. Al-Dalain, and O. H. Alsadoon, "A novel enhanced softmax loss function for brain tumour detection using deep learning," *Journal of Neuroscience Methods*, vol. 330, p. 108520, 2020.
35. J. Kang, Z. Ullah, and J. Gwak, "MRI-based brain tumor classification using ensemble of deep features and machine learning classifiers," *Sensors*, vol. 21, no. 6, p. 2222, 2021.
36. E. Irmak, "Multi-classification of brain tumor MRI images using deep convolutional neural network with fully optimized framework," *Iranian Journal of Science and Technology, Transactions of Electrical Engineering*, vol. 45, no. 3, pp. 1015–1036, 2021.

37. B. Anilkumar, and P. R. Kumar, "Tumor classification using block wise fine tuning and transfer learning of deep neural network and KNN classifier on MR brain images," *International Journal of Emerging Trends in Engineering Research*, vol. 8, no. 2, pp. 574–583, 2020.

38. M. Siar, and M. Teshnehlab, "Brain tumor detection using deep neural network and machine learning algorithm," in *2019 9th International Conference on Computer and Knowledge Engineering (ICCKE)*. IEEE, 2019, pp. 363–368.

39. A. K. Anaraki, M. Ayati, and F. Kazemi, "Magnetic resonance imaging-based brain tumor grades classification and grading via convolutional neural networks and genetic algorithms," *Biocybernetics and Biomedical Engineering*, vol. 39, no. 1, pp. 63–74, 2019.

40. P. T. Selvy, V. Dharani, and A. Indhuja, "Brain tumour detection using deep learning techniques," *International Journal of Scientific Research in Computer Science, Engineering and Information Technology*, vol. 169, p. 175, 2019.

41. R. V. Kurup, V. Sowmya, and K. Soman, "Effect of data pre-processing on brain tumor classification using capsulenet," in *ICICCT 2019—System Reliability, Quality Control, Safety, Maintenance and Management: Applications to Electrical, Electronics and Computer Science and Engineering*. Springer, 2020, pp. 110–119.

42. A. El Boustani, M. Aatila, E. El Bachari, and A. El Oirrak, "MRI brain images classification using convolutional neural networks," in *Advanced Intelligent Systems for Sustainable Development (AI2SD'2019) Volume 4—Advanced Intelligent Systems for Applied Computing Sciences*. Springer, 2020, pp. 308–320.

43. X. Gu, Z. Shen, J. Xue, Y. Fan, and T. Ni, "Brain tumor MR image classification using convolutional dictionary learning with local constraint," *Frontiers in Neuroscience*, vol. 15, p. 679847, 2021.

44. A. Rehman, M. A. Khan, T. Saba, Z. Mehmood, U. Tariq, and N. Ayesha, "Microscopic brain tumor detection and classification using 3D CNN and feature selection architecture," *Microscopy Research and Technique*, vol. 84, no. 1, pp. 133–149, 2021.

45. M. Sajjad, S. Khan, K. Muhammad, W. Wu, A. Ullah, and S. W. Baik, "Multi-grade brain tumor classification using deep CNN with extensive data augmentation," *Journal of Computational Science*, vol. 30, pp. 174–182, 2019.

46. S. Deepak, and P. Ameer, "Automated categorization of brain tumor from MRI using CNN features and SVM," *Journal of Ambient Intelligence and Humanized Computing*, vol. 12, pp. 8357–8369, 2021.

47. R. L. Kumar, J. Kakarla, B. V. Isunuri, and M. Singh, "Multi-class brain tumor classification using residual network and global average pooling," *Multimedia Tools and Applications*, vol. 80, pp. 13 429–13 438, 2021.

48. M. A. Khan, I. Ashraf, M. Alhaisoni, R. Damaševičius, R. Scherer, A. Rehman, and S. A. C. Bukhari, "Multimodal brain tumor classification using deep learning and robust feature selection: A machine learning application for radiologists," *Diagnostics*, vol. 10, no. 8, p. 565, 2020.

49. O. N. Belaid, and M. Loudini, "Classification of brain tumor by combination of pretrained VGG16 CNN," *Journal of Information Technology Management*, vol. 12, no. 2, pp. 13–25, 2020.

50. D. Amrapur, "Computer based diagnosis system for tumor detection & classification: A hybrid approach," *International Journal of Pure and Applied Mathematics*, vol. 118, no. 7, pp. 33–43, 2018.

51. X. Chen, Y. Chen, C. Ma, X. Liu, and X. Tang, "Classification of pancreatic tumors based on MRI images using 3D convolutional neural networks," in *Proceedings of the 2nd International Symposium on Image Computing and Digital Medicine*, New York, NY, 2018, pp. 92–96.

52. M. Havaei, A. Davy, D. Warde-Farley, A. Biard, A. Courville, Y. Bengio, C. Pal, P. M. Jodoin, and H. Larochelle, "Brain tumor segmentation with deep neural networks," *Medical Image Analysis*, vol. 35, pp. 18–31, 2017.

53. S. Sajid, S. Hussain, and A. Sarwar, "Brain tumor detection and segmentation in mr images using deep learning," *Arabian Journal for Science and Engineering*, vol. 44, pp. 9249–9261, 2019.

54. Z. Akkus, A. Galimzianova, A. Hoogi, D. L. Rubin, and B. J. Erickson, "Deep learning for brain MRI segmentation: State of the art and future directions," *Journal of Digital Imaging*, vol. 30, pp. 449–459, 2017.

55. H.-C. Shin, H. R. Roth, M. Gao, L. Lu, Z. Xu, I. Nogues, J. Yao, D. Mollura, and R. M. Summers, "Deep convolutional neural networks for computer-aided detection: CNN architectures, dataset characteristics and transfer learning," *IEEE Transactions on Medical Imaging*, vol. 35, no. 5, pp. 1285–1298, 2016.

56. K. Chang, *et al.*, "Automatic assessment of glioma burden: A deep learning algorithm for fully automated volumetric and bidimensional measurement," *Neuro-Oncology*, vol. 21, no. 11, pp. 1412–1422, 2019.

57. P. Moeskops, M. A. Viergever, A. M. Mendrik, L. S. De Vries, M. J. Benders, and I. Išgum, "Automatic segmentation of MR brain images with a convolutional neural network," *IEEE Transactions on Medical Imaging*, vol. 35, no. 5, pp. 1252–1261, 2016.

58. Z. N. K. Swati, Q. Zhao, M. Kabir, F. Ali, Z. Ali, S. Ahmed, and J. Lu, "Brain tumor classification for MR images using transfer learning and fine-tuning," *Computerized Medical Imaging and Graphics*, vol. 75, pp. 34–46, 2019.

59. E. Avşar and K. Salçin, "Detection and classification of brain tumours from MRI images using faster R-CNN," *Tehnički Glasnik*, vol. 13, no. 4, pp. 337–342, 2019.

60. S. Sarkar, A. Kumar, S. Chakraborty, S. Aich, J.-S. Sim, and H.-C. Kim, "A CNN based approach for the detection of brain tumor using MRI scans," *Test Engineering and Management*, vol. 83, pp. 16 580–16 586, 2020.

61. P. Shanthakumar, and P. G. Kumar, "Computer aided brain tumor detection system using watershed segmentation techniques," *International Journal of Imaging Systems and Technology*, vol. 25, no. 4, pp. 297–301, 2015.

62. Y. Yuehong, Y. Zeng, X. Chen, and Y. Fan, "The internet of things in healthcare: An overview," *Journal of Industrial Information Integration*, vol. 1, pp. 3–13, 2016.

63. E. J. Topol, "High-performance medicine: The convergence of human and artificial intelligence," *Nature Medicine*, vol. 25, no. 1, pp. 44–56, 2019.

64. P. Zhang, D. C. Schmidt, J. White, and G. Lenz, "Blockchain technology use cases in healthcare," *Advances in Computers*, vol. 111, pp. 1–41, 2018.

65. D. M. Hilty, K. Randhawa, M. M. Maheu, A. J. McKean, R. Pantera, M. C. Mishkind, and A. Rizzo, "A review of telepresence, virtual reality, and augmented reality applied to clinical care," *Journal of Technology in Behavioral Science*, vol. 5, pp. 178–205, 2020.

66. X.-W. Chen, and X. Lin, "Big data deep learning: Challenges and perspectives," *IEEE Access*, vol. 2, pp. 514–525, 2014.

67. D. S. Rajeswari, *et al.*, "A review on remote health monitoring sensors and their filtering techniques," *Global Transitions Proceedings*, vol. 2, no. 2, pp. 392–401, 2021.

68. C. Thapa, and S. Camtepe, "Precision health data: Requirements, challenges and existing techniques for data security and privacy," *Computers in Biology and Medicine*, vol. 129, p. 104130, 2021.

69. M. K. Hasan, T. M. Ghazal, R. A. Saeed, B. Pandey, H. Gohel, A. Eshmawi, S. Abdel-Khalek, and H. M. Alkhassawneh, "A review on security threats, vulnerabilities, and counter measures of 5g enabled internet-of-medical-things," *IET Communications*, vol. 16, no. 5, pp. 421–432, 2022.

70. J. Qi, P. Yang, G. Min, O. Amft, F. Dong, and L. Xu, "Advanced internet of things for personalised healthcare systems: A survey," *Pervasive and Mobile Computing*, vol. 41, pp. 132–149, 2017.

71. H. Bhatia, S. N. Panda, and D. Nagpal, "Internet of things and its applications in healthcare-a survey," in *2020 8th International Conference on Reliability, Infocom Technologies and Optimization (Trends and Future Directions)(ICRITO)*. IEEE, 2020, pp. 305–310.

72. J. G. A. Ebenezer, and S. Durga, "Big data analytics in healthcare: A survey," *Journal of Engineering & Applied Sciences*, vol. 10, no. 8, pp. 3645–3650, 2015.

73. N. Kumar, R. K. Kaushal, S. N. Panda, and S. Bhardwaj, "Impact of the internet of things and clinical decision support system in healthcare," in *IoT and WSN Based Smart Cities: A Machine Learning Perspective*. Springer, 2022, pp. 15–26.

74. N. Y. Philip, J. J. Rodrigues, H. Wang, S. J. Fong, and J. Chen, "Internet of things for in-home health monitoring systems: Current advances, challenges and future directions," *IEEE Journal on Selected Areas in Communications*, vol. 39, no. 2, pp. 300–310, 2021.

75. H. Naz, R. Sharma, N. Sharma, and S. Ahuja, "IoT-inspired smart healthcare service for diagnosing remote patients with diabetes," in *Machine Learning for Edge Computing*. CRC Press, 2022, pp. 97–114.

76. M. S. Patel, D. A. Asch, and K. G. Volpp, "Wearable devices as facilitators, not drivers, of health behavior change," *JAMA*, vol. 313, no. 5, pp. 459–460, 2015.

77. P. Verma, R. Tiwari, W.-C. Hong, S. Upadhyay, and Y.-H. Yeh, "Fetch: A deep learning-based fog computing and IoT integrated environment for healthcare monitoring and diagnosis," *IEEE Access*, vol. 10, pp. 12548–12563, 2022.

9 Machine Learning and Its Applications in Healthcare

Mohd. Akram and Pooja Sharma
Rayat-Bahra University, Mohali, India

BACKGROUND

The phrase healthcare means an organization that involves the improvement of health-related offerings in order to meet individuals' clinical needs. Patients, doctors, clinicians, researchers, or medical enterprises are all working to protect and restore health records in medical facilities. Data have been growing steadily in all industries as a result of astounding technological improvements recently, particularly healthcare, facilitating an increasing number of data mining applications.

Yet, as the healthcare system digitizes, medical organizations generate a vast volume of healthcare data [1]. Overall, healthcare data [2] refers to all health-related documents that are digitally preserved. It could include thorough information on the patient's health record, doctor's prescribed notes, and so on. All of this information is massive, multidimensional, and diverse in character. Because of the growing amount of healthcare data, making healthy decisions is becoming more difficult in the modern world.

ML, data mining, and statistical approaches are important topics of research which enhance a person's ability to take good decisions for maximizing outcome of all working domains [3]. When contrasted with the amount of data recorded, rate of human data analysis ability is substantially lower. This is especially important in the healthcare industry, where the quantity of available experts for healthcare analysis of data is quite low.

As a result, computerized (semi-automatic) medical disease diagnosis systems are required to enable doctors and healthcare providers to take educated decisions for all individuals. This increases care quality, enhances diseases' diagnosis, and eventually lowers medical expenses.

Current research is mostly concerned with classification/prediction challenges in healthcare data using ML (supervised) methodologies. There are numerous learning algorithms that may be used in conjunction with data mining approaches to tackle classification issues in the medical field, improving diagnostic precision. Figure 9.1 depicts data-driven knowledge extraction [2] in healthcare.

With the growth of computer technology, many renovation studies are taking place in the digital realm. Information has become an essential component of the digital world. When it comes to data analytics, data acquisition is critical. Data analytics is the study of evidence with the goal of reaching a conclusion (useful insights)

DOI: 10.1201/9781003438205-9

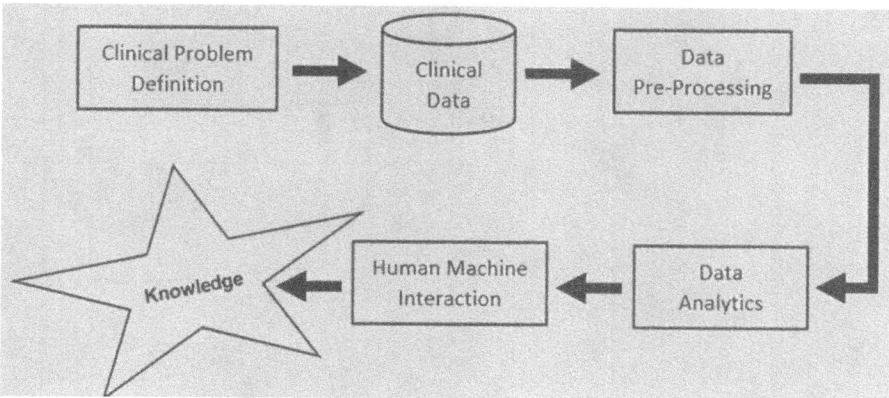

FIGURE 9.1 KDD process [2].

based on data. Data analytics is used in every business. In the healthcare industry, data analytics is more significant than in the financial and other commercial sectors [3]. As a result, efficient use of data analytics in the healthcare area may enhance healthcare quality as well as more crucially promote preventive interventions.

Particularly, according to a recent big data assessment, the estimated value of healthcare information is expected to be around $300 billion [4]. Furthermore, Sun et al. [5] stated in 2012 that the computerized version of medical data from all over the world was estimated to be 500 petabytes, with a potential increase to 25,000 petabytes by 2020. Data scientists in healthcare are responsible for understanding and developing knowledge from healthcare data in order to extract valuable and notable (actionable) information from it.

Many data mining solutions for healthcare sectors have recently been created to aid decision-making. According to [1], the writers examined the evaluation, visualization, and mining of healthcare information as well as how data may be effectively utilized to improve the company's capacity to generate income, regulate risk, and cost.

Raja et al. [2] have discussed and summarized how clinical creativity, electronic healthcare records (EHRs), medical devices, and other healthcare-related elements are transforming medical services. Medical data analytics have proven to be critical in dealing with a variety of difficulties in healthcare fields.

HEALTHCARE MONITORING

The healthcare business is currently facing a significant problem in ensuring compatibility among applications offered by various manufacturers. Multiple applications can be used by each department of hospital management or medical clinic to share organizational and clinical data. Such applications could be technologically advanced. Regrettably, each program may support several communication interfaces, which necessitate ongoing development and maintenance [6]. The development of interoperability among heterogeneous healthcare data systems is critical in

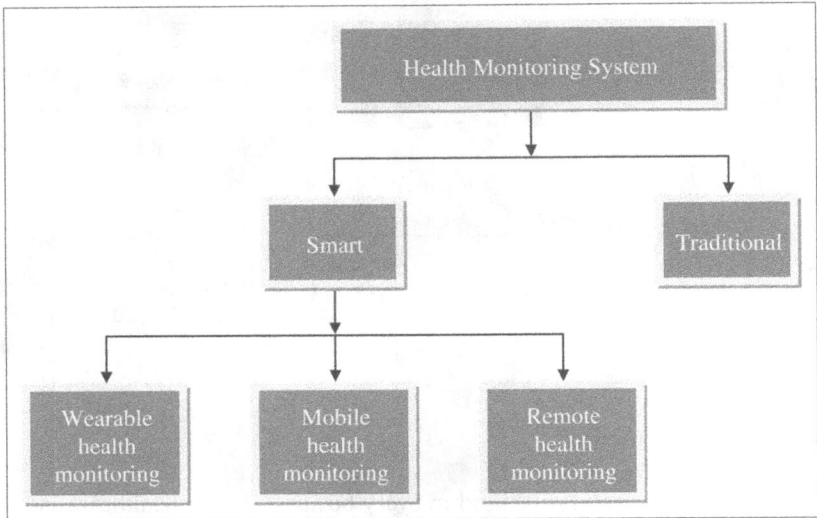

FIGURE 9.2   Classification of healthcare monitoring system [7].

light of the decrease in healthcare costs, as it will lead to better and more effective treatment of patients, management, and care.

In recent years, health monitoring systems have played an important role in lowering hospitalization costs, medical staff load, time, and so on.

Healthcare system is divided into three types. Figure 9.2 shows categories of health monitoring. The mobile health monitoring system is concerned with smartphones, PCs, and other similar devices.

Mobile devices based on smartphones are becoming more and more common. One can take responsibility for their well-being at any point, from anywhere, using mobile health monitoring, and software is easy to use. People are tracked and treated via remote health monitoring systems, which transmit and receive data. This type of technology makes it possible to treat an extensive variety of illnesses or utilize it in homes and hospitals. Biological sensors called wearable health monitoring systems might be worn to take vital signs.

Smart health tracking systems (SHM) are an innovative technology and a novel method for health monitoring. Heart rate, blood pressure, ECG, and temperature are all measured as part of an overall health monitoring system [7].

Over the previous few years, the healthcare system has advanced significantly. Smart health monitoring is essential in hospitals, residential settings, or outdoor settings, as it is enabled by various communication techniques like GPS and radio frequency identification. Figure 9.2 shows the classification of healthcare monitoring system [7].

Increment in healthcare technology is concerned with medical data quality, security, stability, and accuracy of system. With the rate of false alarms increasing, advancements in healthcare are concerned with data quality, security, and its acceptability by medical professionals and patients.

Data quality and safety are all important considerations in a health monitoring system. Remote health monitoring relies heavily on safe data transfer and middleware design, whereas mobile health monitoring relies heavily on battery life, energy consumption with efficacy and confidentiality. Figure 9.3 shows the fundamental design of a smart health monitoring platform. The architecture of most systems is

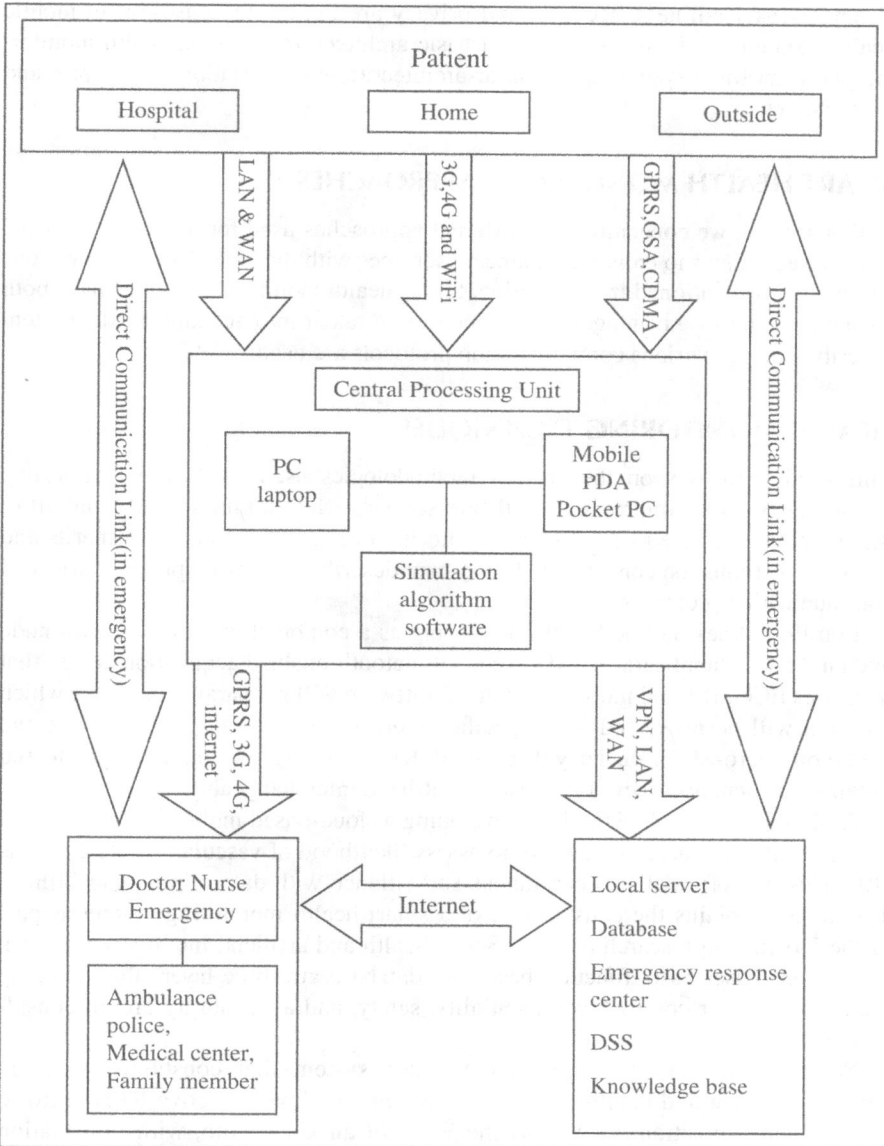

FIGURE 9.3 Design of heath monitoring system [8].

similar, with variations in software [8]. Increment in healthcare technology concern with medical data quality, security, stability, and accuracy of the system. In health monitoring system, data quality, security, and privacy are the major considerations. In wearable health monitoring user comfort, data transmission rate power consumption, the context awareness needs to take into account. Secure data transmission, real-time availability, and middleware design with user-friendliness are major parts of remote health monitoring, whereas power consumption, energy required with efficiency, user-friendliness, security, and privacy are important constraint in mobile health monitoring. Figure 9.3 shows a basic architecture of smart health monitoring platform. Most system uses similar architecture with variation in software and technologies.

SMART HEALTH MONITORING APPROACHES

In this section, we concentrate on different approaches used for health monitoring. Various techniques to enhance healthcare services with the consideration of various parameters is mentioned in this section. Smart health monitoring systems have both advantages and disadvantages which need to be taken into account. In the system described below, various communication protocols are used.

HEALTH MONITORING TECHNIQUES

This section focuses on the various methodologies used for health monitoring. Numerous ways for improving healthcare services while taking into account all of the variables are listed here. Smart monitoring designs for health have merits and demerits that must be considered. The system described below employs a variety of communication protocols.

Lou [9] defines mobile health monitoring as a combination of a 3AHcare node and an Android application. 3AHcare is a Bluetooth-enabled acquisition design that measures ECG, BP, and mobility. Android software will get parameters, after which the data will be processed by a specific algorithm to generate steady waves and stored on micro-SD flash. They discovered that technology is simple to operate and obtains excellent accuracy. This design is utilized interchangeably.

P. Melillo's original method for developing a cloud-based mobile health platform largely employs heart rate viability to assess likelihood of vascular events and falls [10]. This platform delivers continuous surveillance with data mining capabilities. The author explains the construction of a smart health monitoring system as part of the UE-funded research project "Smart health and artificial intelligence for risk estimation." User base, function base, and database are three layers that make up architecture. User comfort, confidentiality, safety, and accessibility are all considered when designing an architecture.

Yadav et al. [11] describe an embedded system that constantly monitors patient's posture and health status. This system employs an active RFID card to trace patient's position, as well as the WSN of an xbee radio, with information being obtained in a local record on the PC. The utilization of an xbee radio raises system's expense.

Izumi et al. [12] describe an electrocardiography processor as well as a wearable healthcare device. To improve the precision of detected heart rate, an immediate heart rate sensor with short-term autocorrelation algorithms is utilized. For heart rate logging applications, the ECG processor chip uses 13.7 A.

A prototype for using an Android application to track a patient's vital signs was created by Kanrar et al. This electronic healthcare system is capable of gathering not just a patient's biological and private data, but also vital signs, and storing that data on a server that houses medical databases. The Gaussian mixture model is used by them. Distinct feature classes are presented in a gentle manner.

Ref. [13] concentrate on remote health monitoring systems with mobile phone and online service offerings that allow widespread and continuous patient health monitoring. A multilayer architecture is built for the workflow, which includes a mobile terminal, a smartphone, and a remote server. Bluetooth protocols are employed for communication, and the Zephyr BioHarness sensor is used as a portable interface. This system has two modes of operation: routine status monitoring or emergent reaction. On an emergency, the indoor localization mechanism on a smartphone is used. It provides consistent performance, but only for immediate tracking of patient status and not for professional analysis. Figure 9.4 shows the overall system architecture of the healthcare monitoring system [10].

Types of Heath Monitoring Systems

Remote Health Monitoring Systems

RHMS refers to the application of electronic information and communication technologies to support and enhance healthcare quality when patients and healthcare professionals are geographically separated. To transmit crucial patient information in real-time to clinicians from a remote location, RHMSs often use cutting-edge information and communication technologies. In addition to RHMSs, mobile phone

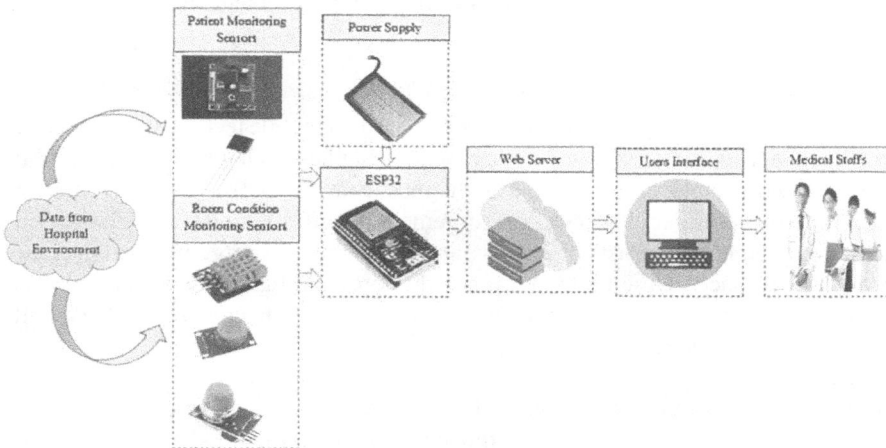

FIGURE 9.4 The overall system architecture of the healthcare monitoring system [10].

systems and wearable monitoring equipment are deployed. The use of this technology has numerous advantages and offers unique options for delivering healthcare through monitoring patients from afar. The section that follows describes such systems.

For effective engagement, increased information exchange reliability, and ideal energy utilization of wearable medical equipment, a reliable, smart, confidential tracking and administration system has been developed [14]. A client-server architecture was used to construct a built-in mobile ECG tracking system. Patient's ECG signals are sent to server via RFID reader or a patient monitor that is normally housed in patient's home [15]. The latter uses fuzzy logic to generate health condition prognoses by determining the connection among different illnesses and symptoms.

A further apparatus has been created that measures five signals. The identification of single fiber action potentials through a homogenous, equivalent isotropic, infinite volume conduction media served as fundamental building component of the sEMG model. Guo et al. [16] monitored the EMG, MMG, SMG, and torque output during the course of nine males. The greatest torque value measured during the isometric contraction in this investigation, known as the maximal voluntary contraction, was determined. The MVC torque was computed by combining two recorded greatest torque values from two trials, with a five-minute break allowed for the two MVC tests.

Review and Critical Analysis

Remote monitoring devices not only track vital signs but also look for anomalies before sending the information to medical staff in real-time. These systems frequently transmit data with a delay as a result of processing real-time data. Data privacy and security, namely, with regard to patient identity and the confidentiality of medical data, pose a serious danger to those systems. There is still an opportunity for enhancement of the system's design and structure to bring it into compliance with ethical or medical standards because these problems have not yet been resolved.

Mobile Health Monitoring Systems

To compile patient data, a multi-agent design based on mobile technology (GSM) and intelligent agents for weight and cardiac monitoring has been designed. In a mobile setting, intelligent agents communicate diagnostic data in bulk and suggest medical solutions. Self-powered devices allow for remote monitoring of BP, pulse oximeters, and ECG. The latter consists of a solar panel setup and an energy-harvesting circuit board that is placed close to 34 W overhead fluorescent lights. Tri-axis accelerometers and low-power sensors are used in this type of system to create extended mobile phone graphical display charts.

It has also been built as an adaptable structure that performs physiological analysis of information for tracking patient health problems. Sensors gathered physiological parameters, which were then analyzed using data mining methods. Mobile devices were used for real-time processing, or critical circumstances might be signaled by a proper alert. With various sampling periods and time windows, a system design

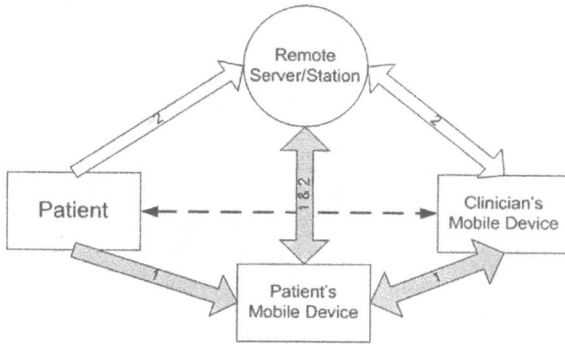

FIGURE 9.5 Data transfer structure for mobile health monitoring [17].

was carried out utilizing the clustering methods basic *K*-means, farthest first, and expectation maximization. It has been reported that using sophisticated algorithms has improved the findings.

Review and Critical Analysis

The ability to quickly obtain health information would be advantageous for many patients and medical personnel, especially in emergency situations, thanks to mobile health monitoring technologies. Although the technology itself is constantly being improved, there are still issues with how well it can be used in therapeutic settings. For example, while raw information is easily transferred from phone, its analysis remains a top priority. This is because data processing may have a significant negative effect on a phone's battery life and can cause a latency in data transfer. As illustrated in Figure 9.5, the data transmission framework for portable health monitoring devices is divided in two transmission kinds and three processes. In type 1, phone gathers patient information, which is then sent to distant server for analysis. Information is transferred either directly to clinician's or via patient's phone. Type 2 includes forwarding the patient information to a remote station for processing before it is forwarded to other equipment. In some circumstances, only the outcomes or notifications will be sent. Both may lead to delays in the output of findings. There will likely be a direct connection among patient and clinician, as shown in Figure 9.5 (refer to dashed line) [7]. Mobile devices' continuous data transfers, including transmitting and receiving, drastically shorten battery life.

ML in Healthcare

ML is a subset of AI that teaches systems how to use data to behave or anticipate without having explicitly programmed. Figure 9.6 demonstrates how the use of ML for medical purposes has chance to completely alter how diseases are recognized, treated, and prevented. Some potential uses of ML for healthcare include the following:

- Diagnosis and treatment: ML approaches are taught to examine medical pictures like CT scans in order to assist in a patient's diagnosis or choose the best course of action.

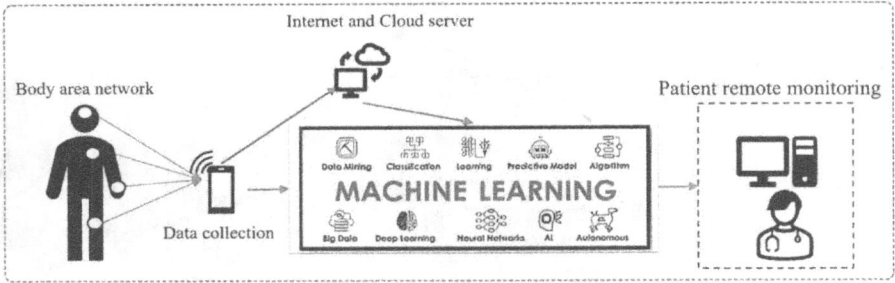

FIGURE 9.6 ML in healthcare [18].

- Personalized medicine: Using a patient's unique traits like genetics and medical history, ML can be used to forecast which medicines are most likely to be beneficial for that patient.
- Clinical decision support: To assist healthcare professionals in making more informed decisions regarding patient care, ML approaches are included into clinical decision support systems.
- Population health management: By analyzing data from huge populations, machine learning may be used to spot sequences that help in creation of public health programmers. In general, applying ML to healthcare has the potential to boost system effectiveness, lower costs, and improve patient outcomes. Figure 9.6 indicates the ML in healthcare [18].

Ensemble Machine Learning for Healthcare Data Analysis

Health-related data about patients are gathered for EHRs from clinics, hospitals, lab tests, and so on. Both unstructured and structured formats can be used for representing this type of information. This information makes it possible to detect diseases early, which can avert serious health issues and lower mortality rates. Making sense of healthcare data requires ML, which is essential. Another significant issue in the healthcare industry is classification of diseases using available information. By utilizing cutting-edge techniques to enhance the accuracy of classification, ML is also utilized to gather healthcare knowledge [19]. By boosting model decision, these categorization models might enhance patient care. Recent research has demonstrated that categorization using a single model yields worse classification accuracy than classification using multiple models. Ensemble learning is the term used to describe these combinations of different models. The benefit of the ensemble is that it combines the top models, which outperform the individual classifiers in terms of performance as shown in Figure 9.7.

MACHINE LEARNING

The fact that ML is a broad, interdisciplinary subject having roots in data, algebra, processing data, information analytics, and so on makes it difficult to develop a new description of it [21]. A particular AI method called machine learning collects data from training sets. We are not instructing machines where to search in this process

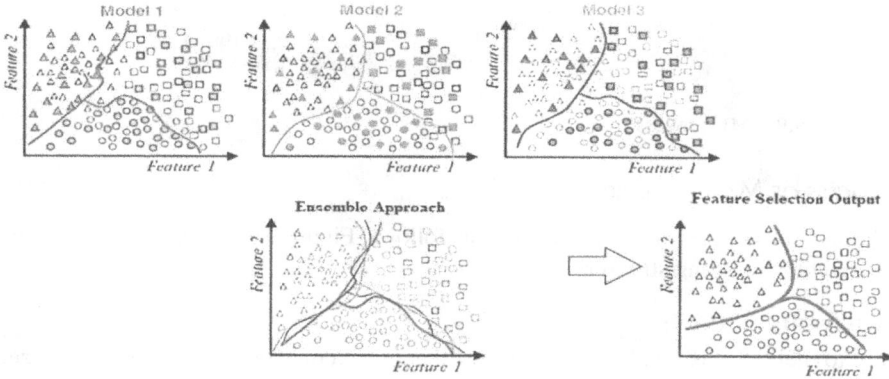

FIGURE 9.7   Representation of feature selection using ensemble approach [20].

of learning because it is at base of the tree and contains countless branches and sub-branches. The main categories of ML [21] are represented in Figure 9.8:

a. **Supervised Machine Learning**
   This knowledge is applied during training, and a method creates a task that matches inputs to associated outputs [22].

b. **Semi-supervised Learning**
   This method allows you to choose the most precise classifier from both labeled and unlabeled data. Using unlabeled data allows for good classification accuracy transmission. This method can only work if a restricted set of underlying assumptions are true.

c. **Machine Learning without Supervision**
   Labels that are used in unsupervised learning acquire no instruction. Unsupervised learning is a form of methodology that employs clustering, fuzzy clustering, hierarchical clustering, K-means clustering, and association rule mining.

d. **Reinforcement Learning**
   For the purpose of this learning, a computer's system provides access to a constantly changing setting. Since the program works around its fault, it receives information in the form of incentives and punishments [23].

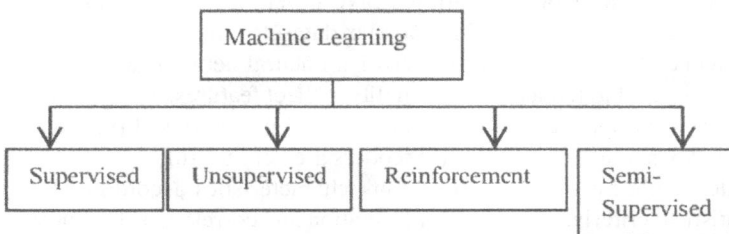

FIGURE 9.8   ML approaches [21].

FIGURE 9.9   ML process.

PROCESS OF MACHINE LEARNING

The specifics of this phase are detailed as follows. Figure 9.9 depicts the fundamental machine learning structure.

Healthcare Datasets

Due to their vastness and complexity, healthcare data may prove difficult to analyze. However, to acquire insights into the health of patients and create targeted efforts to enhance patient care, scientists can employ cutting-edge analytical tools like ML and NLP.

Feature Extractions

Feature extraction, a key aspect of machine learning, is the process of choosing the most pertinent features from a dataset [24]. The process of FE entails turning raw data into features with high pattern recognition capabilities. In this method, the gathered characteristics are regarded to have stronger recognition capacity than the original data. Finding the essential characteristics or qualities from the initial information to use as inputs for an ML algorithm to carry out a certain task is the goal of this approach. PCA, LDA, t-SNE, autoencoders, filter methods [22], and wrapper methods are just a few of the numerous techniques that can be used for feature extraction.

- PCA: In statistical analysis and ML, PCA is a common dimensionality reduction technique. It uses the linear transformation strategy to project high-dimensional information onto a lower-dimensional space in order to identify patterns there. PCA's main goal is to capture the data's most significant variations while reducing noise or redundancy [25].
- T-SNE: The non-linear reduction of dimensionality technique known as t-SNE is particularly good at displaying high-dimensional data. In 2008, Geoffrey Hinton and Laurens van der Maaten founded t-SNE. Its main goal is to preserve the data's local structures, which requires keeping track of the separations among nearby data points when the amount of dimension is reduced. Because of this feature, t-SNE is very good at revealing patterns, clusters, and structures inside complex datasets [26].
- Autoencoder: An unsupervised artificial neural network called an autoencoder is used to reduce dimensionality, collect features, and learn representations. For applications like denoising and unsupervised pre-training for complex neural networks, autoencoders are very useful.
- Filter methods: These methods rank characteristics according to a set of statistical criteria, like mutual information and correlation. The most apparent characteristics are then picked. Techniques like Pearson's correlation and the information gain serve as examples of filter methods [27].

- Wrapper methods: These are feature selection techniques used in data analysis and ML. Their major goal is to identify the best subset of characteristics that improves a certain ML algorithm's efficiency. These are computationally intensive than filter approaches, which rely on intrinsic properties of the data because they directly evaluate possible feature combinations based on the effectiveness of algorithm [28].

SVM

Classification issues can be solved using the traditional ML technique called support vector machine. In a massive data mining context, SVM supports multidomain uses, which is significant. In order to produce trustworthy estimators from a dataset, SVM requires some model features to train data [29]. The idea of SVM maximizes the shortest distance between the nearest point of the sample in Figure 9.10 and the hyperplane using encoders, filter methods, and wrapper methods [29].

The gap that separates the lines with dots is the greatest margin for dividing various classes, and the straight red line denotes the separation hyperplane. When working with a huge dataset, SVM performs better than other pattern recognition methods like Bayesian networks, and so on. The relatively simple nature of SVM's data training is another key benefit. SVM has the drawback of being incredibly slow at ML since a lot of training time is required. Furthermore, the square of the quantity of training instances leads to an increase in the need for memory. SVM is one of the best ML algorithms for recognizing patterns. The majority of SVM applications focus on facial detection, speech recognition, picture recognition, sickness detection,

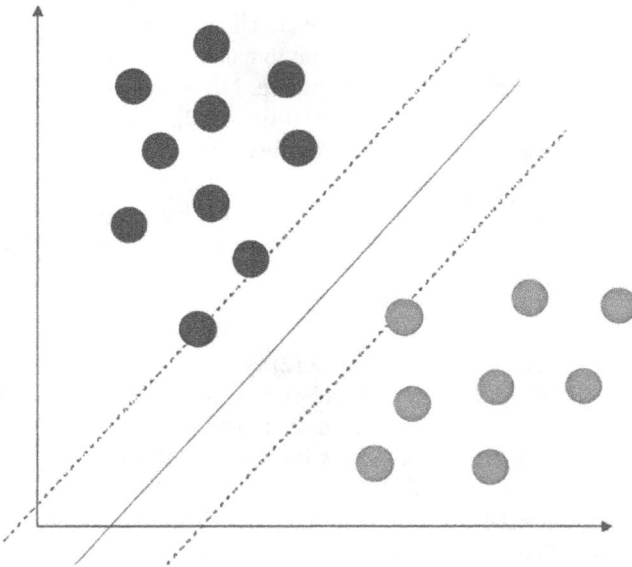

FIGURE 9.10   SVM [29].

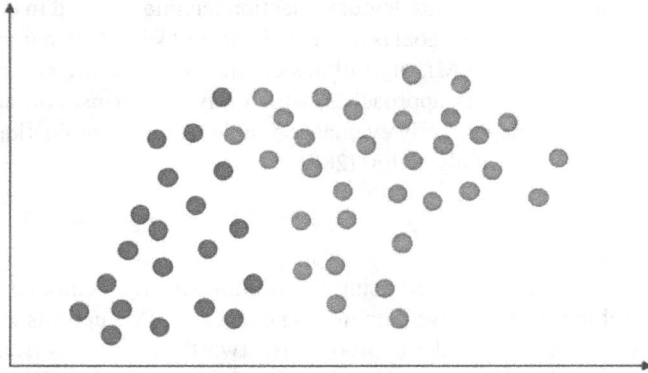

FIGURE 9.11   Naïve Bayes [34].

and prevention. For early heart failure prediction, researchers have adopted an upgraded stacked SVM. Their conclusions showed that the model performed better than expected, with accuracy between 57.85% and 91.83% [30]. In another study, coronary heart disease was diagnosed using fuzzy support vector machines [31].

NAÏVE BAYES

One of the most used algorithms for classification is naive Bayes (NB). NB is the simplest Bayesian network because it only considers one parent node and a small number of independent child nodes. As depicted in Figure 9.11, NB employs the probability classification technique by dividing each attribute's unique likelihood [34]. This straightforward method produces impressive classification outcomes while presuming stability among attributes. This may enable it to handle numerous classification tasks and tiny datasets more effectively. Furthermore, NB is appropriate for incremental training (where it may train further data in real-time) [32]. One study used a NB classifier to analyze skin image data to identify skin diseases, and the findings outperformed other approaches with an accuracy range of 91.2–94.3% [33]. In the medical field, Gupta [34] employed naïve Bayes for feature selection-based heart disease identification, with experimental results obtaining 88.16% accuracy in the test dataset.

DECISION TREES

A decision trees (DT) classifier (Figure 9.12) employs graphical tree data to provide potential options, outcomes, and end values [35]. In order to decide amongst a few possible courses of action, a computer procedure to calculate chances is required. The decision trees approach is built on the training specimens of data and their corresponding class labels. Since the set used for training is iteratively divided into subgroups according to feature values, the information within every subset is cleaner compared to the data in the original set. Each branch node in DT shows evaluation outcome, while leaf nodes show class label. Each internal node in DT provides a test feature. Because it serves to identify class label for a sample

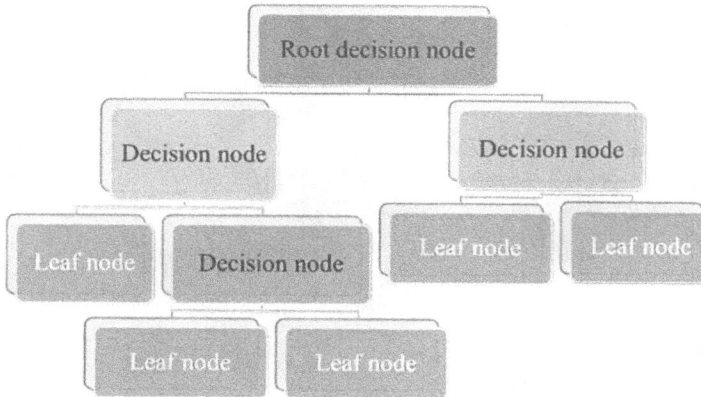

FIGURE 9.12 Decision trees [35].

that is unidentified, the classifier decision tree will be capable to proceed along path from root node to leaf nodes and retain a category label for specimen [35]. The DT method has the benefit of being quick and easy, requiring no parameter's setting or domain knowledge, while handling large dimensional data in the given situation. Each internal node supports several functions that also offer incremental learning, which is immutable.

K-Nearest Neighbor

The most straightforward approach in data mining is K-nearest neighbor (K-NN) classification technique. By giving a value to K, K training sample's nearest neighbor class is assumed to be the same as the classification shown in Figure 9.13 in order to detect an unknown pattern [37]. The classifier takes into account a number of variables, including the chosen K-value and distance measurement, among others. Compared to other machine learning methods, K-NN needs less time to train data. However, the categorization phase calls for greater calculation time. K-NN has the benefit of being simple to comprehend and use for classification. Additionally, it can function well with a dataset that has a large number of class labels.

In a similar vein, the data training phase of machine learning techniques is quicker. The disadvantages of K-NN are its high computing cost, significant unlabeled sample, and the classification phase time delay. In addition to cross-validation, K-NN is computationally costly but lacks concepts to sign K's value. Additionally, if there are many unrelated attributes in data, confusion may develop, resulting in subpar accuracy [36]. Their findings suggest that switching from neural networks to K-NN could improve the accuracy of heart disease diagnosis [37].

Fuzzy Logic

Fuzzy set theory gave rise to fuzzy logic. These numbers fall between 0 and 1. It is a technique that is frequently utilized in engineering applications.

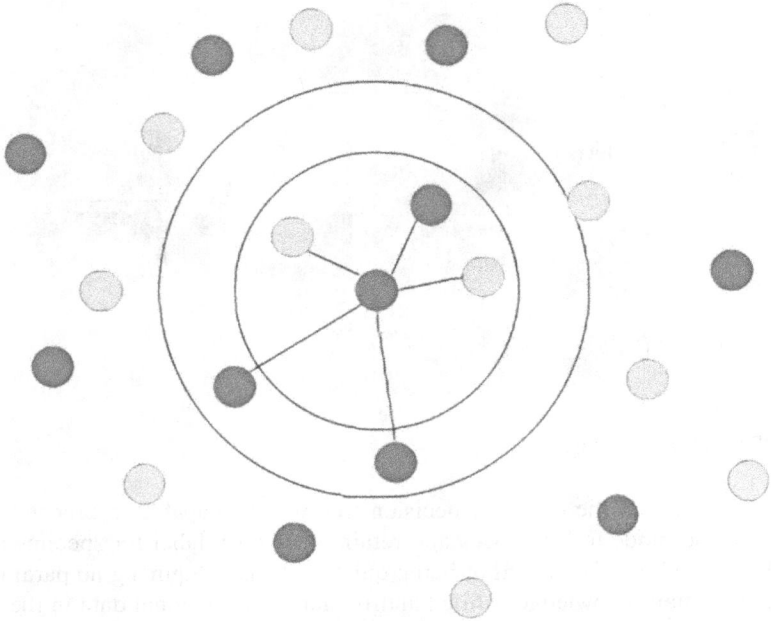

FIGURE 9.13   *K*-NN [37].

CART

The CART stands for Classification and Regression Tree Methodology. The goal variable is expressed as a continuous and categorical variable in classification and regression trees. To forecast values in the tree, these variables are needed.

APPLICATIONS OF ML IN HEALTHCARE

Clinical applications tend to be appropriate to this facility, particularly those that rely on cutting-edge genomics and proteomics measurements. It is commonly used to identify and diagnose a variety of disorders. ML algorithms will develop better treatment strategies for patients by recommending the adoption of efficient medical facilities. The ML algorithms are helpful in locating complex patterns in a wealth of enormous data [38].

ML USED FOR DISEASE PREDICTION

DETECTION OF HEART DISEASE

To increase accuracy of heart disease diagnosis, ML methods are frequently used. Dataset in question is one from UCI repository. Parthiban and Srivatsa proposed an ML strategy for analysis and detection of heart disease [39] using NB and SVM. SVM provides the accuracy of 94.60%, whereas NB produces the accuracy of 74%. Bayes Net and SVM have both been employed by Otoom to forecast coronary heart disease. SVM is 88.3% accurate, whereas Bayes Net is 84% accurate [40].

ANALYSIS OF DIABETIC DISEASES

To increase the accuracy of examination of diabetic diseases, several ML techniques are used. Dataset in question is one from the UCI repository. Iyer proposed NB and decision trees to forecast diabetic problems. NB has an accuracy of 79.56%, while decision trees yield 76.95% [41]. Dash and Sen identified diabetes using ML methods. Accuracy rate provided by Logiboost for CART algorithms is 77.479% [42].

Senturk et al. among other classification algorithms employed SVM, NB, KNN, and DT to identify breast cancer. *K*-NN offers 95.15% accuracy and SVM 96.40% accuracy [43]. Majali et al. used decision trees and common patterns as data mining tools to predict breast cancer [44]. They come to the conclusion that decision trees offer 94% accuracy.

DIAGNOSIS OF THYROID DISORDER

Techniques for machine learning are used to forecast thyroid disorders. SVM and decision trees are utilized for classification, and data from the UCI repository is taken into consideration.

Fuzzy maps and data mining techniques were proposed by Papageorgiou EI and Papandrianos NI as enhanced methods for identifying thyroid illnesses [45]. Table 9.1 provides a summary of the numerous ML approaches used for disease diagnosis.

MATRIX EVALUATION FOR ML

Evaluation Matrix of Supervised Classification Algorithms

Accuracy, sensitivity, and specificity are frequently used to measure effectiveness of supervised systems. Sensitivity measures proportion of real positive data points that are correctly identified as genuine positive, while specificity measures proportion of real negative data points that are correctly identified as true negative. Accuracy measures percentage of prediction rate in model [46]:

$$\text{Accuracy} = \frac{TP+TN}{TP+FN+TN+FP}$$

$$\text{Sensitivity} = \frac{TP}{TP+FN}$$

$$\text{Specificity} = \frac{TN}{TN+FP}$$

TABLE 9.1

Summary of Findings of Healthcare Data Analysis Using Machine Learning Techniques

Reference	Disease Dataset	Machine Learning Algorithms	Performance Parameters	Results	Limitation
Tahmasseb et al. [47]	Breast cancer (mammography data)	Purpose: Prediction used learners: • Decision tree • Logistic regression • Support vector machine • Random forest • Linear discriminant analysis • Adaptive boosting (AdaBoost) • Extreme gradient boosting (XGBoost)	• Classification accuracy	XG boost shows the best result • Accuracy = 0.94	Small-size dataset
Dagli et al. [48]	Lung cancer (mammography data)	Purpose: Prediction used learners: • Logistic regression • Single perception neural network • Multilayer neural network	• Area under curve (AUC) • Classification accuracy precision	Multilayer neural network produces the best result • AUC = 0.75 • CI = 0.693–0.806 • Accuracy = 0.76 • Precision = 0.72	–
Zheng et al. [49]	Diabetes (electronic health records)	Purpose: Prediction used learners: • Support vector machine • K-Nearest neighbor • Logistic regression • Random forest • Decision tree • Naïve Bayes	Classification accuracy • Sensitivity • Specificity • Precision • Area under the curve (AUC)	Support vector machine performs the best result • Accuracy = 96% • Sensitivity = 95% • Specificity = 96% • Precision = 91% • AUC = 0.96	Small-size dataset

(Continued)

TABLE 9.1 (Continued)

Reference	Disease Dataset	Machine Learning Algorithms	Performance Parameters	Results	Limitation
Rahimian et al. [50]	Emergency admission EHR data (risk data)	Purpose: Risk prediction used learners: • Random forest • Cox model • Gradient boosting	• Area under curve (AUC)	Gradient boosting performs the best result. AUC = 0.779	Better techniques can be used
Raja and Sarkar [51]	14 Real-world disease datasets • Hepatitis • Cancer • Heart, and so on	Purpose: Classification and prediction used learners: • C4.5 • RIPPER • Naïve Bayes	Classification accuracy	Naïve Bayes shows the best result. Accuracy increased for all 14 medical datasets	–
Kaur and Malhotra [52]	Stress detection (mobile sensor data)	Purpose: Prediction used learners: • Hybrid approach • Bayesian network • Heuristic technique	• Classification accuracy	Hybrid approach performed well • Accuracy = 92.86% (BP) • Accuracy = 85.71% (HR)	Real-time sensor data not included

Evaluation Matrix of Supervised Regression Algorithms

For regression analysis, the mean absolute error, mean squared error, and root mean squared error are typically employed to assess model effectiveness. MSE is a common metric for comparing expected values to their corresponding actual values. MSE is distinguished by

$$\text{MSE} = \frac{1}{n}\sum_{i=1}^{n}(y_i - y_j)^2$$

where n is the number of data points, y_i is ith data point's actual value, and y_j is jth data point's anticipated value.

Similar to MSE, RMSE calculates square roots of average of squared variations to determine variance among a set of anticipated values and corresponding actual values.

$$\text{RMSE} = \sqrt{\frac{1}{n}\sum_{i=1}^{n}(y_i - y_i)^2}$$

where n is the total amount of data points, y_i is the ith data point's actual value, and y_i is the ith data point's anticipated value.

As a mean of absolute discrepancies among anticipated values and actual values, MAE calculates difference among a collection of expected values and associated actual values. The MAE is written down as

$$\text{MAE} = \frac{1}{n}\sum_{i=1}^{n}(y_i - y_i)^2$$

where n is total amount of data points, y_i is ith data point's actual value, and y_i is ith data point's anticipated value.

MAPE is derived as the average of absolute percentage discrepancies among projected actual values is a metric for evaluating the accuracy of a prediction or a mathematical model.

$$\text{MAPE} = \frac{1}{n}\sum_{i=1}^{n}\left|\frac{(y_i - y_i)}{y_i}\right| \times 100$$

where n is the total amount of data points, y_i is the ith data point's actual value, and y_i is the ith data point's anticipated value. These are often used to assess performance indicators in a variety of industries, particularly when employing algorithms for regression to forecast, time series, and analyze medical data.

EVALUATION MATRIX

As a component of data analysis, assessing the performance of clustering methods is essential. To evaluate the model's quality, certain metrics can be employed. The Euclidean distance may be calculated using SSE (sum of squared errors). Better cluster performance is indicated by a lower SSE value. The variance ratio criterion, commonly known as the Calinski-Harabaz index, is a measure of dispersion within as well as across clusters. The range between 1 and 1 is determined by the silhouette coefficient.

REFERENCES

1. Priyanka, K., and Kulennavar, N., "A survey on big data analytics in health care", Int. J. Comput. Sci. Inf. Technol. 5(4):5865–5868, 2014.
2. Raja, R., Mukherjee, I., and Sarkar, B.K., "A systematic review of healthcare big data", Sci. Programming 2020: 15, 2020, Article ID: 5471849.
3. Bhardwaj, R., and Nambiar, A.R., "A study of machine learning in healthcare", in IEEE 41st Annual Computer Software and Applications Conference, 2017.
4. Manyika, J., Chui, M., Brown, B., Bughin, J., Dobbs, R., Roxburgh, C., and Hung Byers, A., Big Data: The Next Frontier for Innovation, Competition, and Productivity. McKinsey Global Institute, 2011.
5. Sun, J., and Reddy, C.K., "Big data analytics for healthcare", in Proceedings of the 19th ACM SIGKDD International Conference on Knowledge Discovery and Data Mining—KDD '13, Chicago, IL, USA, 2013.
6. Yadav, A., Kaundal, V., Sharma, A., Sharma, P., Kumar, D., and Badoni, P., "Wireless sensor network based patient health monitoring and tracking system", Adv. Intell. Syst. Comput., 479, 2017.
7. Loua, D., Chena, X., and Zhao, Z., "A wireless health monitoring system based on android operating system", in IERI Procedia, International Conference on Electronic Engineering and Computer Science, 208–215, 2013.
8. Hong, Y.-J., Kim, I.J., Ahn, S.C., and GonKim, H., "Mobile Health monitoring system based on activity recognition using accelerometer", Simul. Modul. Theory Pract. 18:446–455, 2010.
9. Lou, D., Chen, X., Zhao, Z., Xuan, Y., Xu, Z., Jin, H., Guo, X., Fang, Z., "A wireless health monitoring system based on android operating system", IERI Proc. 4:208–215, 2013.
10. Melillo, P., Orrico, A., Scala, P., Crispino, F., and Pecchia, L., "Cloud-based smart health monitoring system for automatic cardiovascular and fall risk assessment in hypertensive patients", Journal of Medical Systems, 39(10), 2015.
11. Yadav, A., Kaundal, V., Sharma, A., Sharma, P., Kumar, D., and Badoni, P., "Wireless sensor network based patient health monitoring and tracking system", in Proceeding of International Conference on Intelligent Communication, Control and Devices, 903–917, 2016.
12. Kanrar, S., and Mandal, P.K., "E-health monitoring system enhancement with Gaussian mixture model", Multimedia Tools Appl. 76(8):10801–10823, 2016.
13. Zhang, Y., Liu, H., Su, X., Jiang, P., and Wei, D., "Remote mobile health monitoring system based on smart phone and browser/server structure", J. Healthcare Eng. 6(4):717–738, 2015.
14. Yonglin, R., Pazzi, R.W.N., and Boukerche, A., "Monitoring patients via a secure and mobile healthcare system", IEEE Wireless Commun. 17(1):59–65, 2010.
15. Pantelopoulos, A., and Bourbakis, N.G., "Prognosis—a wearable health-monitoring system for people at risk: methodology and modeling", IEEE Trans. Inf. Technol. Biomed. 14(3):613–621, 2010.

16. Guo, J.-Y., Zheng, Y.-P., Xie, H.-B., and Chen, X., "Continuous monitoring of electromyography (EMG), mechanomyography (MMG), sonomyography (SMG) and torque output during ramp and step isometric contractions", Med. Eng. Phys. 32(9):1032–1042, 2010.

17. Chan, V., Ray, P., and Parameswaran, N., "Mobile e-health monitoring: an agent-based approach", IET Commun. 2(2):223–230, 2008.

18. Dhillon, A., and Singh, A., "Machine learning in healthcare data analysis: a survey", J. Biol. Today's World 8:1–10, 2019.

19. Johnson, K., Soto, J.T., and Benjamin, S., "Artificial intelligence in cardiology: applications, benefits and challenges", Br. J. Cardiol. 25:86–87, 2018.

20. Zhang, Q., Xiao, Y., Dai, W., Suo, J., Wang, C., Shi, J., Zheng, H., "Deep learning based classification of breast tumors with shear-wave elastography," Ultrasonics 72:150–157, 2016.

21. Mitchell, T., Machine Learning, McGraw Hill, p. 2, 1997.

22. Allix, N.M. "Epistemology and knowledge management concepts and practices". J. Knowl. Manage. Pract. 1:1–21, 2003.

23. Kaelbling, L.P., and Moore, A., "Reinforcement learning a survey", J. Artif. Intell. Res. 4:237–285, 1996.

24. Muna, A.H., Moustafa, N., and Sitnikova, E., "Identification of malicious activities in industrial Internet of Things based on deep learning models", J. Inf. Sec. Appl. 41:1–11, 2018.

25. Song, F., Guo, Z., and Mei, D., "Feature selection using principal component analysis", in Proceedings of the 2010 International Conference on System Science, 12–14 November 2010, Vol. 1, pp. 27–30.

26. Li, M.A., Luo, X.Y., and Yang, J.F., "Extracting the nonlinear features of motor imagery EEG using parametric t-SNE", Neurocomputing 218:371–381, 2016.

27. Nagarajan, S.M., Muthukumaran, V., Murugesan, R., Joseph, R.B., Meram, M., and Prathik, A., "Innovative feature selection and classification model for heart disease prediction", J. Reliable Intell. Environ. 8:333–343, 2022.

28. Li, B., Zhang, P., Tian, H., Mi, S.S., Liu, D.S., and Ren, G.Q., "A new feature extraction and selection scheme for hybrid fault diagnosis of gearbox", Expert Syst. Appl. 38:10000–10009, 2011.

29. Kazemi, M., Kazemi, K., Yaghoobi, M.A., and Bazargan, H., "A hybrid method for estimating the process change point using support vector machine and fuzzy statistical clustering", Appl. Soft Comput. 40:507–516, 2016.

30. Ali, L., Niamat, A., Khan, J.A., Golilarz, N.A., Xingzhong, X., Noor, A., Nour, R., and Bukhari, S.A.C., "An optimized stacked support vector machines based expert system for the effective prediction of heart failure", IEEE Access 7:54007–54014, 2019.

31. Nilashi, M., Ahmadi, H., Manaf, A.A., Rashid, T.A., Samad, S., Shahmoradi, L., Aljojo, N., and Akbari, E., "Coronary heart disease diagnosis through self-organizing map and fuzzy support vector machine with incremental updates", Int. J. Fuzzy Syst. 22:1376–1388, 2020.

32. Abikoye, O.C., Omokanye, S.O., and Aro, T.O., "Text classification using data mining techniques: a review", Comput. Inf. Syst. J. 1:1–8, 2018.

33. Balaha, H.M., and Hassan, A.E.S., "Skin cancer diagnosis based on deep transfer learning and sparrow search algorithm", Neural Comput. Appl. 35:815–853, 2023.

34. Gupta, A., Kumar, L., Jain, R., and Nagrath, P., "Heart disease prediction using classification (naive bayes)", in Proceedings of the First International Conference on Computing, Communications, and Cyber-Security (IC4S 2019), Chandigarh, India, 12 October 2019, pp. 561–573.

35. Kotsiantis, S.B., Zaharakis, I., and Pintelas, P., "Supervised machine learning: a review of classification techniques", in Proceedings of the 2007 Conference on Emerging Artificial Intelligence Applications in Computer Engineering: Real Word AI Systems with Applications in EHealth, HCI, Information Retrieval and Pervasive Technologies, pp. 3–24, IOS Press, 2007.

36. Bhavsar, H., and Ganatra, A., "A comparative study of training algorithms for supervised machine learning", Int. J. Soft Comput. Eng. 2:2231–2307, 2012.
37. Shouman, M., Turner, T., and Stocker, R., "Applying k-nearest neighbour in diagnosing heart disease patients", Int. J. Inf. Educ. Technol. 2:220–223, 2012.
38. Sukanya, J., "Applications of big data analytics and machine learning techniques in health care sectors", Int. J. Eng. Comput. Sci. 6:21963–21967, 2017.
39. Parthiban, G., and Srivatsa, S.K., "Applying machine learning methods in diagnosing heart disease for diabetic patients", Int. J. Appl. Inf. Syst. 3:25–30, 2012.
40. Otoom, A.F., Abdallah, E.E., Kilani, Y., and Kefaye, A., "Effective diagnosis and monitoring of heart disease", Int. J. Software Eng. Appl. 9:143–156, 2015.
41. Iyer, A., Jeyalatha, S., and Sumbaly, R., "Diagnosis of diabetes using classification mining techniques", Int. J. Data Mining Knowl. Manage. Process 5:1–14, 2015.
42. Sen, S.K., and Dash, S., "Application of meta learning algorithms for the prediction of diabetes disease", Int. J. Adv. Res. Comput. Sci. Manage. Stud. 2:396–401, 2014.
43. Senturk, Z.K. and Kara, R., "Breast cancer diagnosis via data mining: performance analysis of seven different algorithms", Comput. Sci. Eng. 1:1–10, 2014.
44. Majali, et al., "Data mining techniques for diagnosis and prognosis of cancer", Int. J. Adv. Res. Comput. Commun. Eng. 4(3):613–616, 2015.
45. Papageorgiou, E.I., Papandrianos, N.I., Apostolopoulos, D.J., and Vassilakos, P.J., "Fuzzy cognitive map based decision support system for thyroid diagnosis management", in International Conference on Fuzzy Systems pp. 1204–1211, 2008.
46. Sidey-Gibbons, J.A., and Sidey-Gibbons, C.J., "Machine learning in medicine: a practical introduction", BMC Med. Res. Methodol. 19:1–18, 2019.
47. Tahmasseb, "Impact of machine learning with multiparametric magnetic resonance imaging of the breast for early prediction of response to neoadjuvant chemotherapy and survival outcomes in breast cancer patients", Investigative Radiology 54(2):110, 2019.
48. Dagli, Y., Choksi, S., and Roy, S., "Prediction of two-year survival among patients of non-small cell lung cancer", in Computer Aided Intervention and Diagnostics in Clinical and Medical Images, pp. 169–177, Springer, 2019.
49. Zheng, T., et al., "A machine learning-based framework to identify type 2 diabetes through electronic health records", Int. J. Med. Inf. 97:120–127, 2017.
50. Rahimian, F., et al, "Predicting the risk of emergency admission with machine learning: development and validation using linked electronic health records", PLoS Med. 15(11), e1002695, 2018.
51. Raja, R., and Sarkar, B.K., "An entropy-based hybrid feature selection approach for medical datasets", in Machine Learning, Big Data, and IoT for Medical Informatics, pp. 201–214, Academic Press, 2021.
52. Kaur, P., and Malhotra, S., "Improved SLReduct framework for stress detection using mobile phone-sensing mechanism in wireless sensor network", in Panigrahi, C., Pujari, A., Misra, S., Pati, B., Li, K.C. (Eds.), Progress in Advanced Computing and Intelligent Engineering. Advances in Intelligent Systems and Computing, vol 714. Springer, 2018. https://doi.org/10.1007/978-981-13-0224-4_45

10 Multi-perspective Issues and Role of WSN and Different Technologies in Healthcare Information System

Renu Popli[1], Rajeev Kumar[1], Isha Kansal[1], Kanwal Garg[2], Sumit Kumar Mahana[3], Ashutosh Sharma[4], and Syed Hassan Shah[5]
[1]Chitkara University Institute of Engineering and Technology, Chitkara University, Punjab, India
[2]Department of Computer Science and Applications, Kurukshetra University, Kurukshetra, Haryana
[3]Department of Computer Engineering, National Institute of Technology, Kurukshetra, Haryana
[4]Henan University of Science and Technology, Luoyang, Henan, China
[5]California Sate University, Long Beach, California, USA

INTRODUCTION

The wireless sensor network (WSN) has a wide range of potential applications and is regarded as one of the most significant technologies that has the potential to alter the history of mankind [1]. The real-time transmission, accuracy, and comprehensiveness of this network are its key strengths.

The main parts of a WSN are sensor nodes, which have the following functionalities: collecting data from various sensors; buffering and caching of sensor data; processing of data; self-testing and monitoring; receiving, transmitting, and forwarding of data packets; and coordinating networking tasks. The majority of the time sensor nodes are deployed in large numbers in inaccessible locations and run on batteries. These limitations serve as a guide for the design of sensor nodes in order to achieve the highest power efficiency and lowest unit cost. To collect data from various sensors, sensor nodes are equipped with specialised hardware and software that can interface with different types of sensors. This allows them to gather information such as temperature, humidity, light intensity, and more. Additionally, sensor nodes are designed to communicate wirelessly with each other and with a central base station. This enables them to form a network and share

DOI: 10.1201/9781003438205-10

data seamlessly. The wireless communication protocols used by sensor nodes vary depending on the application and the range of communication required. Some common protocols include Zigbee, Bluetooth Low Energy (BLE), and Wi-Fi. These protocols ensure reliable and efficient transmission of data while minimising power consumption. Moreover, sensor nodes often employ data compression techniques to reduce the amount of data transmitted, further optimising power efficiency. Furthermore, to extend the lifespan of the batteries, sensor nodes incorporate power management strategies such as sleep modes and duty cycling. Sleep modes allow the nodes to conserve energy by temporarily disabling non-essential components when not in use, while duty cycling involves periodically activating the node for data collection and communication tasks. Overall, the design of sensor nodes prioritises power efficiency, cost-effectiveness, and compatibility with various sensors and communication protocols to enable seamless data collection in diverse environments. The collected data are then buffered and cached within the sensor nodes to ensure reliable storage even in the event of intermittent connectivity or power loss. Sensor nodes are crucial components in WSNs, as they enable the collection of data in a power-efficient and cost-effective manner. These nodes are designed to be compatible with different sensors and communication protocols, allowing seamless data collection in various environments. To achieve power efficiency, two common techniques are employed: sleep mode and duty cycling. Sleep mode conserves energy by keeping the node inactive when not in use, while duty cycling involves periodically activating the node for data collection and communication tasks. This design approach ensures that the sensor nodes can operate for extended periods without draining their power sources. Additionally, to ensure reliable storage of the collected data, the sensor nodes have built-in buffering and caching mechanisms. These mechanisms allow for data storage even during intermittent connectivity or power loss situations, guaranteeing that no valuable information is lost. Overall, the design of sensor nodes focuses on optimising power efficiency, cost-effectiveness, and compatibility to enable efficient and reliable data collection in diverse environments [2, 3]. To optimise power efficiency, sensor nodes employ techniques like duty cycling, where they periodically wake up from sleep mode to perform sensing and processing tasks before returning to a low-power state. This helps prolong the battery life of the nodes, allowing them to operate for extended periods without requiring frequent battery replacements. In addition to duty cycling, sensor nodes also utilise energy harvesting techniques to further enhance power efficiency. These techniques involve harnessing energy from the surrounding environment, such as solar power or kinetic energy, to supplement or even replace the need for batteries. By tapping into these alternative energy sources, sensor nodes can operate autonomously for even longer durations, reducing the overall cost of maintenance and battery replacements. Moreover, cost-effectiveness is a crucial aspect of designing sensor nodes for efficient data collection. Manufacturers strive to develop affordable yet reliable components that can be mass-produced without compromising on quality. This ensures that deploying large-scale sensor networks becomes economically feasible, enabling widespread data collection across diverse environments. Additionally, advancements in manufacturing processes and materials have led to reduced production costs while maintaining high-performance standards. These advancements have also allowed for the miniaturisation of sensor nodes, making them smaller and more discreet. This is particularly important in applications where sensor nodes need to be deployed in hard-to-reach or sensitive areas. The compact size of these nodes also allows for easier installation and integration into existing infrastructure. Moreover, manufacturers are constantly innovating to improve the energy efficiency of sensor nodes. By developing low-power components and optimising power management techniques, sensor nodes can operate for extended periods without the need for frequent battery replacements. This not only reduces maintenance costs but also minimises the environmental impact. Furthermore, manufacturers are incorporating wireless communication capabilities into sensor nodes, enabling seamless data transmission over long distances. This eliminates the need for physical connections and simplifies network deployment and management. In conclusion, the continuous advancements in manufacturing processes, materials, energy efficiency, and wireless communication have revolutionised the design of sensor nodes, making them more affordable, reliable, and versatile than ever before. Furthermore, compatibility plays a vital role in facilitating efficient and reliable data collection. Sensor nodes are designed to be compatible with various communication protocols and networking technologies, allowing seamless integration into existing infrastructure. This compatibility enables easy deployment and integration of sensor networks. Self-testing and monitoring mechanisms are integrated into sensor nodes to ensure their proper functioning. These mechanisms can detect issues such as sensor failures or communication errors and trigger appropriate actions like sending alerts or initiating self-repair processes. Sensor nodes also play a crucial role in data transmission and forwarding.

The use of wireless sensors can be advantageous for both the health sciences and the healthcare system. Intel has conducted research on the issues that affect senior citizens. Wireless sensors can be used to monitor and manage cognitive disorders in their early stages, which may prevent Alzheimer's disease [4]. Such an experiment is being conducted by Intel in Las Vegas and Portland (proactive health research). The nodes can be used to remind senior citizens, record recent actions (such as taking medication or having a visitor last), reveal the person's true behaviour, or identify a developing issue. By tracking the frequency with which patients touch particular objects, Intel and the University of Washington have used RFID tags in related research to examine patient behaviour and customs. The application has a display that will assist the "care provider" in gathering information about the incapacitated person covertly and without offending them. Finally, notes can be employed to research young children's behaviour. For instance, the study described in [5] sought to analyse children's behaviour by tracking sensors hidden inside toys. Retinal prostheses are used to restore human vision in another medical application. In order to support a function, sensors are implanted into human organs. These sensors need to be able to communicate wirelessly with an external computer system, which performs the sophisticated processing. This computation cannot be done onboard the sensors due to energy constraints. Other suggested uses for wireless sensors in healthcare include keeping tabs on doctors and patients or monitoring drug use in hospitals, as shown in Figure 10.1.

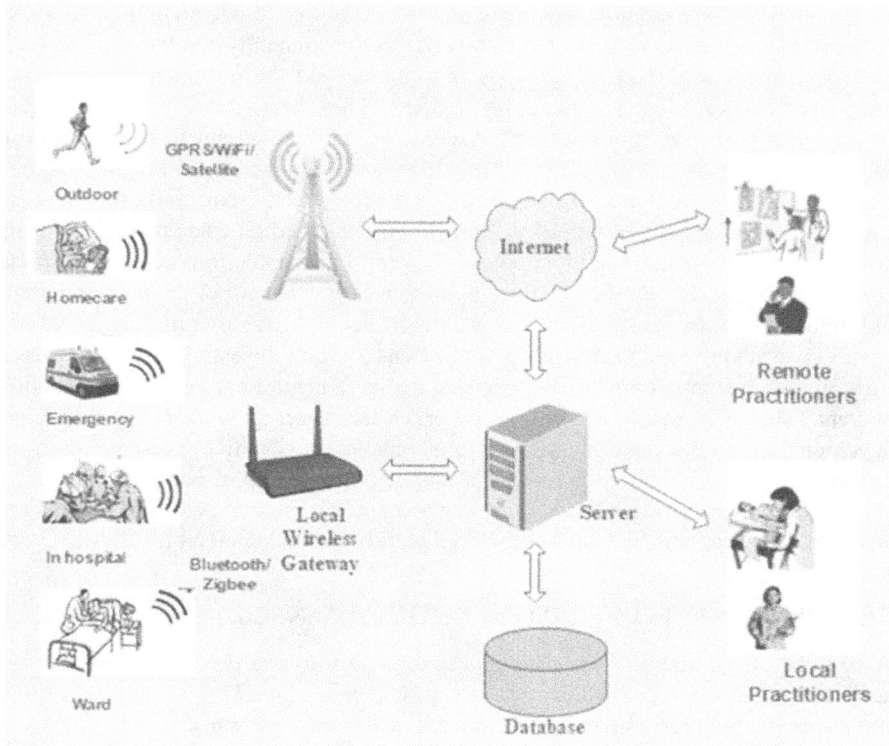

FIGURE 10.1   WSN for healthcare services.

Body sensor network systems can help people by providing healthcare services such medical monitoring, memory enhancement, home appliance control, access to medical data, and communication in emergency scenarios [6]. Continuous monitoring using wearable and implantable body sensor networks will enhance early disease and emergency condition identification in vulnerable individuals and provide a range of healthcare services for persons with various degrees of physical and cognitive limitations. These systems will help families where both parents must work to provide their infants and children with high-quality care, not just the elderly and the chronically ill. Furthermore, these systems can also revolutionise the field of sports medicine by providing real-time data on athletes' performance and health, allowing for personalised training programmes and injury prevention strategies. Additionally, the integration of artificial intelligence and machine learning algorithms into these systems will enable predictive analytics and personalised medicine, leading to more accurate diagnoses and targeted treatments. Moreover, with the advancement of telemedicine, these systems can facilitate remote consultations and virtual healthcare services, especially in rural areas where access to medical facilities is limited. Ultimately, the widespread adoption of such advanced healthcare systems has the potential to improve overall public health outcomes and enhance the quality of life for individuals across various demographics.

The COVID-19 pandemic has made us more receptive to telemedicine and more at-home health monitoring, and it has probably permanently altered the way we receive medical care. Our personal pulse oximeter and the no-touch thermometer are currently more than just desirable gadgets. They have genuine clinical applications in a potential "point-of-use" system. We can get screened at the point of use instead of going to the clinic. This shift towards remote healthcare has not only increased convenience but also improved accessibility, especially for those in rural areas or with limited mobility. Additionally, telemedicine has proven to be an effective tool for mental health support, allowing individuals to receive therapy or counselling from the comfort of their own homes. Furthermore, the pandemic has highlighted the importance of prioritising public health infrastructure and preparedness. Governments and healthcare organisations are now investing more in disease surveillance systems, early warning mechanisms, and rapid response protocols to prevent future outbreaks. Moreover, the crisis has spurred innovation in vaccine development and distribution, leading to advancements in mRNA technology and collaborative efforts on a global scale. As we navigate the post-pandemic world, it is crucial that we maintain the lessons learned and continue to prioritise healthcare advancements that enhance both individual well-being and societal resilience.

REQUIREMENTS FOR WIRELESS MEDICAL SENSORS

Wearability, dependability, security, and interoperability are the four primary characteristics that wireless medical sensors should meet [7]. This section describes some specific features that are required for wireless medical sensors:

a. *Wearability*: Wireless medical sensors must be portable and light in order to accomplish non-invasive and undetectable continuous health monitoring. The size and weight of batteries play a major role in determining the size and weight of sensors. However, a battery's capacity is inversely correlated with its size. We may anticipate that as technology continues to evolve, integrated circuits and batteries will become smaller, making it easier for designers to increase the comfort of users and the wearability of medical sensors. As technology advances, the development of smaller integrated circuits and batteries holds immense potential for enhancing user comfort and the wearability of medical sensors. The portability and lightweight nature of these sensors are crucial for achieving non-invasive and undetectable continuous health monitoring. The size and weight of the batteries have a significant impact on the overall dimensions of the sensors. Yet there is an inverse correlation between a battery's capacity and its size. Fortunately, ongoing technological advancements are expected to address this challenge by enabling the miniaturisation of integrated circuits and batteries. This progress will empower designers to create more compact and lightweight medical sensors, ultimately revolutionising the field of continuous health monitoring.

b. *Reliable communication*: For medical applications that rely on WBANs, reliable communication between body sensors is crucial. Various medical sensors require varying sample rates for communication, ranging from less than 1 Hz to 1,000 Hz. Moving beyond telemetry and processing the sensor signal is one way to increase reliability. For instance, feature extraction on the sensor can be used to convey only information about an event rather than the raw electrocardiogram (ECG) data that would otherwise be sent. Additionally, to lower the high demands on the communication channel, the decreased communication needs result in lower overall energy costs and longer battery life. For the best system architecture, communication and processing must be carefully balanced. This can be achieved by implementing a distributed system where some processing is done locally on the sensor itself while more complex computations are offloaded to a central processing unit. By distributing the workload, the communication bandwidth can be optimised, reducing latency and ensuring efficient data transmission. Furthermore, intelligent algorithms can be employed to prioritise and compress data before transmission, further minimising energy consumption. This balanced approach not only enhances reliability but also maximises the lifespan of the device, making it an ideal solution for long-term monitoring applications. Additionally, incorporating error detection and correction mechanisms into the communication protocol adds another layer of reliability to ensure accurate data transmission. Overall, by carefully considering the trade-offs between communication and processing in the system architecture, a robust and energy-efficient solution can be achieved for telemetry applications.

c. *Security*: The security of the entire WBAN system is another crucial problem [8]. A WBAN-based telemedicine system has three tiers, each of which has a security issue. Wireless medical sensors should provide data integrity and conform to the minimum privacy standards set forth by law for all medical devices. Even though key establishment, authentication, and data integrity are challenging problems in the resource-constrained environment of medical sensors, they are manageable thanks to the typical WBAN's small number of nodes and communication ranges. Furthermore, the second tier of the WBAN-based telemedicine system involves the transmission and storage of sensitive patient data. Ensuring the confidentiality and privacy of this data is paramount to protecting patients' personal information from unauthorised access. Robust encryption algorithms and secure storage protocols must be implemented to safeguard against potential breaches. Additionally, the third tier focuses on the communication between the WBAN and the healthcare provider's infrastructure. This introduces another security concern as it requires a secure and reliable channel for transmitting data while also preventing any malicious attacks or unauthorised modifications. Implementing secure communication protocols, such as SSL or TLS, can help mitigate these risks and ensure the integrity of the transmitted data. Overall, addressing these security challenges in each tier of a WBAN-based telemedicine system is essential to guarantee patient safety and maintain trust in remote healthcare services.

d. *Interoperability*: Depending on the user's state of health, wireless medical sensors should make it simple for users to construct a reliable WWBAN. Standards defining how wireless medical sensors interact will promote vendor competition and eventually result in more user-friendly devices [9]. Additionally, interoperability is crucial for the seamless integration of wireless medical sensors into the healthcare system. By establishing standards that govern how these sensors interact, users can easily create a robust WWBAN tailored to their specific health needs. This not only encourages competition among vendors but also paves the way for the development of more user-friendly devices. With interoperability in place, healthcare professionals can access and analyse data from various sensors with ease, enabling them to make informed decisions and provide personalised care to patients. Moreover, it allows for efficient data sharing between different healthcare providers, leading to improved coordination and better overall patient outcomes. In this resource-constrained environment, where reliability and data integrity are paramount, the establishment of interoperability standards is a significant step towards harnessing the full potential of wireless medical sensors to revolutionise healthcare delivery.

THE ROLE OF DIFFERENT TECHNOLOGICAL MECHANISMS IN HEALTHCARE

The role of different technological mechanisms in healthcare has been transformative, revolutionising patient care, medical research, and healthcare management. In this section, some key technological mechanisms and their roles in healthcare are highlighted. Table 10.1 outlines the applications of WSNs and different technologies in healthcare information systems. It provides a concise overview of the key features and benefits of each technology in the healthcare domain, along with relevant references to support the information.

Apart from the above-mentioned technologies, the role of various wireless technologies in healthcare is shown in Figure 10.2.

RFID APPLICATIONS IN HEALTHCARE: ENHANCING PATIENT SAFETY

Hospitals experience over 42.7 million adverse events annually as a result of medical and human errors [25], underscoring the critical need to improve patient safety. RFID has a tremendous potential to raise the standard of patient care by reducing human error during interactions with medical personnel [26]. However, there hasn't been much use of RFID technology in hospitals because of the high prices and difficulties proving a return on investment. In the literature, a variety of RFID applications for healthcare have been suggested. These applications include monitoring and tracking of patients, monitoring and tracking of pharmaceuticals, monitoring and tracking of blood transfusions, monitoring and tracking of equipment and assets, and collecting sensor-derived data. The four applications that fall under the category of patient safety are patient identification, patient tracking, patient monitoring, and patient drug compliance. These

TABLE 10.1

Role of Technologies in Healthcare Information Systems

Technology	Application in Healthcare Information Systems	Benefits	References
WSN	Real-time patient monitoring and tracking	Continuous data collection	[10]
WSN	Remote patient care and telemedicine	Remote healthcare services	[11]
WSN	Environmental monitoring in healthcare facilities	Reduced manual intervention	[12]
WSN	Fall detection and emergency response	Enhanced patient safety	[13]
WSN	Real-time patient monitoring	Continuous data collection	[14]
WSN	Remote patient tracking and health data collection	Remote healthcare services	[15]
AI	Medical diagnosis and decision support	Enhanced accuracy and precision	[16]
AI	Predictive analytics for patient outcomes	Faster diagnosis and treatment planning	[17]
AI	Medical diagnosis and imaging analysis	Enhanced accuracy and precision	[18]
AI	Predictive analytics for patient outcomes	Faster diagnosis and treatment planning	[19]
Wearable devices	Continuous health monitoring and fitness tracking	Patient empowerment and self-management	[20]
Wearable devices	Remote patient engagement and telemedicine	Patient self-management	[21]
IoT devices	Remote patient monitoring and health tracking	Improved patient engagement	[22]
IoT devices	Medication adherence monitoring	Personalised treatment plans	[23]
IoT devices	Home-based medical devices and wearables	Early detection of health issues	[24]

WSN: Wireless sensor network; IoT: Internet of Things; AI: artificial intelligence.

applications aim to enhance patient safety and improve overall healthcare outcomes. Patient identification systems utilising RFID technology can ensure accurate and efficient identification of patients, reducing the risk of medical errors, and ensuring proper care is provided to the right individual. RFID-based patient tracking systems enable healthcare providers to monitor the movement and location of patients within a facility, ensuring their safety and timely response in case of emergencies. Patient monitoring applications utilise RFID tags to collect real-time data on vital signs, allowing healthcare professionals to closely monitor patients' health status and intervene promptly if necessary. Additionally, RFID technology can be employed to promote patient drug compliance by tracking medication administration and reminding patients about their prescribed doses. These innovative applications not only streamline healthcare

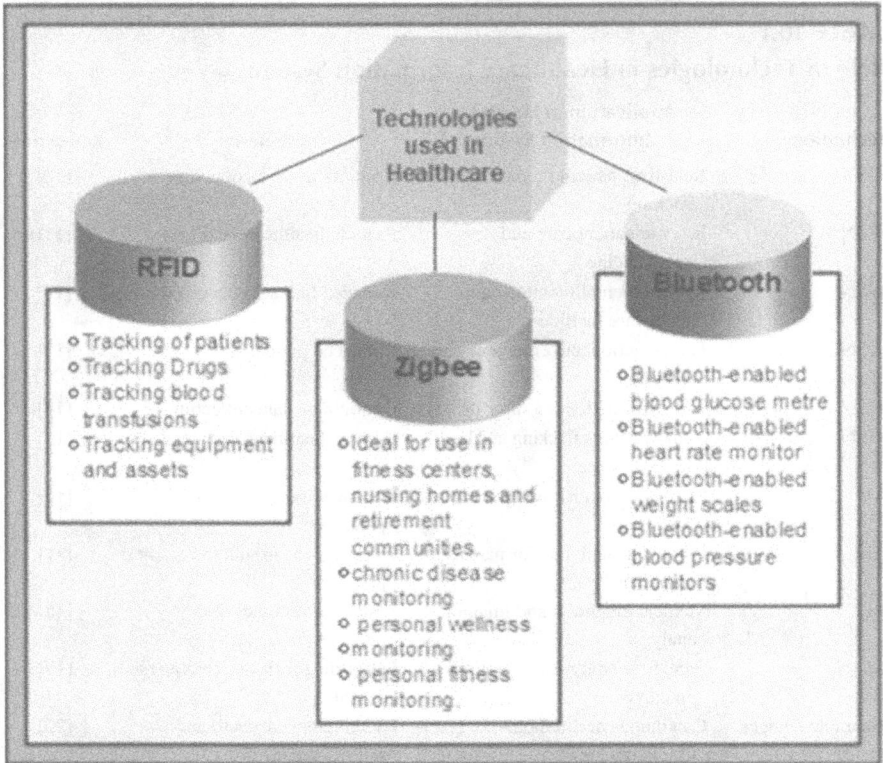

FIGURE 10.2   Wireless technologies used in healthcare.

processes but also significantly contribute to improving patient safety and the quality-of-care delivery [27].

Despite the many benefits of employing RFID to improve patient safety, there are still a number of difficulties to its implementation in the healthcare sector, including issues with tracking, identifying, and monitoring patients as well as assuring medicine compliance [28]. Three categories can be used to group these difficulties: There are three types of problems: technological, organisational, and financial; security, privacy, and data management challenges.

HEALTHCARE APPLICATIONS USING ZIGBEE SENSOR-BASED MONITORING

The wireless communication technology ZigBee, which is based on the wireless standard 802.15.4, is expanded by the WPAN. There are several notable features of this technology [29]. The ZigBee transmission range is between 10 and 100 m, which is sufficient for the requirements of mobility equipment. The first point is the short range. When no data exchange is necessary, the second option is low power, in which case the node enters a very low-power sleep state. A healthcare system that provides less expensive and wiser management tools can be developed using ZigBee technology to improve nursing care for chronic diseases. Special management is necessary for devices connected to the ZigBee personal area network. The authors of Ref. [30] suggested a dynamic medical management system for the ZigBee network. In order to collect biological or environmental data, this approach would utilise wireless personal area networks to link the signal sensor device to the mobile system. The authors of Ref. [31] created a way to quickly expand their sensor information network-based ZigBee home network in the meantime. The patient won't require special computer settings for managing their diabetes. Senior users can greatly benefit from pathology and documenting data such as weight, blood pressure, and other data. Additionally, this device enables real-time data analysis and remote monitoring, giving medical experts important information about the patient's status. The sensor gadget can easily transfer biological and environmental data to the mobile system by exploiting wireless personal area networks, doing away with the requirement for clumsy connected connections. This not only increases convenience but also guarantees the patient's freedom of movement. The addition of ZigBee technology to the home network further broadens the scope of sensor data, enabling a thorough comprehension of the patient's health status. With this novel strategy, diabetics may easily control their disease by keeping track of important statistics like weight, blood pressure, and other pertinent information. The older customers of this user-friendly solution, who may have low technical expertise or mobility issues, will particularly benefit from it. Overall, by utilising wireless connectivity and cutting-edge sensor technology to empower patients and enhance their quality of life, this system revolutionises diabetes control.

BLUETOOTH TECHNOLOGY-BASED HEALTHCARE MONITORING APPLICATIONS

Bluetooth technology has revolutionised the healthcare industry by enabling wireless communication between various devices. It has been widely used in healthcare monitoring applications to collect and transmit health data for remote monitoring and analysis [32]. Here are some examples of Bluetooth-based healthcare monitoring applications: One such application is the Bluetooth-enabled blood glucose metre [33], which allows diabetic patients to easily monitor their blood sugar levels without the need for frequent doctor visits. Another example is the Bluetooth-enabled heart rate monitor, which can be worn by patients during exercise or daily activities to track their heart rate and detect any abnormalities. Additionally, Bluetooth technology has been used in medication adherence monitoring devices, which remind patients to take their medication at the right time and track their compliance. Other applications include Bluetooth-enabled weight scales and blood pressure monitors, which allow patients to track their progress and share data with healthcare providers. Overall, Bluetooth technology has greatly improved healthcare monitoring and made it more convenient for patients to manage their health from home. In addition to medication adherence monitoring devices, Bluetooth technology has found its way into various other healthcare applications. For instance, it has been integrated into smartwatches and fitness trackers, enabling individuals to monitor their physical activity levels and sleep patterns. These data can be synced with mobile apps, providing users with valuable insights into their overall well-being. Moreover, Bluetooth-enabled glucose metres have revolutionised diabetes management by allowing patients to effortlessly track their blood sugar levels and share the information with their healthcare team for better treatment decisions. Furthermore, Bluetooth technology has been utilised in wearable heart rate monitors, empowering individuals to keep a close eye on their cardiovascular health during exercise or daily activities. The seamless connectivity that Bluetooth offers has significantly changed how patients interact with their healthcare providers and take charge of their own health outcomes. With continuous advancements in this field, the future holds immense potential for further innovations in Bluetooth-enabled healthcare devices.

The BLE-Based Health Monitoring System has revolutionised the way we monitor our health [34]. With its low power consumption and wireless connectivity, Bluetooth technology has made it possible for patients to track their health data in real-time and share it with healthcare providers. This has greatly improved patient compliance and reduced the need for frequent hospital visits. The system is also being used to monitor chronic conditions such as diabetes, where patients can track their blood glucose levels and receive alerts when they need to take medication or adjust their diet. Additionally, Bluetooth-enabled weight scales and blood pressure monitors have made it easier for patients to manage their weight and blood pressure from home. The data collected by these devices can be shared with healthcare providers, who can then provide personalised advice and treatment plans based on the patient's individual needs. Overall, Bluetooth technology has revolutionised healthcare monitoring by making it more convenient, accessible, and effective for patients of all ages and backgrounds.

This author [35] discusses the use of BLE technology for remote patient monitoring, highlighting its advantages, challenges, and potential applications in healthcare. Bluetooth technology and its applications in healthcare are rapidly growing fields with the potential to revolutionise the way healthcare is delivered. The advantages of Bluetooth-enabled medical devices are numerous, including increased patient convenience, improved accuracy and reliability of data, and reduced healthcare costs. However, challenges such as data security and privacy concerns must be addressed to ensure widespread adoption. Despite these challenges, the potential applications of Bluetooth technology in healthcare are vast, ranging from remote patient monitoring to telemedicine consultations. With continued innovation and development, Bluetooth-enabled medical devices have the potential to transform healthcare delivery and improve patient outcomes.

The authors in Ref. [36] provided an overview of Bluetooth technology and its applications in healthcare, covering topics like wearable devices, remote monitoring, and patient safety. Bluetooth-enabled inhalers for asthma management are a promising innovation in telemedicine consultations. The use of Bluetooth technology in healthcare has opened up new possibilities for remote patient monitoring and management. One promising innovation is the development of Bluetooth-enabled inhalers for asthma management [37]. These inhalers can track medication usage and provide real-time feedback to patients and healthcare providers, allowing for personalised treatment plans and improved medication adherence. Wearable devices such as smartwatches and fitness trackers can also be integrated with Bluetooth technology to monitor vital signs like heart rate, blood pressure, and oxygen saturation levels. These data can be transmitted to healthcare providers for analysis, enabling early detection of potential health issues and timely interventions. Patient safety is also improved with the use of Bluetooth-enabled medical devices, as they reduce the risk of errors associated with manual data entry [38, 39]. Overall, the integration of Bluetooth technology in healthcare has the potential to revolutionise patient care by providing more personalised, efficient, and effective treatment options through telemedicine consultations.

SOME REAL PROJECTS AND RELATED WORKS

The study of wireless healthcare systems has received a lot of attention. Wearable health technology has been the subject of several recent projects. Governmental agencies and various private organisations have taken up these initiatives. These projects span a wide range of healthcare topics, including cancer diagnosis, geriatric patient monitoring, glucose-level monitoring, stress monitoring, and ECG monitoring. One notable project in the field of wireless healthcare systems is the development of wearable devices for cancer diagnosis. These devices utilise advanced sensors to detect early signs of cancer, allowing for timely intervention and treatment. Another project focuses on geriatric patient monitoring, where wearable technology is used to track vital signs and provide real-time alerts in case of emergencies. Additionally, there are projects dedicated to glucose level monitoring, enabling individuals with diabetes to conveniently monitor their blood sugar levels without invasive procedures. Stress monitoring is another area of interest, with wearable devices measuring physiological responses to stressors and providing personalised stress management strategies. Lastly, ECG monitoring projects aim to create wearable devices that can continuously monitor heart activity, allowing for early detection of cardiac abnormalities. These initiatives demonstrate the vast potential of wireless healthcare systems to improve patient care and overall well-being through innovative technological solutions. Here are some of the most significant indoor and outdoor application projects now underway around the globe.

PROJECTS AND APPLICATIONS IN REAL LIFE

HealthGear is a product of Microsoft Research [40]. It consists of a cell phone Bluetooth-connected to a group of physiological sensors. It effectively performs the same functions as a wearable, real-time health system to monitor and analyse physiological data [41]. HealthGear has revolutionised the way we monitor our health and well-being. By seamlessly integrating a cell phone with physiological sensors, this innovative technology provides real-time insights into our body's vital signs. Gone are the days of bulky medical equipment; HealthGear offers a compact and convenient solution for tracking and analysing physiological information. Whether you're an athlete looking to optimise performance or an individual managing a chronic condition, this wearable health system empowers you to take control of your own health. With its Bluetooth connectivity, HealthGear ensures seamless data transfer and allows for easy integration with other health monitoring devices. The possibilities are endless, as this technology opens up new avenues for personalised healthcare and remote patient monitoring. From tracking heart rate variability to monitoring sleep patterns, HealthGear is paving the way for a future where technology plays a crucial role in enhancing patient care and overall well-being. As these indoor and outdoor application projects continue to evolve globally, we can expect even more groundbreaking advancements in the field of healthcare technology.

The European Commission has financed the mobile healthcare initiative MobiHealth [42]. It intends to create cutting-edge approaches to individualised healthcare and remote patient monitoring. MobiHealth is revolutionising the delivery of healthcare services with the help of cutting-edge mobile technologies. By leveraging the power of smartphones and wearable devices, patients can now conveniently monitor their vital signs and share real-time data with healthcare professionals. This not only enhances the efficiency of medical consultations but also empowers individuals to take control of their own health. Moreover, MobiHealth's focus on personalised healthcare ensures that treatment plans are tailored to each patient's unique needs, leading to better outcomes and improved quality of life. With the European Commission's backing, MobiHealth is set to drive significant advancements in mobile healthcare technology, ultimately transforming the way we approach patient care. By utilising UMTS and GPRS networks, it enables patients to be completely mobile while receiving ongoing health monitoring. The management of patients with arrhythmic heart disease and the follow-up observation of post-operative care in patients who have undergone surgery are two areas that are being taken into consideration.

A study called CodeBlue [43] was conducted at Harvard University in the United States. It incorporates wireless devices and sensor nodes into a disaster response environment. It is made to function with a variety of wireless devices and network densities. PCs can be combined in CodeBlue, ranging from tiny sensor motes to more potent devices like PDSs. The CodeBlue study at Harvard University has revolutionised disaster response by integrating wireless devices and sensor nodes. This innovative system is designed to adapt to various wireless devices and network densities, allowing for seamless communication in critical situations. From small sensor motes to powerful devices like PDSs, CodeBlue enables effective coordination and data collection. This advancement holds immense potential for improving post-operative care for patients who have undergone surgery, as it allows for real-time monitoring of vital signs and disease progression. By leveraging this technology, healthcare professionals can ensure timely interventions and personalised care, minimising the risk of complications and optimising patient outcomes. The integration of CodeBlue into healthcare settings marks a significant step forward in enhancing patient safety and well-being.

For context-aware computing research, the wearable sensor and notification platform eWatch [44] was created. It can be worn as a wristwatch, making it easily accessible, immediately visible, and appropriate in social settings. The eWatch detects and records light, motion, sound, and temperature while providing haptic, aural, and visual alerts. This comprehensive data collection allows healthcare professionals to monitor patients' vital signs and detect any abnormalities or potential health risks in real-time. The integration of eWatch into healthcare settings enables continuous monitoring and timely interventions, ultimately improving patient outcomes. By providing context-aware alerts, the eWatch ensures that healthcare providers are promptly notified of any critical changes in a patient's condition, allowing for immediate action to be taken. This proactive approach not only enhances patient safety but also contributes to better overall patient well-being. Moreover, the eWatch's discreet design and functionality make it suitable for use in social settings, minimising any potential discomfort or stigma associated with traditional medical devices. With its ability to capture a wide range of environmental data and deliver alerts through various sensory channels, the eWatch revolutionises patient monitoring by combining convenience, accuracy, and effectiveness.

The Vital Jacket mobile device [45] is an intelligent wearable garment that can continuously monitor ECG waves and heart rate for a variety of fitness, high-performance sports, security, and medical applications. Here, data can be sent to a PDA over Bluetooth while also being saved on a memory card. This allows for real-time monitoring and analysis of vital signs, providing valuable insights into an individual's health and performance. The Vital Jacket's seamless integration with other devices and platforms enables healthcare professionals to remotely access and review patient data, facilitating timely interventions and personalised care. Additionally, its compact design and comfortable fit make it suitable for long-term wear, ensuring uninterrupted monitoring throughout the day. With its advanced technology and comprehensive features, the Vital Jacket sets a new standard in wearable healthcare devices, empowering individuals to take control of their well-being while enhancing the efficiency of medical professionals.

All of these initiatives seek to offer reasonably priced ongoing monitoring of a person's health-related conditions. The cost-effectiveness and power consumption of these devices are the main areas of study. Although these gadgets serve a novel purpose, they have significant social implications in terms of security, privacy, and legal considerations. Some of these apps, for instance, rely significantly on Bluetooth-like technologies. Eavesdropping and service denial are two threats to security that these technologies pose. They must also address worries about potential health risks from implanted devices. In the following sections, we will outline the key issues that are connected to them.

MULTI-PERSPECTIVE ISSUES IN WSN WITH HEALTHCARE

There are various concerns associated with the use of WSNs in healthcare. These concerns include privacy and confidentiality, data integrity and security, interoperability, ethical and legal considerations, and potential health risks. Privacy and confidentiality are crucial in healthcare settings as they involve sensitive personal information. Data breaches or unauthorised access to this information can have severe consequences for individuals. Ensuring data integrity and security is also vital to prevent tampering or manipulation of patient data, which could lead to incorrect diagnoses or treatments. Interoperability is another challenge, as different WSN devices may use different protocols or standards, making it difficult to exchange information seamlessly. Ethical considerations arise when using WSN in healthcare, such as ensuring informed consent from patients and addressing issues of equity in access to these technologies. Finally, potential health risks from implanted devices must be carefully evaluated to minimise any adverse effects on patients' well-being. Overall, a multi-perspective approach is necessary to address these complex issues and ensure the safe and effective use of WSN in healthcare. These issues are described in the following:

SECURITY CONCERNS

The possibility of a security breach exists even in sensor network applications used in healthcare [46, 47]. Threats and attacks can be broadly divided into passive and active kinds. While the data packets are being routed, passive attacks could happen. Attackers may alter packet destinations or introduce erratic routing [7, 9]. By listening in on the wireless connection medium, the attackers can also steal health information in this situation. Threats that are active provide a greater threat than those that are passive. Eavesdropping could allow criminals to determine the user's location. Threats to one's life could result from this. The typical design of sensor devices leaves limited room for external security mechanisms, making them vulnerable to physical tampering This makes the devices more vulnerable and creates more difficult security issues. Eavesdropping can also be used to intercept crucial data flows from WBAN networks via GPRS or other networks. The authors have described a few types of attacks on health monitoring in more detail, including jamming attacks, eavesdropping on medical data, altering medical data, fabricating alarms on medical data, denial of service attacks, user location and activity tracking, and modification of medical data.

PRIVACY CONCERNS

Regarding healthcare applications, privacy in WSNs has always been an issue. Data transmission from a patient via wireless technology can seriously jeopardise that person's right to privacy [48]. They have emphasised that there is a risk of a public backlash that will lead to mistrust and prevent the technology from being used for the many beneficial applications where it can offer significant benefits if the privacy-related issues are not honestly debated in a reasoned and open manner. People may be prevented from enjoying the full benefits of the system if misuse or privacy concerns prevent them from doing so, regardless of whether the data were received with the person's consent or without it because the system required it (e.g., emergency data from a patient). Furthermore, the potential misuse or privacy concerns surrounding this technology should not overshadow the numerous beneficial applications it can offer. It is crucial that we engage in honest and open debates to address these issues effectively. By doing so, we can ensure that people are not deprived of the system's full benefits. Whether it is receiving emergency data from a patient or any other consent-based information, we must find a balance that safeguards privacy while still allowing for the utilisation of this technology's advantages. Only through reasoned discussions can we strike the right balance and unlock the immense potential it holds for improving various aspects of our lives. Let us not allow fear or apprehension to hinder progress but rather foster an environment where privacy concerns are addressed responsibly, enabling us to harness the true power of this transformative technology.

Legal Concerns

Legal difficulties must arise since the data in WSNs for healthcare systems will be sent to various people in various locations [49]. Who will be responsible and liable for the data obtained from a person? This is the most obvious query. Other questions that may come up include who will be in charge of enforcing regulations and who qualifies for access rights to utilise these data. Who will be liable if there is abuse? A person's data generated from them may be used unlawfully. This could cause issues for the person whose data are being made public. He or she might experience discrimination and other issues in life, such as blackmail. Someone who has certain healthcare-related secrets that are known may experience prejudice at work. Furthermore, the potential consequences of unauthorised data usage extend beyond individual privacy concerns. With the increasing reliance on data-driven decision-making, it becomes crucial to address the question of who will be responsible for enforcing regulations in this domain. Establishing a governing body or regulatory authority becomes imperative to ensure compliance and prevent misuse. Additionally, determining access rights to this data is another critical aspect. It is essential to define clear criteria and guidelines for who qualifies to access and utilise such sensitive information. Striking a balance between enabling innovation and protecting individuals' privacy rights will be paramount in this process. Moreover, liability issues must be addressed to hold accountable those responsible for any abuse or unlawful use of personal data. Implementing robust legal frameworks can help establish accountability mechanisms that deter potential wrongdoers and provide recourse for victims. The repercussions of unauthorised data usage can be far-reaching, leading to various adverse outcomes for individuals whose private information is exposed.

The use of internal sensor implants in humans is a significant additional problem. These gadgets may become an unwelcome intrusion on a person's privacy, which will place psychological strain on that person. The cost of maintaining these gadgets could result in additional financial hardship for both the individual and the government. The average person may become afraid as a result of this at any time. It would be challenging to employ sensor networks on a large scale if these issues weren't effectively handled. One possible solution to address the concerns surrounding sensor implants in humans is to prioritise privacy and security measures. Implementing strict regulations and encryption protocols can help ensure that individuals' personal information remains confidential and protected from unauthorised access. Additionally, providing transparent information about data collection practises and allowing individuals to have control over their own data can help alleviate privacy concerns. Moreover, developing affordable and accessible maintenance plans for these gadgets can help mitigate the financial burden on both individuals and the government. By offering subsidies or incentives for regular maintenance, the cost of upkeep can be reduced, making it more feasible for people to adopt these technologies. Furthermore, fostering public awareness and education about the benefits of sensor networks can help alleviate fear and resistance. By emphasising the potential positive impacts on healthcare, safety, and overall quality of life, individuals may become more open to embracing this technology. Ultimately, addressing privacy concerns, reducing financial burdens, and promoting awareness are crucial steps in effectively implementing sensor networks on a large scale while minimising potential drawbacks.

CONCLUSION AND FUTURE SCOPE

WSN technologies are crucial to improving healthcare quality of life. Wearable monitoring systems provide continuous physiological data and information on individual health, reducing healthcare costs and enhancing quality of life. Furthermore, the integration of WSN technologies with artificial intelligence and machine learning algorithms holds immense potential for the early detection and prediction of diseases. By analysing the vast amount of data collected from wearable devices, healthcare professionals can identify patterns and trends that may indicate the onset of a health condition. This proactive approach not only allows for timely intervention but also enables personalised treatment plans tailored to each individual's needs. Additionally, advancements in WSN technologies can revolutionise remote patient monitoring, enabling healthcare providers to remotely monitor patients' vital signs and provide real-time feedback and support. This not only reduces the burden on hospitals but also allows patients to receive quality care from the comfort of their homes. Moreover, the widespread adoption of WSN technologies in healthcare can contribute to the development of smart cities and communities. By integrating sensor networks with existing infrastructure, cities can become more efficient in managing resources such as energy, water, and transportation systems. This holistic approach to urban planning promotes sustainability, reduces environmental impact, and enhances the overall quality of life. This chapter discusses devices and techniques for monitoring blood pressure, glucose levels, cardiac activity, and respiratory activity, as well as on-body propagation issues. Wearable and implantable body sensor networks face challenges and open research problems, but they offer significant benefits. For instance, wearable devices like smartwatches and fitness trackers have revolutionised the way individuals monitor their health. These devices can track various vital signs such as heart rate, sleep patterns, and physical activity levels. They provide real-time feedback and enable users to make informed decisions about their well-being. Additionally, implantable body sensor networks have shown great potential in the field of healthcare. These tiny sensors can be placed inside the body to monitor specific conditions or deliver targeted treatments. They have the ability to collect data continuously and transmit it wirelessly to healthcare professionals for analysis. However, there are challenges that need to be addressed, such as ensuring data privacy and security, optimising power consumption, and developing reliable communication protocols. Despite these obstacles, wearable and implantable body sensor networks hold immense promise for improving healthcare outcomes and empowering individuals to take control of their own health. By integrating these technologies into urban planning initiatives, cities can create smarter and more sustainable environments that prioritise the well-being of their residents while minimising resource consumption.

The use of gas sensors in healthcare is still in its infancy [50]. They do, however, have potential in the medical industry and in environmental sensing. Checking a person's "out breath"—the gas content of their breath—can be useful in determining their risk of developing diabetes and maybe other ailments. Numerous devices actually take advantage of this aspect, but they aren't as optical as poking your finger to check your blood sugar or diabetes. Therefore, our gasoline sensors are another aspect of this whole breath sensor thing. Gasoline sensors have the potential to revolutionise the way we detect and monitor pollutants in the environment. These sensors, although still in their early stages of development, hold promise for applications in the medical industry and environmental sensing. By analysing the gas content of a person's breath, these sensors can provide valuable insights into their health status, including the risk of developing diabetes and other ailments. While there are already devices that utilise this concept, they lack the convenience and non-invasiveness of traditional methods like blood sugar testing. This is where gasoline sensors come into play, offering a more optical and efficient approach to breath analysis. With further advancements, these sensors could significantly contribute to improving healthcare outcomes and environmental monitoring efforts.

REFERENCES

1. Neves, Paulo, Michal Stachyra, and Joel Rodrigues, "Application of wireless sensor networks to healthcare promotion," Journal of Communications Software and Systems, vol. 4, no. 3, pp. 181–190, 2008.
2. Tan, M., "Security issues of wireless sensor networks in healthcare applications," BT Technology, vol. 24, no. 6, pp. 138–144, 2006.
3. Al Ameen, Moshaddique, Jingwei Liu, and Kyungsup Kwak, "Security and privacy issues in wireless sensor networks for healthcare applications," Journal of Medical Systems, vol. 36, pp. 93–101, 2012.
4. Arampatzis, Th, John Lygeros, and Stamatis Manesis, "A survey of applications of wireless sensors and wireless sensor networks," In Proceedings of the 2005 IEEE International Symposium on Mediterranean Conference on Control and Automation Intelligent Control, pp. 719–724, IEEE, 2005.
5. Darwish, Ashraf, and Aboul Ella Hassanien, "Wearable and implantable wireless sensor network solutions for healthcare monitoring," Sensors, vol. 11, no. 6, pp. 5561–5595, 2011.
6. Stanford, V., "Using pervasive computing to deliver elder care," IEEE Pervasive Computing, vol. 1, no. 1, pp. 10–13, 2002.
7. Rani, Shalli, Deepika Koundal, Fnm Kavita, Muhammad Fazal Ijaz, Mohamed Elhoseny, and Mohammed I. Alghamdi, "An optimized framework for WSN routing in the context of industry 4.0," Sensors, vol. 21, no. 19, p. 6474, 2021.
8. Bahache, Anwar Noureddine, Noureddine Chikouche, and Fares Mezrag, "Authentication schemes for healthcare applications using wireless medical sensor networks: A survey," SN Computer Science, vol. 3, no. 5, p. 382, 2022.
9. Rathore, Pramod Singh, Jyotir Moy Chatterjee, Abhishek Kumar, and Radhakrishnan Sujatha, "Energy-efficient cluster head selection through relay approach for WSN," The Journal of Supercomputing, vol. 77, pp. 7649–7675, 2021.
10. Cook, D., and S. Das, Smart Environments: Technology, Protocols and Applications, Wiley, 2005.
11. Xu, Y., et al., "Wireless Sensor Networks for Healthcare," Wiley, 2013.

12. Hossain, A., and M. Alam, Wireless Sensor Networks for Healthcare: A Survey, Elsevier, 2014.
13. Hu, C., L. Xu, and L. Chan, Wireless Sensor Networks in Healthcare, Springer, 2009.
14 Ko J, Lu C, Srivastava MB, Stankovic JA, Terzis A, Welsh M. "Wireless Sensor Networks for Healthcare," in Proceedings of the IEEE, vol. 98, no. 11, pp. 1947–1960, Nov. 2010, doi: 10.1109/JPROC.2010.2065210.
15 Alemdar, H. and Ersoy, C., "Wireless sensor networks for healthcare: A survey," Computer Networks, vol. 54, no. 15, pp. 2688–2710, 2010.
16. Amann, J., Blasimme, A., Vayena, E., Frey, D., Madai, V.I. and Precise4Q Consortium, " "Explainability for artificial intelligence in healthcare: a multidisciplinary perspective BMC Medical Informatics and Decision Making, vol. 20, pp. 1–9, 2020
17. Buczak, Anna L., and Erhan Guven, "A survey of data mining and machine learning methods for cybersecurity intrusion detection," IEEE Communications Surveys & Tutorials, pp. 1153–1176, 2015.
18. Esteva, A., Kuprel, B., Novoa, R.A., Ko, J., Swetter, S.M., Blau, H.M. and Thrun, S., "Dermatologist-level classification of skin cancer with deep neural networks," Nature, vol. 542, no. 7639, pp. 115–118, 2017.
19. Rajkomar, A., Oren, E., Chen, K., Dai, A.M., Hajaj, N., Hardt, M., Liu, P.J., Liu, X., Marcus, J., Sun, M. and Sundberg, P., "Scalable and accurate deep learning with electronic health records," NPJ Digital Medicine, vol. 1, pp. 18, 2018.
20. Patel, S., Park, H., Bonato, P., Chan, L. Rodgers, M., "A review of wearable sensors and systems with application in rehabilitation," Journal of Neuroengineering and Rehabilitation, vol. 9, no. 1, pp. 1–17, 2012.
21. Bonato, P. "Wearable sensors and systems," *IEEE Engineering in Medicine and Biology Magazine*, vol. 29, no. 3, pp. 25–36, 2010.
22. Ganti, R. K., and R. Jayaraman, Internet of Things: Healthcare Applications, Springer, 2014.
23. Ray, P. K., and B. Ray, Internet of Things for Healthcare Applications, CRC Press, 2017.
24. Shakshuki, E., Internet of Things and Big Data Analytics for Smart and Connected Health, Springer, 2018.
25. Ajami, Sima, and Ahmad Rajabzadeh, "Radio frequency identification (RFID) technology and patient safety," Journal of Research in Medical Sciences, vol. 18, no. 9, p. 809, 2013.
26. Haddara, Moutaz, and Anna Staaby, "RFID applications and adoptions in healthcare: A review on patient safety," Procedia Computer Science, vol. 138, pp. 80–88, 2018.
27. Haddara, Moutaz, and Anne Staaby, "RFID applications for patient safety in the healthcare sector," In Quality of Healthcare in the Aftermath of the COVID-19 Pandemic, pp. 155–179. IGI Global, 2022.
28. Ting, S. L., Siu Keung Kwok, Albert H. C. Tsang, and Wing Bun Lee, "Critical elements and lessons learnt from the implementation of an RFID-enabled healthcare management system in a medical organization," Journal of Medical Systems, vol. 35, pp. 657–669, 2011.
29. Huang, Yueh-Min, Meng-Yen Hsieh, Han-Chieh Chao, Shu-Hui Hung, and Jong Hyuk Park, "Pervasive, secure access to a hierarchical sensor-based healthcare monitoring architecture in wireless heterogeneous networks," IEEE Journal on Selected Areas in Communications, vol. 27, no. 4, pp. 400–411, 2009.
30. Sharma, Rinki, Shreyas K. Gupta, K. K. Suhas, and G. Srikanth Kashyap, "Performance analysis of ZigBee based wireless sensor network for remote patient monitoring," In 2014 Fourth International Conference on Communication Systems and Network Technologies, pp. 58–62, IEEE, 2014.
31. Laine, Teemu H., Chaewoo Lee, and Haejung Suk, "Mobile gateway for ubiquitous health care system using ZigBee and Bluetooth," In 2014 Eighth International Conference on Innovative Mobile and Internet Services in Ubiquitous Computing, pp. 139–145, IEEE, 2014.

32. Omre, Alf Helge, and Steven Keeping, "Bluetooth low energy: Wireless connectivity for medical monitoring," Journal of Diabetes Science and Technology, vol. 4, no. 2, pp. 457–463, 2010.
33. Ly, Trang T., Jennifer E. Layne, Lauren M. Huyett, David Nazzaro, and Jason B. O'Connor, "Novel Bluetooth-enabled tubeless insulin pump: Innovating pump therapy for patients in the digital age," Journal of Diabetes Science and Technology, vol. 13, no. 1, pp. 20–26, 2019.
34. Lin, Zhe-Min, Cheng-Hung Chang, Nai-Kuan Chou, and Yuan-Hsiang Lin, "Bluetooth low energy (BLE) based blood pressure monitoring system," In 2014 International Conference on Intelligent Green Building and Smart Grid (IGBSG), pp. 1–4, IEEE, 2014.
35. Kumar, B. G. Anil, K. C. Bhagyalakshmi, K. Lavanya, and K. H. Gowranga, "A Bluetooth low energy based beacon system for smart short range surveillance," In 2016 IEEE International Conference on Recent Trends in Electronics, Information & Communication Technology (RTEICT), pp. 1181–1184, IEEE, 2016.
36. De Raeve, Nick, Adnan Shahid, Matthias De Schepper, Eli De Poorter, Ingrid Moerman, Jo Verhaevert, Patrick Van Torre, and Hendrik Rogier, "Bluetooth-low-energy-based fall detection and warning system for elderly people in nursing homes," Journal of Sensors, vol. 2022, pp. 1–14, 2022.
37. Ferrante, Giuliana, Amelia Licari, Gian Luigi Marseglia, and Stefania La Grutta, "Digital health interventions in children with asthma," Clinical & Experimental Allergy, vol. 51, no. 2, pp. 212–220, 2021.
38. Zhang, Qiaoyang, and Zhiyao Liang, "Security analysis of Bluetooth low energy based smart wristbands," In 2017 2nd International Conference on Frontiers of Sensors Technologies (ICFST), pp. 421–425, IEEE, 2017.
39. Filippoupolitis, Avgoustinos, William Oliff, and George Loukas, "Bluetooth low energy based occupancy detection for emergency management," In 2016 15th international conference on ubiquitous computing and communications and 2016 International Symposium on Cyberspace and Security (IUCC-CSS), pp. 31–38, IEEE, 2016.
40. Oliver, Nuria, and Fernando Flores-Mangas, "HealthGear: A real-time wearable system for monitoring and analyzing physiological signals," In International Workshop on Wearable and Implantable Body Sensor Networks (BSN'06), pp. 4, IEEE, 2006.
41. Oliver, Nuria, and Fernando Flores-Mangas, "Healthgear: Automatic sleep apnea detection and monitoring with a mobile phone," Journal of Communications, vol. 2, no. 2, pp. 1–9, 2007.
42. Van Halteren, Aart, Richard Bults, Katarzyna Wac, Dimitri Konstantas, Ing Widya, Nicolay Dokovsky, George Koprinkov, Val Jones, and Rainer Herzog, "Mobile patient monitoring: The MobiHealth system," The Journal on Information Technology in Healthcare, vol. 2, no. 5, pp. 365–373, 2004.
43. Malan, David J., Thaddeus Fulford-Jones, Matt Welsh, and Steve Moulton. "Codeblue: An ad hoc sensor network infrastructure for emergency medical care," In International Workshop on Wearable and Implantable Body Sensor Networks, 2004.
44. Maurer, Uwe, Anthony Rowe, Asim Smailagic, and Daniel P. Siewiorek, "eWatch: A wearable sensor and notification platform," In International Workshop on Wearable and Implantable Body Sensor Networks (BSN'06), pp. 4, IEEE, 2006.
45. Cunha, João P. Silva, Bernardo Cunha, António Sousa Pereira, William Xavier, Nuno Ferreira, and Luis Meireles, "Vital-Jacket®: A wearable wireless vital signs monitor for patients' mobility in cardiology and sports," In 2010 4th International Conference on Pervasive Computing Technologies for Healthcare, pp. 1–2, IEEE, 2010.
46. Agrawal, Vivek, "Security and privacy issues in wireless sensor networks for healthcare," In Internet of Things. User-Centric IoT: First International Summit, IoT360 2014, Rome, Italy, October 27–28, 2014, Revised Selected Papers, Part I, pp. 223–228. Springer International Publishing, 2015.

47. Prakash, Shiva, "An overview of healthcare perspective based security issues in Wireless Sensor Networks," In 2016 3rd International Conference on Computing for Sustainable Global Development (INDIACom), pp. 870–875, IEEE, 2016.
48. Narayan, Vipul, and A. K. Daniel, "Design consideration and issues in wireless sensor network deployment," Invertis Journal of Science & Technology, vol. 13, pp. 101–109, 2020.
49. Al Ameen, Moshaddique, and Kyung Sup Kwak, "Social issues in wireless sensor networks with healthcare perspective," International Arab Journal of Information Technology, vol. 8, no. 1, pp. 52–58, 2011.
50. Chen, Xiaohu, Michelle Leishman, and Darren Bagnall, and Noushin Nasiri, "Nanostructured gas sensors: From air quality and environmental monitoring to healthcare and medical applications," Nanomaterials (Basel), vol. 11, no. 8, p. 1927, 2021.

11 Exploring the Applications of Wireless Sensor Networks and the Internet of Things in Agricultural and Livestock Production and Management

Anirudh K. C.[1], Muskan Dixit[2], and Shilpa Karat[3]
[1]ICAR-National Dairy Research Institute, Karnal, Haryana, India
[2]Chitkara University Institute of Engineering and
Technology, Chitkara University, Punjab, India
[3]College of Agriculture, Kerala Agricultural
University, Thrissur, Kerala, India

INTRODUCTION

Agricultural production is a prerequisite for raw materials for the majority of manufactured products and thus for an enhanced gross domestic product, better food security, employment generation, capital formation, market size extension, and many other aspects related to an economy. To enhance the growth rate of the agricultural sector, both in quantity and quality, there is a need to emphasize and inculcate emerging technologies, establish institutions and agencies to instigate growth, and mobilize agricultural services along with equipping and accrediting the farmers to strengthen the farming value chain. Technological advances are making the field of agriculture smart and intelligent. The worldwide network of devices that are connected mutually, called the Internet of Things (IoT), is one such advancement. It combines universal communications, prevalent computing, and ambient intelligence (Figure 11.1). IoT is a vision where "things," especially everyday objects, such as all home appliances, furniture, wearables, vehicles, roads, and other devices are readable, recognizable, locatable, addressable, and/or controllable through the Internet. This will provide the basis for many new applications, such as energy monitoring, transport safety systems, or building security. This vision will evolve with time, especially with the advanced synergies among identification technologies, wireless

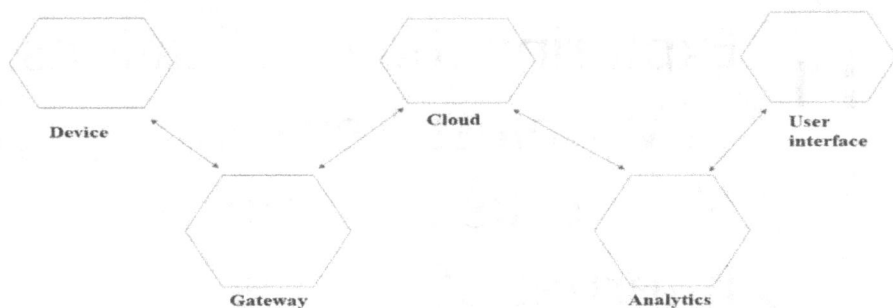

FIGURE 11.1   Graphical representation of the IoT. (Generated by the author.)

sensor networks, intelligent devices, and nanotechnology [1]. Pieces of evidence point out that regenerative and resource-conserving technologies and practices can bring about both environmental and economic benefits for farmers, communities, and nations [2]. Digitalization takes place through different technologies such as IoT, artificial intelligence (AI), machine learning (ML), blockchain, graphic information systems, and many more which have different applications. These digital innovations confer real-time benefits to both consumers and producers in agriculture. Technology improves the land use pattern, farming practices, seed selection and use, recovery of unused land, identification of better cropping patterns, and ensuring seed germination and viability [3].

Four interconnected hexagons represent the connection between device, gateway, cloud, analytics, and user interface.

IoT is an integrated model which evolved from machine-to-machine communication that connects objects/devices to different networks and large databases. This development converts simple standalone objects into a dynamic system of connected things that are intelligent enough to sense their environment and initiate actions by themselves. IoT is drawn on the functionality offered by various existing technologies to realize the vision of a fully interactive and responsive network environment [4]. Some prominent cloud platforms which support IoT devices are Thingworx8, Google Cloud, Cisco, Microsoft Azure, AWS, and Oracle. The integration of cloud platforms is an important factor in developing an effective IoT system [5].

Application of IoT in the field of agriculture includes sensor-based green-house management systems (Smart Greenhouse), livestock monitoring to track the location and health, wireless sensors to monitor climate conditions, remote sensing for crop assessment, weather conditions, and soil quality, sensor camera-based computer imaging for pest and disease detection, quality control, automated input application, irrigation monitoring, sorting, grading, and other post-harvest management. Innovative use of technologies such as radio frequency identification (RFID), near-field communication, ZigBee, and Bluetooth is contributing to creating a value proposition for stakeholders of IoT. IoT will connect the world's objects in a both sensory and intelligent manner by combining technological developments in item identification which is also termed as tagging things, embedded systems, or the thinking things, feeling things which include connected and wireless sensor networks, and lastly nanotechnology which is aptly termed as shrinking things. In 2005, Wal-Mart and the US Department of Defence demanded that their major contractors and suppliers mark their shipments with RFID tags for inventory control. The explosion of the RFID market in 2005 marked the dawn of the thinking about the IoT [6].

The review of literature shows the potential of emerging technologies, namely, IoT and AI in enhancing production, productivity, and efficiency (Table 11.1). They also would aid in real-time management and automated decision-making leading to prompt action. It is evident from the literature that the scope of

TABLE 11.1
Review of Literature

Author(s)	Year	Context	Major Findings
Bali et al. [7]	2023	IoT generates large volumes of data. It is necessary to schedule tasks based on their urgency. The authors try to design a priority aware task scheduling (PaTS) algorithm for sensor networks to schedule priority-aware tasks to offload data on edge and cloud servers	The authors develop an algorithm with overall improvement; average queue delay, computation time, and energy obtained for 200 tasks are 17.2%, 7.08%, and 11.4%, respectively, significantly higher than the existing benchmarks
Ali et al. [8]	2023	The study analyses the improvement in efficiency with the use of IoT and machine learning techniques over traditional practices	Neural Net Work models can predict yield with an accuracy of 92%. The emerging technology including IoT and AI can reduce energy consumption by 8% and irrigation expenses by up to 25%. They are also able to predict common diseases with an accuracy of 90%
Kour et al. [9]	2022	The chapter investigates major agronomic factors that have a direct or indirect impact on the growth of saffron and in determining their optimal value. The chapter also suggests an IoT-based system using renewable energy sources for the cultivation of saffron	The study concludes that the optimum level of agronomic factors differs significantly from existing literature and suggests a model for the application of IoT in saffron cultivation considering the major limitations such as latency and cost
Kour et al. [10]	2022	The chapter provides a novel approach to sensor selection for saffron cultivation in an IoT-based environment. The authors try to create an IoT-based smart hydro-phonic saffron cultivation system and is tested using the AquaCrop simulator	The chapter establishes sensor configurations, performance metrics, and a sensor-based saffron cultivation model. They conclude that IoT-based cultivation practices provide better results compared to natural cultivation

(Continued)

TABLE 11.1 *(Continued)*

Author(s)	Year	Context	Major Findings
Boursianis et al. [11]	2022	The authors conduct a survey of the literature on the application of IoT and unmanned aerial vehicles in agriculture for fertilization, irrigation, pest and disease monitoring, plant growth management, weed management, pesticide application, and plant phenotyping	IoT and unmanned aerial vehicles have the potential to transform traditional agriculture into future-ready intelligent agriculture
Junior et al. [12]	2022	Even though IoT and other intelligent techniques produce large volumes of data, all these data might not be important. The authors try to use machine learning to reduce data and make cloud upload easier	The study concludes that with efficient data reduction algorithms, there can be a substantial reduction in the volume of data that is to be stored and transmitted using IoT, with a reduction possible of up to 6% of the present transfers
Farooq et al. [13]	2022	The chapter tries to review and analyse research published on the application of IoT in livestock farming	The application of IoT and AI has enhanced the quality of animal health monitoring, field and herd monitoring, stress monitoring supply chain supervision, and automation of input supply. When applied on a large scale, data security should be given due consideration
Maginga et al. [14]	2022	Disease detection using IoT relies mostly on visible symptoms from images captured using drones or cameras. This is possible only after symptoms are shown by the plants. An attempt is done to detect disease in the pre-visual symptomatic stage by detecting the presence of volatile organic compounds or changes in nutrient consumption patterns	Previsual symptomatic stage detection of diseases provided with autonomous alert systems to warn the farmers will be more effective in disease control. Automated chatbots for information dissemination can help reach farmers with lesser grades of education
Nagasubramanian et al. [15]	2021	The chapter tries to develop an IoT and AI-based system for the early-stage detection of crop diseases and to compare various machine learning algorithms for doing the same	The results suggest that observable data collected through sensors and cameras aid well in early-stage crop disease identification and thus play a crucial role in decision-making
Mate [16]	2021	The author tries to develop an IoT-based smart irrigation system to monitor and balance soil nutrient status along with ensuring efficient use of water based on crop and climate	A system with a microcontroller and soil parameter sensors was developed with a programmer device. The program could be installed on any Android device and would help in automated irrigation and nutrient management

(Continued)

TABLE 11.1 *(Continued)*

Author(s)	Year	Context	Major Findings
Gao et al. [17]	2020	The authors research on the applicability of IoT and unmanned aerial vehicles for ground (micro) and air (macro) detection of pest and disease spread. The images captured by drones are used to analyse the spread of disease in a particular area	A long-term pest-environment model is developed. The research also suggests methods to enhance the energy efficiency of farms by using self-adjusting solar panels and coordinating drone movement with wind direction
Unold et al. [18]	2020	The authors develop a disease monitoring cum disease detection system using IoT and AI for dairy animals. They concentrate on sensors to detect temperature and heart rate	Disease detection with an accuracy of 90% is possible using IoT and AI
Muangprathub et al. [19]	2019	The study aims to design and develop a control system using node sensors in the crop field with data management via smartphone and a web application for the automation of irrigation. Irrigation can be initiated automatically based on data from the sensors or manually using the mobile phone application	The model helped in maintaining optimum moisture content in the soil and reduced cost and increased productivity
Vyas et al. [20]	2019	The authors employ IoT and various machine learning algorithms for the detection of foot and mouth disease (FMD) and mastitis in dairy animals	Application of IoT and ML in the detection of disease will have a drastic effect on the prevention of the disease and reduce the lousy quality of milk and thus reducing processing costs
Raj and Jayanthi [21]	2018	Large-scale poultry farms require strenuous monitoring to check the spread of diseases among the animals. The authors try to develop an embedded system that helps in the detection and isolation of affected birds	A combined system of audio, image, thermal, and motion sensing would allow in identifying affected birds and removing them from the herd to check the spread of contagious diseases
Culman et al. [22]	2017	Plant leaf images can be used to detect nutrient deficiencies in them and manage nutrient supply accordingly. An IoT-based application, connected to the cloud, would store and process historical data for in-situ identification of nutrients required	The authors created a model using Microsoft Azure which can be scaled up to include a large number of devices dedicated to the purpose

(Continued)

TABLE 11.1 *(Continued)*

Author(s)	Year	Context	Major Findings
bin Ismail and Thamrin [23]	2017	An attempt is made to develop an automated control system for vertical farming. The major objective is to develop a model that uses sensors, handheld devices, and laptops for controlling soil moisture and water content	The model lets the users supervise vertical farms remotely and make sure that moisture is adequately maintained. Higher production due to optimum water availability and lesser supervision cost is ensured
Zhang et al. [24]	2017	Fertilization and irrigation are mostly done following general recommendations without giving due consideration to terrain, stage of free growth, soil difference, and species characteristics. An IoT-based system is tested for specific nutrient and irrigation management based on the above-mentioned characteristics	The application of emerging technologies including ZigBee led to a commendable reduction in labour costs and environmental degradation due to agricultural chemicals
Mohanraj et al. [25]	2016	The authors try to develop an e-agriculture application based on the knowledge base and monitoring modules. They try to include information on all major aspects from production to marketing and monitoring is done through various sensors	The new model has enhanced the efficiency of production with lesser requirement of water and labour cost was reduced significantly

application of IoT and AI in agriculture and allied sectors is unlimited and that any model can be developed depending upon the creativity and availability of resources, from basic automation of irrigation to disease detection and on the spot application of nutrients and chemicals using drone mapping. Detection of diseases, in crops and livestock, before the exhibition of visible symptoms, by analysing emissions and other minute morphological changes, which is nearly impossible for humans to detect will help in preventing commendable financial loss if adopted on a large scale. Ensuring adequate cyber security and lowering the cost of installation of IoT-based systems would further boost its application among the farmers. In the case of India, agricultural production is highly scattered and farmers are resource-poor. For them to adopt such practices, it is necessary that the investment is made affordable. Unless they provide higher rates of return on investment and are made financially accessible to small-scale farmers, their advantages over traditional systems may be reaped only by a small proportion of resource-rich, large-scale commercial farmers. Thus, with better sensors, efficient algorithms, and a perfect mix of interconnected devices, AI enables IoT systems will be the future of agriculture.

WORKING OF IoT

A hesitant, unassuming, and lucrative system of item identification is critical for connecting commonplace objects and devices to substantial databases and networks, and undeniably to the internet or the network of networks. This is very crucial for the collection and processing of data from the devices. Radiofrequency identification (RFID) offers this functionality. Data collection will benefit from the ability to detect changes in the physical status of things, using sensor technologies. Embedded intelligence in the devices or things themselves can to a great extent augment the power of the network by decentralizing information processing capabilities to the edges of the network. Lastly, advances in miniaturization and nanotechnology imply that smaller things will have the capacity to communicate and connect. An amalgamation of all this progress will generate the IoT that will connect the world's objects intelligently.

ARCHITECTURE OF IoT

In the realm of IoT solutions, there exist four distinct models that are used to explain their architecture. These models serve as a framework for understanding the various components and interactions that make up an IoT system.

The IoT is generally defined as a network, which connects everything with the Internet by RFID, sensors, global positioning systems, laser scanners, and other information sensing devices in harmony with the approved protocol and exchange information to realize intelligent identification, location tracking, monitoring, and management. The network can realize the automatic identification of objects and locations, and track, monitor, and trigger the corresponding event. It makes use of RFID technology for scanning and reading tags on items for information sharing. The IoT architecture (Figure 11.2) is split up into four layers, namely, the perception layer (data acquisition layer), network layer, middleware layer, and application layer.

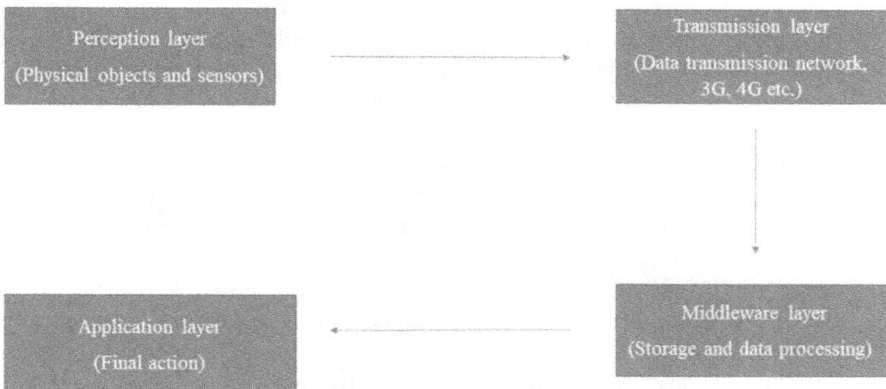

FIGURE 11.2 Architecture of IoT. (Generated by the author.)

When we view the image, we see the connection between the perception layer and the transmission layer, which is represented by an arrow, followed by an arrow indicating the middleware layer and a final arrow indicating the application layer.

A data acquisition layer, self-organizing sensor networks, and middleware make up the sensing layer.

1. Acquisition sublayer consists of barcode sensors, satellite positioning, RFID, and multimedia information acquisition technology. This is used for and helps in data acquisition and access to a variety of data and even information in the form of audio, video, and other multimedia formats that mark the real physical world.
2. Self-organizing sensor network includes digital link coding, modulation, and demodulation technology based on data transmission, networking, traffic management, and routing technology based on self-interaction, organization, and coordination among the nodes.
3. Embedded middleware, sensor network middleware, and application in the sensing layer are included in the middleware to resolve the issue of equipment and data management, configuration, standardization, fault detection, and data conversion. For effectively reducing the data redundancy and improving the quality of data, one might need to employ data fusion, compression, clustering, and recognition along with other information processing technology.

The IoT systems for agriculture are composed of the following layers:

- Perception layer or the sensor layer which collects real-world information and detects certain spatial parameters on various aspects of agriculture and converts it into formats that the processing systems can understand. These sensors or actuators accept the data and then the processed data are emitted over the network.
- Network layer that collects and processes the data from the sensor layer. It binds the intelligent objects to smart objects to distribute it for analysis.
- Data-processing layer analyses and processes the data before sending it to a data centre. This layer has two stages, one is the data accumulation stage and another is the data abstraction stage. In the data accumulation stage, the real-time data are captured through Application Programming Interface and thus help in meeting the requirements of real-time applications. This is a conveyance between event-based data generation and query-based data consumption. The Data Abstraction stage combines data from IoT-based, non-IOT-based, ERM, ERP, and CRM systems; it reconciles multiple data formats and aggregates the data in one accessible format.
- Application layer that applies the wisdom received from processing the information obtained. It formats the data, handles its presentation, and serves as the bridge between what the IoT device is doing.

APPLICATION IN AGRICULTURE AND LIVESTOCK MANAGEMENT

Precision Farming

This system of farming aims at enhanced production and better decision-making with the help of sensors and data analysis. From soil testing to real-time monitoring of physical variables such as temperature and humidity to detection and early warning of pests and disease, advanced sensors and data processing techniques using AI are used. The system has useful applications in Farm as well as Green House. The central system is the heart of this architecture, as it does half of the work. It is responsible for data transfers between nodes and central servers and database management as well as integration with the outer world. Central system is made of three components that are communication server, the database, and the web server. The control cabinet delivers easy access to the nodes, which consist of sensors that detect temperature, soil moisture, water level, and rain. The data collected by the nodes are relocated to the central system through the central server. Warnings and notifications of any sort are then communicated with the farmer through the provided gateway. The central system even notifies an agronomist in case of extreme variations [26].

Video Sensing in Precision Farming

The idea of video sensing for fertilizer/chemical application was proposed in 2015. The objective of their work was to enhance energy efficiency by reducing fertilizer use in agriculture. Drones and filed sensors including video sensors are used to capture the field situation and crop geo-positioning is done (Figure 11.3). Drones are again used to spray chemicals on previously marked geo-positions. Optimized Link State Routing Protocol is used to establish the network between drones and the central system. Multi-layer perception algorithm is mostly engaged in such assignments [27].

The image shows a face representing the user interface. It is connected to Cloud servers, which in turn are connected to video sensors placed between the planters

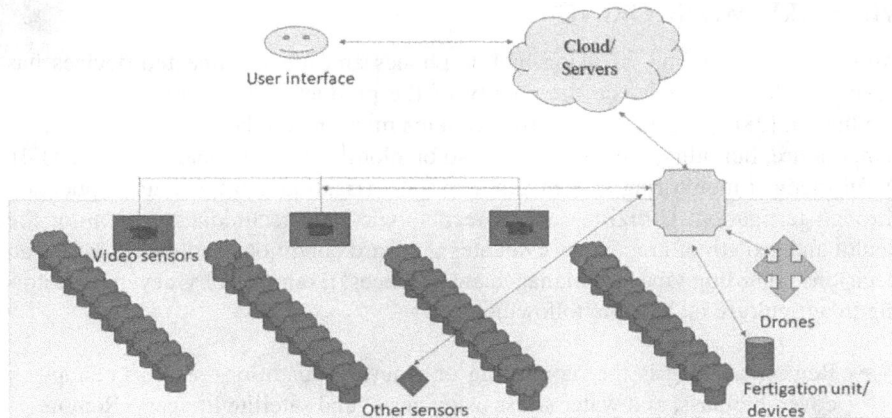

FIGURE 11.3 Precision farming using video sensors and drones. (Generated by the author.)

in the farm. The server is also connected to drones, represented by four arrows. Additionally, there are other sensors connected in the field, and a cylindrical shape represents the fertigation devices.

VERTICAL FARMING

Vertical farming is the cultivation of plants in vertically layered strata using controlled environment agriculture techniques such as artificial illumination, temperature control, and nutrient management. This technology enables farmers to cultivate crops indoors, in urban areas and in areas where traditional farming is not feasible due to factors such as limited land availability or unfavourable weather. In a vertical farm, crops are cultivated in layers or cages that are layered vertically, using artificial illumination and climate control. The crops are grown hydroponically or aeroponically, which means they are grown in a soilless medium, such as nutrient-rich water or air, with controlled irrigation and the addition of the necessary nutrients. Typically, LED illumination is used in vertical farming because it provides the required light spectrum for plant growth and development. Temperature and humidity levels are also precisely controlled in a vertical farm to ensure optimal growing conditions for the plants. To promote plant growth and increase crop yields, carbon dioxide levels are also monitored and regulated. Vertical farming enables farmers to cultivate a vast array of crops, including verdant greens, herbs, fruits, and vegetables, regardless of the external weather conditions, all year round. As crops can be produced and sold locally, this technology also reduces the need for transportation and the associated carbon emissions, thereby increasing the availability of fresh produce in urban areas. Vertical farming has the potential to revolutionize agriculture by rendering it more efficient, sustainable, and accessible to urban populations. By utilizing techniques for agriculture in a controlled environment, vertical farming enables farmers to cultivate crops more efficiently and sustainably, thereby reducing the environmental impact of agriculture and enhancing food security and accessibility.

VINEYARD MONITORING

Application of IoT and AI using mobile phones and other connected devices has been introduced to enhance the quality of the product and to reduce the cost of production [28]. Vineyard monitoring systems monitor weather parameters such as temperature, humidity, wind speed, and so on along with plant health (Figure 11.4). A full-fledged monitoring system can automate irrigation and fertilizer application through fertigation. Utilizing various technologies and techniques to monitor the health and growth of grapevines evaluates vineyard conditions and makes informed decisions regarding vineyard management practices. Examples of vineyard monitoring in agriculture include the following:

- Remote sensing is the monitoring of vineyard conditions such as canopy cover, biomass, and water stress using aerial and satellite imagery. Remote sensing can provide a comprehensive view of the vineyard and assist in identifying areas that may require additional care.

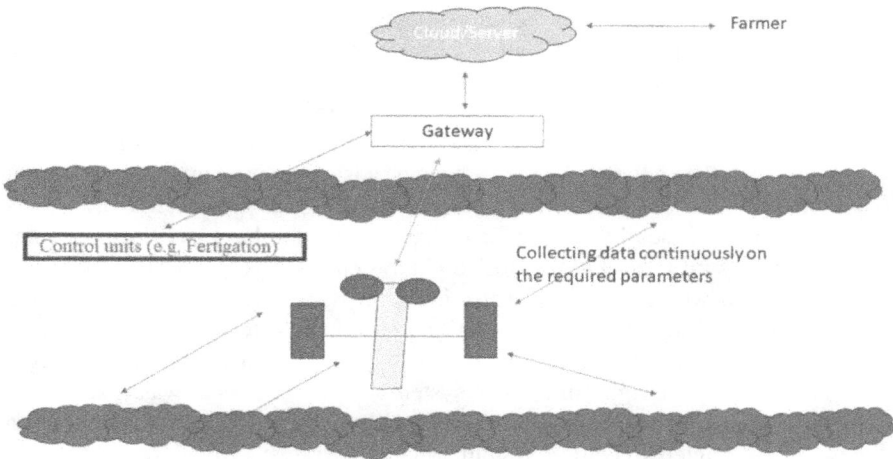

FIGURE 11.4   Vineyard monitoring using IoT. (Generated by the author.)

- These sensors are embedded in the soil to measure soil moisture, soil temperature, and nutrient levels. These data can be used to optimize irrigation and fertilizer application and detect any soil-related issues.
- These instruments are used to measure meteorological conditions such as temperature, humidity, and precipitation at weather stations. This information is crucial for determining the likelihood of frost damage, disease outbreaks, and other weather-related hazards to the vineyard.
- Pest and disease monitoring entails the use of traps, visual inspections, and other techniques to monitor the vineyard for pests and diseases. Eliminating the need for pesticides and preventing the spread of disease can be accomplished through early detection.
- Monitoring the yield involves quantifying grape yields and evaluating grape quality. These data can be used to optimize vineyard management practices and make informed harvesting and marketing decisions.

The cloud server, which takes the form of a cloud, is monitored by a farmer. The cloud is connected to a gateway. Afterwards, there are two parallel vineyards that are like green clouds. In the centre is where all the data collection are conducted. The gateway is connected to the control units, which are also located in the centre.

WATER MANAGEMENT

Water is a very precious but rapidly depleting resource. The most efficient use of water thus has a major role in ensuring sustainability. Agriculture is one of the sectors that require water as a major input, constituting 70% of its use [29], and no production is possible without direct irrigation or water availability. Efficient utilization of water would lead to enhanced production along with reduced costs.

AI-enabled IoT systems are now being employed to sense the availability of water for plants to automatically start and stop irrigation as per the requirements [30]. For irrigation and water management, the sensors are embedded in soil which continuously measure the soil moisture levels and communicate with the central system. Measurements are taken at specified intervals also if continuous monitoring is not necessary. The advanced systems have provisions to inform the farmer about irrigation and field conditions using text messages. The gravimetric method is used to calibrate the sensors. Some farms have claimed to have cut down on water requirements by up to 75% using IoT. Such systems are very useful in drought-prone areas and where irrigation water is not provided free of cost. Automated irrigation systems provide producers with a precise and effective method of managing water resources and increasing crop yields. By providing crops with the right amount of water at the right time, these systems can reduce water waste, save time and labour, and enhance crop health and quality. Here is a detailed explanation of how an automated irrigation system usually operates: in numerous locations across a field, sensors are embedded in the soil. These sensors also measure soil temperature, humidity, and solar radiation. The sensors transmit data to a central controller, which analyses the information and calculates the amount of water required for each section of the field. The controller activates the irrigation system and modifies the water flow rate and duration based on the data from the sensors to satisfy the requirements of the crops in each area of the field. Sprinklers or trickle lines deliver water to the vegetation as part of the irrigation system. As soil moisture levels rise, the sensors detect the change and send the controller updated data. The controller then modifies the water flow rate and duration to provide crops with the optimal amount of water. Depending on the requirements of the crops and the availability of water, the system can be programmed to operate at specific times of day and night.

NUTRIENT MANAGEMENT

IoT-enabled nutrient management is usually carried out using drone technology employing Normalized Difference Vegetation Index to measure the nutrient requirements. The system works based on the absorption and reflection of light for photosynthesis by the vegetation under consideration. Light sensors are used to create an image in the infrared range. When these images are processed, a technician or a farmer can analyse plant vigour and disease conditions using them. Similar set of sensors is employed to detect pest population growth. They also used the system to detect bud necrosis virus in the ground nut. A naïve bias classifier was used to predict the dynamics of pests and diseases (Figure 11.5).

The field is represented at the bottom of the diagram as a green rectangle. Above the field, there is the central system, which is connected to the drone data capture and spray system through arrows.

The mechanism of drone-based nutrient application entails using drones to fly over a field and administer mist or granular nutrients to the vegetation. The drone is equipped with a canister or receptacle that contains the nutrient solution or granules, as well as a spray or dispersal mechanism that applies the nutrients to the vegetation

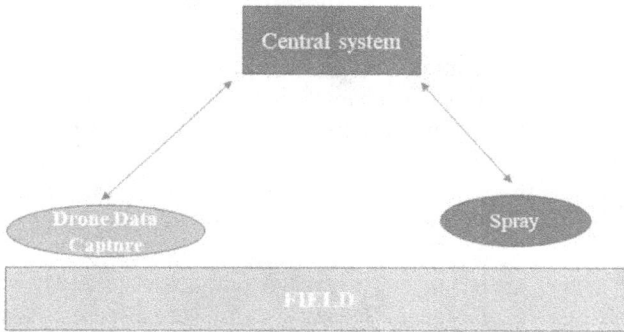

FIGURE 11.5 System to detect pests and diseases. (Generated by the author.)

as it soars over the field. A remote operator controls the drone, or it can be programmed to follow a predetermined flight path using GPS. Based on real-time data such as crop growth stage and weather conditions, the operator or software can modify the rate and pattern of nutrient application. This enables the precise and efficient administration of nutrients, maximizing crop yields while minimizing environmental impacts. In agriculture, nutrient management can be implemented through a central system that monitors and regulates the application of fertilizers and pesticides. Based on real-time data such as weather conditions, soil moisture levels, and crop growth stages, the central system can be programmed to modify the quantity and timing of nutrient applications. This ensures that crops receive the optimal quantity of nutrients at the right time, thereby minimizing waste and reducing the risk of environmental contamination.

FOOD SUPPLY CHAIN MANAGEMENT

In the traditional method, data on agricultural products are collected manually and with the help of bar codes. The chances of errors and missing out are very high when these processes are carried out. The application of IoT for process automation of supply chain tracking can ensure traceability along with the quality of the products marketed. The system uses a combination of cloud computing and RFID tags. Monitoring includes all stages from production to sales. The benefit of IoT systems is that they can even track and ensure storage and logistics environment including temperature and moisture content in the environment as well as the commodity under consideration. The whole process starts with RFID tagging of the lots/products/commodity at the production stage. The coding contains information such as product name, manufacturer, grade, place of origin, net weight, batch number, production date, shelf-life, and so on. During logistics, installing a GPS positioning system on the vehicles enables the managers to know the location and state of the vehicles that transport the agricultural products and adjust driving direction timely in an emergency. At the same time, installing a wireless data acquisition system on the vehicles can not only learn the basic information and quantity of the goods but also detect and prevent lost and stolen.

GENETIC ENGINEERING

Genetic engineering is a technique in which scientists manipulate the DNA of plants and animals to generate traits or characteristics of their choosing. This process entails the insertion, deletion, or alteration of genes in an organism's DNA sequence. Genetic engineering is used in agriculture and livestock management to increase the resilience and productivity of crops and livestock. Here are a few examples of how genetic engineering is revolutionizing agriculture and livestock management:

- Using genetic engineering, it is possible to produce crops and livestock with enhanced disease resistance. Scientists can insert disease-resistant genes into the DNA sequences of plants and animals. These genes can be derived from other species or synthesized artificially in a laboratory.
- Using genetic engineering, it is also possible to create plants that are resistant to parasites. Scientists can insert genes that code for insecticidal proteins into the DNA sequence of plants. These proteins can destroy or repel insects that might otherwise cause crop injury.
- Tolerance to environmental stresses genetic engineering can also be used to develop cultivars that are more tolerant to environmental stresses such as drought, heat, and frost. Scientists can insert genes into the DNA sequence of plants that code for proteins that assist the plant in withstanding these stresses.
- Utilizing genetic engineering, it is possible to produce crops with enhanced nutritional value. Scientists can insert genes that translate for nutrients such as vitamins, minerals, or proteins into the DNA sequence of plants.
- Utilizing biotechnology instruments such as restriction enzymes, DNA ligases, and polymerase chain reaction to manipulate DNA sequences is the mechanism underlying genetic engineering. These instruments allow scientists to insert, eradicate, or modify genes in an organism's DNA sequence.
- To incorporate a new gene into the DNA of an organism, scientists typically use a vector, such as a plasmid, which is a small, circular DNA molecule that can replicate independently of the organism's chromosomal DNA. The vector is then introduced into the organism, where it can integrate into the chromosomal DNA.

Once the gene has been inserted, the organism is able to manifest it, resulting in the desired trait or characteristic. A gene that codes for an insecticidal protein, for instance, can be expressed by a plant to produce a pest-resistant crop.

LIVESTOCK MANAGEMENT

Each year, farmers lose significant amounts of profit due to animal illnesses. There are many ways that IoT-enabled livestock management solutions allow farmers to promote better livestock health:

- Connected sensors in livestock wearables allow farmers to monitor heart rate, blood pressure, respiratory rate, temperature, digestion, and other vitals.

- Data streamed to the cloud directly from wearables allow farmers to identify and address issues like illness and feeding problems before they significantly impact the herd's health.
- Farmers can use IoT solutions to monitor livestock reproductive cycles and the calving process to promote safer and more successful outcomes.
- IoT sensors can be used to track an animal's location, which can be helpful in locating sick animals as well as establishing and optimizing grazing patterns.

DISEASE DETECTION USING PRECISION DAIRY FARMING TECHNOLOGIES

The dairy industry is fast growing with increased demand for milk and milk products. With an increase in the number of animals maintained, management becomes complicated, and the time devoted to each individual decreases [31–33]. The decision-making process in dairy farming must consider issues beyond economic profit such as animal well-being, sustainable production, food safety, quality, and so on [34, 35]. Unfortunately, on-farm decisions are riddled with complexities, many of which the effects must be estimated or guessed at by the producers. One way to counteract these problems is by using automated monitoring systems [36].

Precision livestock farming applies precision agriculture principles to animals, focusing on individual animal production and environmental impact [37]. One goal of precision livestock farming is to develop inline systems that monitor animals objectively, continuously, and automatically, without adding stress to the animals [34]. Precision dairy farming (PDF) is the use of technologies to measure physiological, behavioural, and production indicators on individual animals to improve management and farm performance. IoT can support farmers with wearable sensor devices to keep them aware of the status of every cow (Figure 11.6). The sensor-based system can effectively and correctly detect the illness of the cow before it

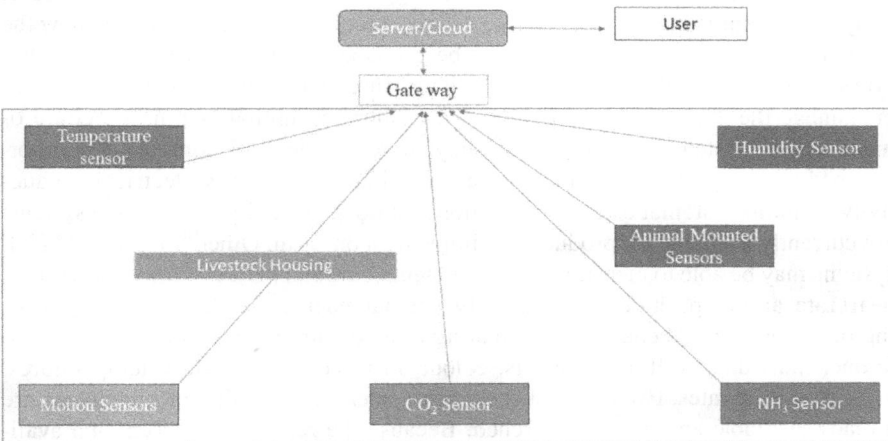

FIGURE 11.6 Livestock management (generated by the author).

affects milk production. The farm owner can place the sensor on the cow's neck, tail, or leg for acquiring real-time data to examine numerous factors like the cow's behaviour, activity, health, feed consumption, milk production, and fertility management. These wearable sensors can spot cow illnesses and diseases such as mastitis or any other disease that can reduce milk production. Smart farming is not just monitoring; it is also related to proper business management. Gathering raw data like insemination moment, pregnancy, or calving time, these data are used for valuable information and for applying some probability and statistics to plan future work and manage dairy farms properly.

The goals of PDF are maximizing individual animal potential, early disease detection, and maximizing preventive care instead of medical treatments. Perceived benefits of PDF technologies include increased efficiency, reduced costs, improved milk quality, minimized environmental impacts, and improved animal health and well-being. Additionally, information from PDF technologies could potentially be incorporated into genetic evaluations for traits targeted at improving subsequent generations' health, well-being, and longevity [32]. Thus, the application of IoT leads to both process and product innovation in the dairy sector. IoT can be applied for herd movement detection and geo-fencing, automated milking, and automated food and water supply. Data from sensors collected over time can be analysed using dynamic linear models to forecast the milk yields per individual cow (milked using robots).

The traditional practices use the experience of the producer to detect the problems/diseases affecting the animals, mainly through examining visible symptoms and signs. Some of the diseases and cases may go unnoticed and unattended in this method. It is here that the importance of precision farming comes into play (Figure 11.6). Real-time data on animal behaviour and status are made available using PDF. However, to achieve success using precision livestock farming processes, three conditions apply. First, animal variables should be monitored continuously, and the data should be analysed consistently. The definition of "continuously" depends on the animal variable of interest, like weight, activity, drinking and feeding behaviour, feed intake, body temperature, and so on. Second, a reliable prediction or expectation on how the animal will respond to the change must be available constantly. Lastly, this prediction should be coupled with the technology measurements in an algorithm to monitor or manage the animals automatically, and to monitor animal health or welfare or make desired system changes. Daily milk yield recording, milk component monitoring, pedometers, automatic temperature recording devices, milk electrical conductivity monitors, automatic oestrus detection monitors, and daily body weight systems are currently available for producers to implement on-farm. Other "theoretical" PDF systems may be able to measure: jaw movements, ruminal pH, reticular contractions, heart rate, animal positioning and activity, vaginal mucus, electrical resistance, feeding behaviour, lying behaviour, ruminating time, odour, glucose, acoustics, progesterone, individual milk components, colour, infrared udder surface temperatures, and respiration rates. Excitingly, just six years later, many of these technologies are already available and being researched. Because the rapid development and availability of new PDF systems continue to grow, they are becoming more feasible for producers to implement in their own herds.

The user's Cloud server is connected to a gateway, which in turn is connected to various sensors. These sensors include a temperature sensor, livestock housing sensor, motion sensor, CO_2 sensor, humidity sensor, animal mounted sensor, and NH_3 sensor.

CONCLUSION

Agriculture is a labour and input-intensive venture, and to enhance profits and sustainability, efficiency should be improved at all stages of production, starting from the selection of crops or livestock to final marketing and sales. Manual supervision and detection of physical variables that have a direct effect on the production and quality of the product from production to logistics to storage are tiresome and the chances of manual error are very high. Employing IoT-based systems for monitoring and process automation can lead to higher efficiency and accuracy as discussed in the chapter. Process automation and site-specific resolutions are the major advantages of IoT. The sophisticated sensors used can precisely detect even minute changes in conditions, which farmers may fail to as they depend on solely experience. With IoT in place, treatment can start even before visible symptoms are observed leading to higher profits and lesser costs. This leads to cost reduction, input savings, higher quality, and higher yields along with better sustainability which are essential features of any venture including agriculture.

REFERENCES

1. Malavade, V. N., & Akulwar, P. K. (2016). Role of IoT in agriculture. *IOSR Journal of Computer Engineering*, 1(1), 2278–0661.
2. Rehman, A., Jingdong, L., Khatoon, R., Hussain, I., & Iqbal, M. S. (2016). Modern agricultural technology adoption its importance, role and usage for the improvement of agriculture. *Life Science Journal*, 14(2), 70–74.
3. Ashraf, S. N. (2022). *Understanding Usability and Applications of Appropriate Technology in Indian Agriculture Sector with reference to Entrepreneurship* (Doctoral dissertation, Entrepreneurship Development Institute of India).
4. Zhao, J. C., Zhang, J. F., Feng, Y., & Guo, J. X. (July 2010). The study and application of the IoT technology in agriculture. In *2010 3rd International Conference on Computer Science and Information Technology*, vol. 2 (pp. 462–465).
5. Ramachandran, V., Ramalakshmi, R., Kavin, B. P., Hussain, I., Almaliki, A. H., Almaliki, A. A., Elnaggar, A. Y., & Hussein, E. E. (2022). Exploiting IoT and its enabled technologies for irrigation needs in agriculture. *Water*, 14(5), 719.
6. Ruan, N., & Hori, Y. (2012). DoS attack-tolerant Tesla-based broadcast authentication protocol in Internet of Things. In *2012 International Conference on Selected Topics in Mobile and Wireless Networking* (pp. 60–65).
7. Bali, M. S., Gupta, K., Gupta, D., Srivastava, G., Juneja, S., & Nauman, A. (2023). An effective technique to schedule priority aware tasks to offload data at edge and Cloud servers. *Measurement: Sensors*, 26, 100670.
8. Ali, A., Hussain, T., Tantashutikun, N., Hussain, N., & Cocetta, G. (2023). Application of smart techniques, Internet of Things and data mining for resource use efficient and sustainable crop production. *Agriculture*, 13(2), 397.
9. Kour, K., Gupta, D., Gupta, K., Juneja, S., Kaur, M., Alharbi, A. H., & Lee, H. N. (2022). Controlling agronomic variables of saffron crop using IoT for sustainable agriculture. *Sustainability*, 14(9), 5607.

10. Kour, K., Gupta, D., Gupta, K., Anand, D., Elkamchouchi, D. H., Pérez-Oleaga, C. M., & Goyal, N. (2022). Monitoring ambient parameters in the IoT precision agriculture scenario: An approach to sensor selection and hydroponic saffron cultivation. *Sensors, 22*(22), 8905.
11. Boursianis, A. D., Papadopoulou, M. S., Diamantoulakis, P., Liopa-Tsakalidi, A., Barouchas, P., Salahas, G., ... Goudos, S. K. (2022). Internet of things (IoT) and agricultural unmanned aerial vehicles (UAVs) in smart farming: A comprehensive review. *Internet of Things, 18*, 100187.
12. Junior, F. M. R., Bianchi, R. A., Prati, R. C., Kolehmainen, K., Soininen, J. P., & Kamienski, C. A. (2022). Data reduction based on machine learning algorithms for fog computing in IoT smart agriculture. *Biosystems Engineering, 223*, 142–158.
13. Farooq, M. S., Sohail, O. O., Abid, A., & Rasheed, S. (2022). A survey on the role of IoT in agriculture for the implementation of smart livestock environment. *IEEE Access, 10*, 9483–9505.
14. Maginga, T., Nsenga, J., Bakunzibake, P., & Masabo, E. (September 2022). Smallholder farmer-centric integration of IoT and Chatbot for early Maize diseases detection and management in pre-visual symptoms phase. In *2022 IEEE Global Humanitarian Technology Conference (GHTC)* (pp. 369–372).
15. Nagasubramanian, G., Sakthivel, R. K., Patan, R., Sankayya, M., Daneshmand, M., & Gandomi, A. H. (2021). Ensemble classification and IoT-based pattern recognition for crop disease monitoring system. *IEEE Internet of Things Journal, 8*(16), 12847–12854.
16. Mate, S. (2021). Internet of things (IoT) based irrigation and soil nutrient management system. *i-Manager's Journal on Embedded Systems, 9*(2), 22.
17. Gao, D., Sun, Q., Hu, B., & Zhang, S. (2020). A framework for agricultural pest and disease monitoring based on Internet of Things and unmanned aerial vehicles. *Sensors, 20*(5), 1487.
18. Unold, O., Nikodem, M., Piasecki, M., Szyc, K., Maciejewski, H., Bawiec, M., & Zdunek, M. (2020, June). IoT-based cow health monitoring system. In *Computational Science—ICCS 2020: 20th International Conference*, Amsterdam, The Netherlands, June 3–5, 2020, Proceedings, Part V (pp. 344–356). Springer International Publishing.
19. Muangprathub, J., Boonnam, N., Kajornkasirat, S., Lekbangpong, N., Wanichsombat, A., & Nillaor, P. (2019). IoT and agriculture data analysis for smart farm. *Computers and Electronics in Agriculture, 156*, 467–474.
20. Vyas, S., Shukla, V., & Doshi, N. (2019). FMD and mastitis disease detection in cows using Internet of Things (IoT). *Procedia Computer Science, 160*, 728–733.
21. Raj, A. A. G., & Jayanthi, J. G. (2018, March). IoT-based real-time poultry monitoring and health status identification. In *2018 11th International Symposium on Mechatronics and its Applications (ISMA)* (pp. 1–7). IEEE.
22. Culman, M. A., Gomez, J. A., Talavera, J., Quiroz, L. A., Tobon, L. E., Aranda, J. M., ... Bayona, C. J. (2017, April). A novel application for identification of nutrient deficiencies in oil palm using the Internet of things. In *2017 5th IEEE International Conference on Mobile Cloud Computing, Services, and Engineering (MobileCloud)* (pp. 169–172).
23. bin Ismail, M. I. H., & Thamrin, N. M. (2017, November). IoT implementation for indoor vertical farming watering system. In *2017 International Conference on Electrical, Electronics and System Engineering (ICEESE)* (pp. 89–94).
24. Zhang, X., Zhang, J., Li, L., Zhang, Y., & Yang, G. (2017). Monitoring citrus soil moisture and nutrients using an IoT-based system. *Sensors, 17*(3), 447.
25. Mohanraj, I., Ashokumar, K., & Naren, J. (2016). Field monitoring and automation using IoT in agriculture domain. *Procedia Computer Science, 93*, 931–939.
26. Ojha, T., Misra, S., & Raghuwanshi, N. S. (2015). Wireless sensor networks for agriculture: The state-of-the-art in practice and future challenges. *Computers and Electronics in Agriculture, 118*, 66–84.

27. Bradski, G., & Kaehler, A. 2008. *Learning OpenCV: Computer Vision with the OpenCV Library*. O'Reilly Media, Inc.
28. Burrell, D. N., Courtney-Dattola, A., Burton, S. L., Nobles, C., Springs, D., & Dawson, M. E. (2020). Improving the quality of "The Internet of Things" instruction in technology management, cybersecurity, and computer science. *International Journal of Information and Communication Technology Education, 16*(2), 59–70.
29. FAO (2016). *AQUASTAT: Water Uses*. Available online: http://www.fao.org/nr/water/aquastat/water_use (accessed on 5 January 2019).
30. Kamienski, C., Soininen, J. P., Taumberger, M., Dantas, R., Toscano, A., Salmon Cinotti, T., & Torre Neto, A. (2019). Smart water management platform: IoT-based precision irrigation for agriculture. *Sensors, 19*(2), 276.
31. Schulze, C., Spilke, J., & Lehner, W. (2007). Data modeling for precision dairy farming within the competitive field of operational and analytical tasks. *Computers and Electronics in Agriculture, 59*(1–2), 39–55.
32. Bewley, J. M., Gray, A. W., Hogeveen, H., Kenyon, S. J., Eicher, S. D., & Schutz, M. M. (2010). Stochastic simulation using @Risk for dairy business investment decisions. *Agricultural Finance Review, 70*(1), 97–125.
33. Brandt, M., Haeussermann, A., & Hartung, E. (2010). Invited review: Technical solutions for analysis of milk constituents and abnormal milk. *Journal of Dairy Science, 93*(2), 427–436.
34. Berckmans, D. (2006). Automatic online monitoring of animals by precision livestock farming. *Livestock Production and Society, 287*, 27–30.
35. Schukken, Y. H., Barkema, H. W., Lam, T. J. G. M., & Zadoks, R. N. (2008). Improving udder health on well-managed farms: Mitigating the perfect storm. *Mastitis Control from Science to Practice*, 1st ed., Wageningen Academic Publishers, 21–35.
36. Chagunda, M. G. G., Friggens, N. C., Rasmussen, M. D., & Larsen, T. (2006). A model for detection of individual cow mastitis based on an indicator measured in milk. *Journal of Dairy Science, 89*(8), 2980–2998.
37. Laca, E. A. (2009). Precision livestock production: Tools and concepts. *Revista Brasileira de Zootecnia, 38*, 123–132.

12 Wireless Sensor Networks and IoT Revolutionizing Healthcare

Advancements, Applications, and Future Directions

Preeti Saini[1], Rakesh Ahuja[1], and Vyasa Sai[2]
[1]Chitkara University Institute of Engineering and
Technology, Chitkara University, Punjab, India
[2]Intel Corp., Santa Clara, California, USA

INTRODUCTION

The development of techniques and models for computer systems to learn from data and make predictions or judgments without explicit programming is the field of machine learning (ML), a subfield of artificial intelligence (AI). ML has enormous potential to transform healthcare by enhancing patient outcomes and medical practice. ML algorithms (MLalgo) can help in the diagnosis of diseases, the planning of treatments, and the maintenance of one's health by analysing enormous volumes of data and spotting complicated patterns. The application of ML in healthcare does, however, present a unique set of difficulties, including issues with data protection, the interpretability of models, and ethical considerations.

Healthcare is not an exception to how numerous industries have been impacted by technological advancements in the age of digital transformation. The Internet of Things (IoT) and wireless sensor networks (WSNs) have come together as a disruptive force that is reshaping the healthcare industry. This combination has created previously unimaginable prospects for the precise, effective, and widely accessible monitoring, diagnosis, and treatment of patients. WSNs and IoT have ushered in a new era of personalized, patient-centric care that holds enormous promise for transforming healthcare in ways previously considered unthinkable by integrating medical devices, wearable sensors, and smart healthcare systems. The relationship between WSNs, the IoT, and ML in a healthcare system is depicted in Figure 12.1.

Figure 12.1 illustrates how WSNs and IoT serve as the backbone for data gathering and communication, while ML provides healthcare systems with powerful data analysis tools to improve patient care, diagnosis, and treatment plans. The "Healthcare

DOI: 10.1201/9781003438205-12

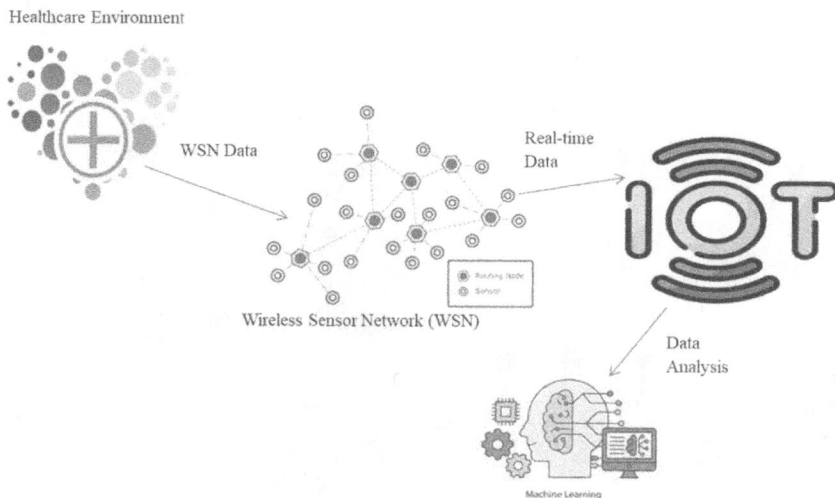

Healthcare Environment

WSN Data

Real-time Data

Wireless Sensor Network (WSN)

Data Analysis

Machine Learning

FIGURE 12.1 Relationship between WSNs, IoT, and ML in a healthcare system.

Environment" refers to the broad environment in which these technologies are included to enhance patient care and medical procedures. The "Wireless Sensor Network (WSN)" is made up of interconnected sensor nodes that gather data from wearable sensors and medical devices in real-time. The WSN is linked to the internet by the "Internet of Things (IoT)," allowing for easy data transfer and remote access to knowledge. With the help of advanced algorithms, "Machine Learning (ML)" analyses the data gathered from the IoT and WSN to find trends, spot abnormalities, and offer predictive insights for better healthcare decision-making. By delivering personalized healthcare solutions and real-time analytics for better patient outcomes and overall healthcare efficiency, these technologies collectively transform the healthcare system.

The diagram in Figure 12.2 illustrates the concept of a smart hospital with interconnected IoT devices and WSNs. It offers a fundamental understanding of the integration of IoT and WSNs into a smart hospital environment.

The "Smart Hospital" is the focal point, illustrating how IoT and WSNs are combined in a healthcare environment. The "Hospital Infrastructure" is linked to a number of gadgets, such as "Medical Equipment" and "Patient Wearables," demonstrating the presence of IoT devices all across the hospital. The interconnection and data exchange between various IoT-enabled components are represented by the "IoT Connectivity" that binds the devices together. The "Data Collection and Monitoring System" illustrates how data travel from IoT devices to the main monitoring system. The processing and analysis of gathered data using ML algorithms to produce valuable insights is represented by the "Data Analysis and ML Algorithms" section.

IoT AND WIRELESS SENSOR NETWORKS' EMERGENCE IN HEALTHCARE

Networks of connected sensor nodes that can communicate with one another wirelessly are known as WSNs. These sensor nodes, which are frequently included into

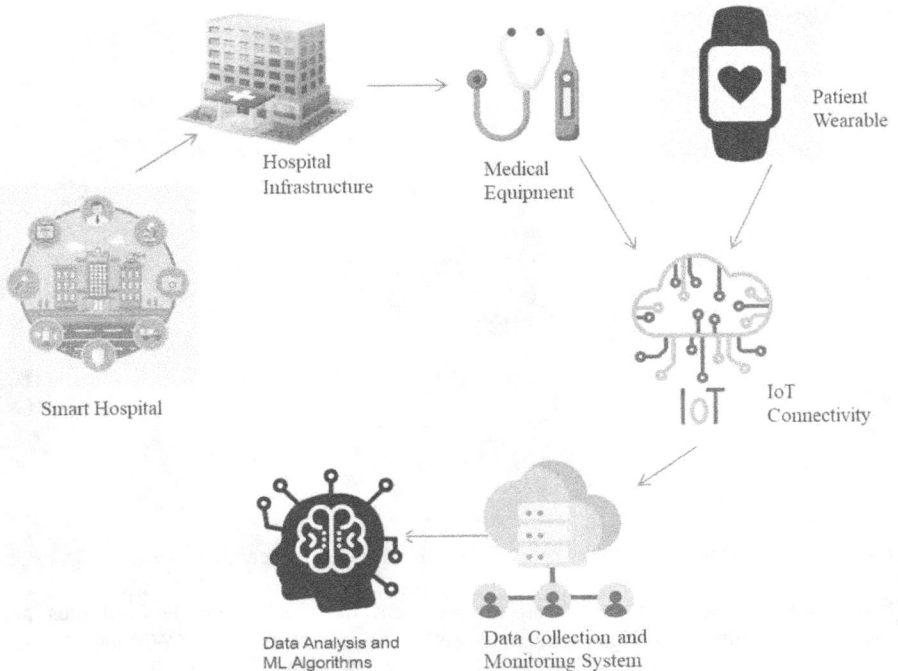

FIGURE 12.2 Smart Hospital environment.

medical equipment or wearable technology, may gather, process, and transmit data in real-time. By linking these sensor networks to the internet and enabling seamless data sharing and remote access to information, the IoT expands the possibilities of WSNs.

A wide range of cutting-edge applications, from telemedicine and remote patient monitoring to smart hospital infrastructure and wearable health monitors, have been made possible by the combination of WSNs and IoT in the healthcare industry. Medical professionals now have access to useful insights that allow the early detection of health issues, effective treatment regimens, and improved disease management through continuous monitoring of vital signs, disease-specific indicators, and patient behaviours.

ADVANCEMENTS ENABLED BY WSNS AND IoT IN HEALTHCARE

The synergy between WSNs and IoT has driven several groundbreaking advancements in healthcare. One significant breakthrough is remote patient monitoring, which has reached unprecedented levels of accessibility and accuracy. With WSNs and IoT, healthcare professionals can now monitor patients' conditions in real-time, even outside traditional clinical settings. This innovation has greatly improved the care of patients with chronic illnesses, elderly individuals, and those residing in remote or underserved areas. Furthermore, the real-time data streams generated by WSNs and IoT have paved the way for predictive analytics and AI in healthcare. ML algorithms can analyse vast amounts of patient data to identify patterns, detect anomalies, and predict potential health risks. This empowers medical professionals with actionable insights, enabling them to deliver personalized treatment plans based on individual patient profiles, ultimately improving patient outcomes and treatment efficacy.

DIVERSE APPLICATIONS OF WSNS AND IoT IN HEALTHCARE

The scope of WSNs and IoT applications in healthcare is vast and continues to expand. Wearable health devices, such as smart watches and fitness trackers, have become ubiquitous and offer real-time monitoring of vital signs like heart rate, blood pressure, and sleep patterns. Beyond consumer wearable, implantable sensors are being developed to provide continuous monitoring of chronic conditions like diabetes and cardiovascular diseases. In healthcare facilities, IoT-enabled smart healthcare systems streamline workflow, automate equipment maintenance, and enhance patient safety through real-time tracking of medical assets and personnel. Additionally, telemedicine solutions leveraging WSNs and IoT facilitate remote consultations, leading to reduced healthcare costs and improved accessibility to medical services, particularly for individuals living in remote or underserved regions.

OVERVIEW OF MACHINE LEARNING

A branch of artificial intelligence (AI) called ML is concerned with creating algorithms that let computers learn from data and make predictions or judgments without having to be explicitly programmed. A subset of ML, which is a subset of AI, is deep learning. As a result, all ML algorithms, including deep learning (DL) algorithms as depicted in Figure 12.3, fall under the umbrella term of AI. Though deep learning has recently attracted a lot of attention and success, it's important to realize that AI comprises a wide range of additional techniques in addition to ML and deep learning. The broad topic of AI studies how to make machines intelligent. A subset of AI known as ML focuses on the techniques that let computers learn from data. DL is a specialized subset of ML that uses deep neural networks for representation

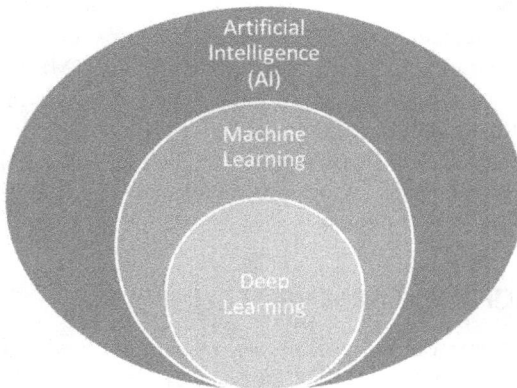

FIGURE 12.3 Machine learning and artificial intelligence.

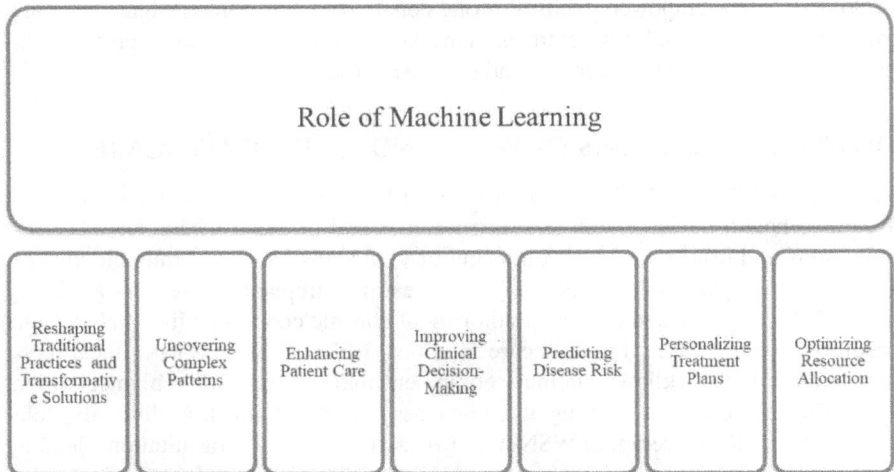

FIGURE 12.4 Role of machine learning in healthcare.

learning and has excelled in a number of AI applications. It makes use of statistical approaches to find trends, draw lessons from past performance, and continuously enhance efficiency. Some of the popular techniques in ML include supervised learning, unsupervised learning, and reinforcement learning.

THE ROLE OF MACHINE LEARNING IN HEALTHCARE

ML has recently become a major player in the healthcare industry, transforming established procedures and providing game-changing innovations. It has created new opportunities for improving patient care and clinical decision-making because of its capability to collect and analyse enormous volumes of healthcare data as well as its ability to spot complicated patterns [1–5]. ML algorithms have a variety of uses, including the ability to personalize treatment programmes, predict illness risk, and find tiny patterns in medical imagery [6]. The fight against COVID 19 has also benefited from genetic engineering applications using ML [7]. ML has the potential to advance precision medicine and enhance patient outcomes, serving as a catalyst for change in the healthcare industry. It aids in medical imaging interpretation, disease prediction, drug discovery, and personalized medicine. It also enhances electronic health record (EHR) management, supports patient engagement, and aids clinical decision-making. Figure 12.4 depicts the role of ML in healthcare.

DIFFERENT MACHINE LEARNING TECHNIQUES

There are several common ML techniques utilized in healthcare. The diagram Figure 12.5 illustrates the different ML techniques used in healthcare and provides examples for each technique. These ML techniques, when applied in healthcare, enable healthcare providers to gain insights from large datasets, make accurate

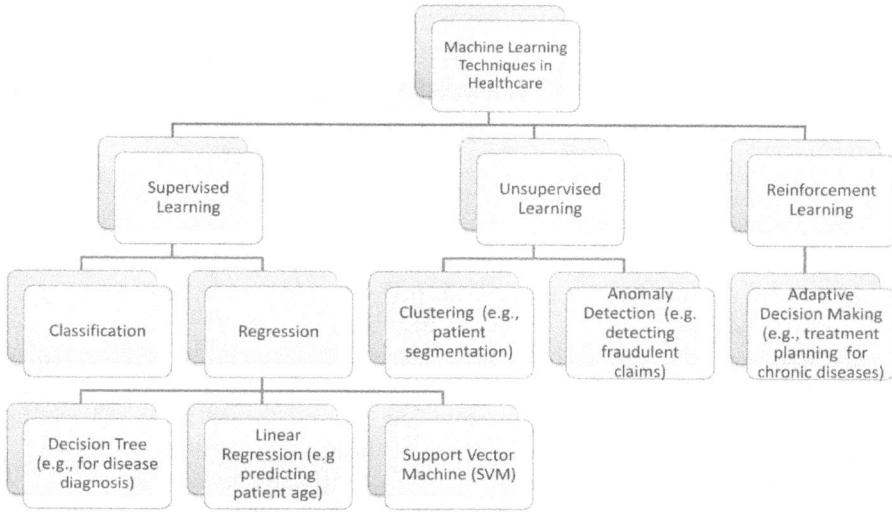

FIGURE 12.5   Machine learning techniques in healthcare.

predictions, personalize treatments, and optimize decision-making processes, ulti-mately leading to improved patient care and outcome.

Supervised learning: A model is trained using labelled data using supervised learning, where the algorithm learns to link input features to desired output labels. Algorithms for supervised learning can be utilized in the healthcare industry to per-form tasks including disease classification, patient outcome prediction, and high-risk individual identification [8, 9]. Although supervised learning excels at making pre-cise predictions, it is limited by the size of the labelled datasets it can use and may have trouble generalizing to new data. Two techniques are mainly used for super-vised learning as described below:

a. Classification: This technique entails categorizing or classifying input data. As an illustration, consider the use of a decision tree algorithm to identify a disease from its symptoms and test findings.
b. Regression: Predicting a continuous numerical value is involved. One illus-tration is the use of linear regression to estimate a patient's age from a variety of health markers.

GENERAL SUPERVISED LEARNING FLOW DIAGRAM

The basic steps followed for supervised learning are illustrated in Figure 12.6. The steps include:

a. **Input data**: This is the baseline dataset that includes the features (inputs) and the target labels (outputs) that go with them.
b. **Preprocessing**: To prepare the data for training, preprocessing involves cleaning, converting, and normalizing the data.

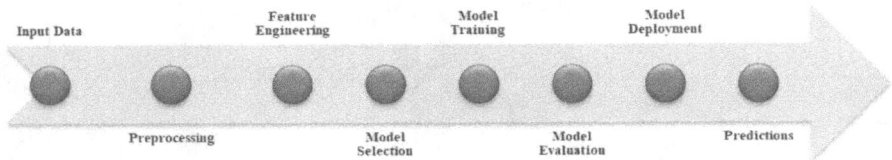

FIGURE 12.6   Basic flow diagram for supervised learning.

c. **Feature engineering**: To improve the performance of the model, this stage involves choosing or developing pertinent features from the raw data.
d. **Model selection**: Depending on the nature and requirements of the problem, select the right ML algorithm or model architecture.
e. **Model training**: Use the preprocessed data to train the chosen model.
f. **Model evaluation**: Utilize performance indicators like accuracy, precision, recall, and so on to gauge the trained model's effectiveness.
g. **Model deployment**: Use the model to generate predictions on fresh, unforeseen data if it performs well.
h. **Predictions**: With new input data, the deployed model is now prepared to generate predictions.

Examples of supervised learning include regression, object detection, fraud detection, and other applications that can all benefit from supervised learning. Depending on the problem and dataset, the specific implementation could change. In many different applications, including speech and image recognition, natural language processing (NLP), and others, supervised learning is used to learn patterns and connections between inputs and outputs using labelled samples.

A. Image classification
   A ML algorithm is trained to recognize the items or categories present in a picture or image, which is a key task in computer vision. The objective is to select a label or class from a predefined list of categories to apply to an input image. There are numerous fields where image classification is used, including:

   i. Identification of items in photos for robots, self-driving cars, and surveillance systems.
   ii. Medical imaging is the process of identifying illnesses and ailments using images from X-rays, MRIs, and CT scans.
   iii. Plant disease classification, crop health monitoring, and yield estimation are all aspects of agriculture.
   iv. Retail: Analysing customer behaviour and inventory levels by recognizing products.
   v. In order to enhance text-based data, NLP extracts information from images.
       The architecture of the model, the amount and quality of the training dataset, and the accessibility of computational resources for complicated deep

learning model training all have a significant role in the success of picture classification. Image categorization has attained impressive performance across a wide range of applications thanks to developments in deep learning and the availability of large-scale datasets. The process of image classification using convolutional neural networks (CNN) typically involves the following steps:

a. The dataset contains labelled images of different classes (e.g., cats and dogs).
b. Images are preprocessed, and features are automatically extracted using a CNN.
c. The CNN is trained on the preprocessed images to learn hierarchical features and patterns.
d. The trained model is evaluated on a separate test set to measure its performance.
e. If the model performs well, it can be deployed to classify new, unseen images.

B. Sentiment analysis.

A NLP technique called sentiment analysis, commonly referred to as opinion mining, tries to ascertain the sentiment or emotional tone expressed in a text. Finding out whether a text conveys a good, negative, neutral, or occasionally a more nuanced emotion like happiness, sadness, anger, and so on is the aim of sentiment analysis. Applications for sentiment analysis can be found in many fields, such as

i. Monitoring social media platforms such as Facebook, Twitter, and Instagram to analyse public opinion, sentiment patterns, and brand perception.
ii. Market research: Analysing consumer sentiment to learn more about product preferences and buying patterns.
iii. Client service is the process of reviewing client feedback in order to find and fix problems, improve products, and improve the customer experience.
iv. Political analysis is the study of political rhetoric in speeches, debates, and online discussions to determine how the public feels about candidates and policies.
v. Online reviews: Examining consumer feedback to gauge satisfaction levels and pinpoint areas for development.

The process of sentiment analysis using NLP involves the following steps:

a. The dataset contains labelled text data with sentiments (e.g., positive or negative reviews).
b. Text data are preprocessed and converted into numerical representations (e.g., word embeddings).
c. A ML algorithm (e.g., Naïve Bayes, support vector machine) is trained on the preprocessed text data.
d. The trained model is evaluated on a separate test set to assess its sentiment prediction accuracy.
e. If the model performs well, it can be deployed to analyse sentiments in new, unseen text data.

UNSUPERVISED LEARNING

Unsupervised learning seeks to identify structures or patterns in unlabelled data without direct instruction. When it comes to tasks like clustering patient subgroups, spotting anomalies, or finding hidden patterns in medical datasets, this technique is useful in the healthcare industry [9]. The accuracy of the patterns found may vary, and interpretations of the results may be more subjective when using unsupervised learning algorithms to assist uncover insights and direct future investigation. Two techniques are mainly used for unsupervised learning as described below:

a. Using a method called **clustering**, related data points are grouped together based on shared traits. To identify unique patient segments for individualized treatment, one example is to group patients based on their health information.
b. **Detecting anomalies** in data entails finding unusual or uncommon patterns. One example is spotting suspicious trends in claims data to identify fraudulent insurance claims.

GENERAL UNSUPERVISED LEARNING FLOW DIAGRAM

Unsupervised learning is frequently applied to exploratory data analysis to better comprehend the underlying data structure. Unsupervised learning can provide valuable information for later supervised learning tasks or other decision-making procedures. The basic steps followed for unsupervised learning are illustrated in Figure 12.7. The steps include:

a. **Input data**: This is the first dataset that only contains features without any associated target labels.
b. **Preprocessing:** Data preparation comprises preparing the data for training by cleaning, converting, and normalizing it.
c. **Feature engineering**: To improve the performance of the model, this stage involves choosing or developing pertinent features from the raw data.
d. **Model selection**: Depending on the needs and the nature of the problem, select the suitable unsupervised learning method or technique.
e. **Model training**: Train the chosen unsupervised learning model using the data that has been processed.

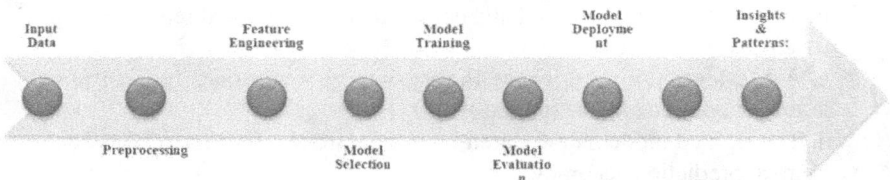

FIGURE 12.7   Basic flow diagram for unsupervised learning.

    f. **Model evaluation**: Utilize relevant metrics for unsupervised learning tasks to assess the trained model's performance, such as silhouette score and inertia. Not all unsupervised learning approaches will be suitable for this step.

    g. **Model deployment**: Use the unsupervised model for further use or analysis if it offers insightful information.

    h. **Clustering or dimension reduction**: Common unsupervised learning techniques include clustering and dimension reduction to group similar data points.

    i. **Insights and patterns**: In order to get insights and identify patterns in the data, the next step entails evaluating the clustering or dimension reduction results.

In this type of learning, the model is developed using data that has not been labelled, that is, the input data do not contain corresponding target labels. Finding patterns, correlations, and structures within the data are the aim of unsupervised learning.

Examples of unsupervised learning are valuable in various applications, such as clustering similar data points, reducing data complexity, and exploratory data analysis. Some of them are as follows:

A. *K*-means clustering

    a. Unlabelled data points, such as client purchase histories, are present in the dataset.

    b. Preprocessing of data points may include feature extraction if necessary.

    c. Based on similarity, the K-means method divides the data points into K groups.

    d. K clusters are the outcome, each of which contains data points that are comparable to one another.

B. Principal Component Analysis

    a. Unlabelled data points with various features can be found in the dataset.

    b. To enhance the quality of the data, feature engineering may be used to preprocess the data points.

    c. Principal component analysis is used to decrease the data's dimensionality while keeping its variance.

    d. The outcome is a representation of the original data in a lower dimensional space that is decreased in dimension.

Without the need for labelled target outputs, unsupervised learning is used in both scenarios to analyse the data and understand its structure and intended results.

Reinforcement learning: Reinforcement learning is the process of teaching an agent to make decisions sequentially in response to rewards or penalties in the environment. Reinforcement learning, albeit less frequently utilized in healthcare, offers the potential to improve treatment schedules and resource allocation, for example, in personalized medicine or dynamic patient monitoring [10]. Determining appropriate reward structures, handling the exploration-exploitation trade-off, and guaranteeing patient safety in medical decision-making are the hurdles, though. The common technique is used for reinforcement learning as described below:

A. **Adaptive decision-making**: Making decisions sequentially in order to maximize long-term benefits is the focus of the adaptive decision-making technique. One illustration is the use of reinforcement learning algorithms to create treatment plans for chronic diseases that change over time in response to patient reactions.

GENERAL REINFORCEMENT LEARNING FLOW DIAGRAM

In situations when the best course of action must be discovered by trial and error rather than being known beforehand, reinforcement learning is very helpful. Exploration and exploitation approaches are combined by the RL agent to strike a balance between trying out novel actions (exploration) and making use of the most well-known actions at the moment (exploitation) in order to maximize long-term rewards. A general flow chart for reinforcement learning activities is shown in Figure 12.8. The steps include:

a. **Environment**: The environment, with which the RL agent interacts, responds to the agent's activities by rewarding or punishing it.
b. **RL algorithm**: This is the fundamental reinforcement learning method that the agent uses to modify its policy and make choices based on input from the environment.
c. **Policy learning**: To choose its behaviours based on the status of the environment at any given time, the agent learns a policy, which is a plan or set of guidelines.
d. **Agent training**: In order to improve decision-making, the agent is trained by frequently interacting with the environment, tracking rewards, and revising its policy.
e. **Policy evaluation**: The training policy is tested to determine how well it performs in the environment.
f. **Trained agent**: Once trained, the agent can use the learnt policy to make judgments in the environment.
g. **Deployment**: The trained agent is used in simulations or real-world events to carry out tasks and make decisions.
h. **Decision-making**: In order to attain its objectives while maximizing the cumulative reward over time, the agent employs its learned policy to make decisions and take actions in the environment.

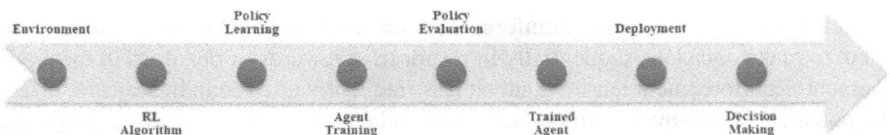

FIGURE 12.8   Basic flow diagram for reinforcement learning.

The objective is to maximize the cumulative benefit over time as the agent receives feedback in the form of rewards or penalties based on its behaviour.

A. Q-Learning for autonomous navigation

Q-learning is a model-free reinforcement learning algorithm used for solving Markov decision processes. It enables an agent to learn the optimal action-value function (Q-function) that estimates the expected cumulative reward for each state-action pair.

a. The agent navigates a grid-like world to reach a goal position while avoiding obstacles.
b. Positive rewards are given for reaching the goal, and negative rewards for hitting obstacles or making inefficient moves.
c. Q-Learning updates the Q values (expected cumulative rewards) for state-action pairs based on observed rewards and the agent's policy.
d. Over time, the agent learns the optimal policy that maximizes cumulative rewards, guiding it to efficiently reach the goal.

B. Deep Q-Network for Playing Atari Games

a. The agent plays an Atari game by observing the game screen as input.
b. It aims to learn effective actions (e.g., move left, right, jump) to achieve high game performance (e.g., points gained, enemies defeated).
c. A Deep Q-Network (DQN) approximates Q values for state-action pairs using a deep neural network.
d. The DQN is trained using experience replay, where past experiences are stored and used for batch training.
e. Over time, the DQN learns to make optimal decisions that maximize cumulative rewards in the game, allowing it to improve its gameplay.

In both cases, reinforcement learning enables the agents to pick up new information from their interactions with the environment and modify their plans of action in order to succeed. In the first case, the Q-values are updated using Q-learning, while a DQN is used in the second example to approximate Q-values for more complex circumstances, such as playing Atari games. These algorithms have proven to be outstanding in their capacities to handle difficult game situations and solve a variety of real-world challenges.

HEALTHCARE APPLICATIONS OF MACHINE LEARNING

ML has numerous applications in healthcare as described in Figure 12.9. This includes the following:

a. **Disease diagnosis and early detection:** ML algorithms can analyse medical records, genetic data, and clinical symptoms to aid in the early detection and accurate diagnosis of diseases [11–13]. For example, they can help identify patterns indicative of cancer, cardiovascular conditions, or neurological disorders, leading to timely interventions and improved patient outcomes.

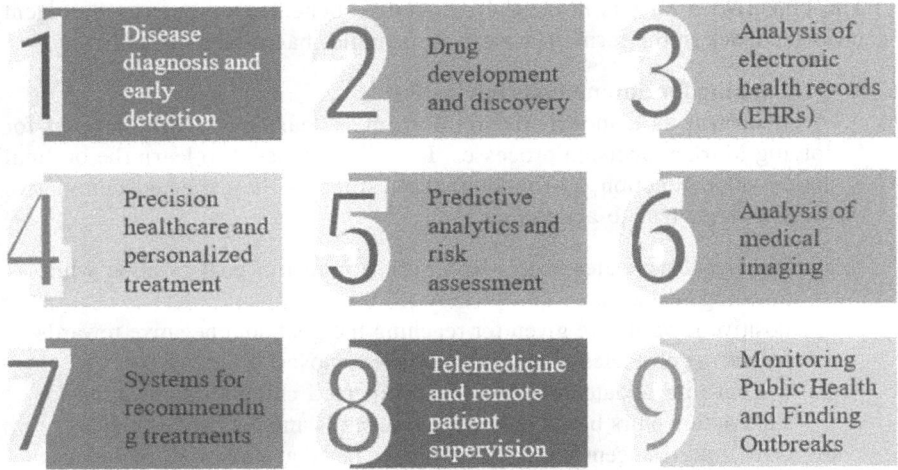

FIGURE 12.9 Machine learning applications in healthcare.

b. **Drug development and discovery:** Drug research and development are greatly accelerated by ML techniques. ML models are able to recognize prospective drug candidates, forecast their effectiveness, and enhance the drug discovery process by analysing large-scale molecular databases, including chemical structures and biological interactions. This may result in the discovery of novel treatments, quicker discovery of therapeutic targets, and more effective pharmacological compound development.

c. **Analysis of EHRs:** The examination of enormous volumes of EHRs is made possible by ML algorithms, which yield insightful data [14]. They can spot trends, correlations, and patterns in patient data, enhancing population health management, resource allocation, and clinical decision support.

d. **Precision healthcare and personalized treatment:** By incorporating individualized patient data, such as genomic details, lifestyle characteristics, and treatment histories, ML enables the development of personalized medicine techniques.

   It makes it possible to identify patient subgroups, forecast treatment outcomes, and tailor interventions based on certain patient traits. This personalized strategy enhances treatment results, reduces side effects, and maximizes the use of healthcare resources.

e. **Predictive analytics and risk assessment:** ML models can leverage patient data, including demographic information, medical history, and biomarkers, to predict the likelihood of future health events. By stratifying patients into risk groups, healthcare providers can focus resources and interventions on those who are most likely to benefit, leading to proactive and targeted care management.

f. **Analysis of medical imaging:** The interpretation and analysis of images from modalities like MRI, CT scans, and X-rays are made easier with the use of ML techniques, which are widely utilized in the field of medical imaging [15]. They can support abnormality detection, organ or tumour segmentation, and disease progression prediction, enabling more accurate and effective diagnosis and treatment planning.

g. **Systems for recommending treatments:** By examining patient characteristics, treatment results, and clinical guidelines, ML algorithms can aid in directing treatment decisions. They can offer the best possible treatment alternatives, dosage modifications, or individualized therapy approaches, supporting medical professionals in making defensible decisions and enhancing the effectiveness of treatment.

h. **Telemedicine and remote patient supervision:** Remote patient monitoring and telemedicine are made possible by ML algorithms that analyses continuous streams of patient data, such as vital signs, sensor data, or patient-reported outcomes. These algorithms are able to identify changes in health status, forecast worsening, and notify healthcare professionals, allowing for prompt interventions and lowering hospital readmission rates. By facilitating remote healthcare delivery, improving diagnostic accuracy, and supplying decision support tools, ML can also improve virtual consultations.

i. **Monitoring public health and finding outbreaks:** ML techniques can contribute to public health surveillance by analysing vast amounts of data from various sources, such as social media, health records, and environmental sensors [16]. These models can detect disease outbreaks, monitor population health trends, and support early warning systems for infectious diseases. By providing real-time insights, ML helps public health agencies in proactive planning, resource allocation, and rapid response to public health emergencies.

ML can be used to improve healthcare delivery and cost reductions, such as skin cancer diagnosis, sepsis prediction, breast cancer detection, and cardiovascular event prediction. Below is a table showing the some healthcare applications using ML algorithms and their accuracy rate (Table 12.1).

INCORPORATING MACHINE LEARNING INTO HEALTHCARE

The process or steps involved in applying ML techniques in healthcare settings as shown in flow diagram in Figure 12.10. It outlines the sequential procedures involved in implementing ML in healthcare. In addition to providing a better understanding of the general procedure, it will also show the connections between the various stages. The major phases are described as below.

a. **Data collection:** Data collection includes patient feedback, genomic data, medical imaging, and EHRs.

b. **Data preprocessing:** Handle missing values, remove noise, and standardize the format of the collected data.

c. **Feature engineering:** The preprocessed data can be used to extract relevant features that can be used by ML algorithms.

TABLE 12.1

Healthcare Applications Using Machine Learning Algorithms

Sl. No.	Healthcare Application	Study/Source	Key Finding
1	Skin Cancer Diagnosis [17]	*Nature Medicine*	A deep learning algorithm outperformed human dermatologists, achieving a 95% accuracy rate compared to dermatologists' 86.6% accuracy rate
2	Early Sepsis Prediction [18]	*Journal of the American Medical Association (JAMA)*	MLalgo analysing electronic health records predicted sepsis up to 48 hours before clinical recognition, reducing mortality rates by 53%
3	Breast Cancer [19] Diagnosis	*PLoS ONE*	MLalgo applied to medical imaging data diagnosed breast cancer with a sensitivity of 90.2%, outperforming conventional approaches (78.2%), leading to earlier detection and improved survival rates
4	Cardiovascular Event Prediction [20]	*The Lancet Digital Health*	MLalgo analysing electrocardiogram (ECG) data accurately predicted cardiovascular events with an AUC of 0.86, surpassing conventional risk scoring systems
5	Potential of Machine Learning in Healthcare [21]	World Health Organization (WHO)	The WHO acknowledges that machine learning has the potential to enhance healthcare by assisting in disease detection, anticipating outbreaks, and optimizing treatment strategies
6	Cost Savings with Machine Learning [22]	National Academy of Medicine report	Machine learning in healthcare could result in yearly savings of up to $100 billion in the United States by enhancing diagnosis accuracy, reducing medical errors, and optimizing resource utilization
7	Medical Image Analysis [23]	*Journal of the American Medical Association (JAMA)*	MLalgo analysing medical images, such as X-rays and MRIs, have shown promising results in accurately diagnosing various conditions, assisting radiologists, and improving patient outcomes
8	Disease Outbreak Prediction [24]	Centers for Disease Control and Prevention (CDC)	MLalgo can analyse vast amounts of data, including social media posts and public health records, to detect early signs of disease outbreaks and support timely public health interventions
9	Drug Discovery and Development [25]	*Frontiers in Pharmacology*	MLalgo can accelerate the process of drug discovery and development by predicting drug-target interactions, identifying potential therapeutic compounds, and optimizing drug design
10	Personalized Treatment Recommendations [26]	*Journal of the American Medical Association (JAMA)*	MLalgo can analyse patient data, including genetics, medical history, and treatment outcomes, to provide personalized treatment recommendations, leading to improved patient care

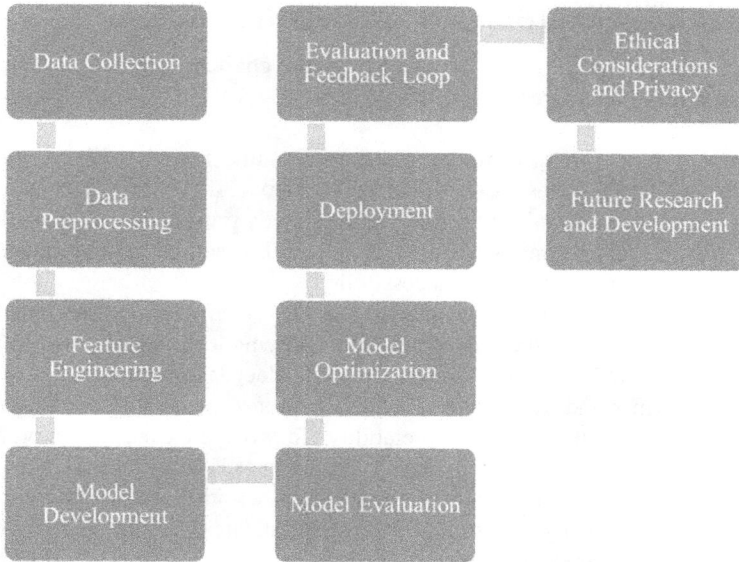

FIGURE 12.10   Steps involved in applying machine learning techniques in healthcare settings.

d. **Model development:** Based on the current healthcare task, choose the best ML algorithms, then train the chosen models using labelled data or unsupervised learning methods.

e. **Model evaluation:** The effectiveness of trained models can be assessed based on metrics such as accuracy, precision, recall, and F1 score. Verify the models using hold-out datasets or cross-validation.

f. **Model optimization**: Improve the performance of the models by adjusting hyper parameters, exploring different model architectures, or using regularization techniques.

g. **Deployment:** For use in the actual world, integrate the trained and improved models into the healthcare system. Include the models in tele-medicine platforms, remote monitoring systems, or clinical decision support systems.

h. **Evaluation and feedback loop:** Monitor and evaluate models in real-world healthcare settings. Refine models based on feedback from healthcare professionals and patients.

i. **Ethical considerations and privacy**: Make sure ethical guidelines are followed, patient privacy laws are followed, and data protection laws are followed. Develop data storage and consent management methods, as well as ML methods that preserve privacy.

j. **Future research and development**: Developing new ML techniques, such as explainable AI and federated learning, to deal with current problems and limitations. In order to improve prediction abilities, integrate additional data sources such as wearable devices and social determinants of health.

CHALLENGES AND ETHICAL CONSIDERATIONS

Implementing ML in healthcare comes with several challenges and ethical considerations that need to be addressed as highlighted in Figure 12.11.

a. **Data quality and availability**: Datasets that are of high calibre, variety, and representativeness are crucial to ML models. However, biases, data fragmentation, and missing values are frequently present in healthcare data. To obtain accurate and trustworthy results, it is essential to guarantee data quality, standardization, and accessibility.

b. **Interpretability and explainability**: It can be difficult to interpret ML models' decisions and comprehend the underlying logic because they might be sophisticated and "black boxes," especially deep learning models. Gaining the confidence and acceptance of healthcare professionals, patients, and regulatory bodies depends on interpretability. To provide clear and comprehensible models, research on explainable AI strategies is required.

c. **Privacy and security**: Healthcare data contain sensitive and private information, such as patient health records. Protecting patient privacy and ensuring data security are paramount. ML algorithms must be designed with robust privacy protection mechanisms, secure data storage, and appropriate data sharing protocols to maintain confidentiality while enabling effective collaboration.

d. **Bias and fairness**: ML algorithms can inherit biases present in the data they are trained on, potentially leading to discriminatory outcomes. Fairness considerations are vital to ensure equitable healthcare delivery. Regular audits, bias detection methods, and fairness-aware training techniques should be employed to minimize and mitigate bias and promote fairness in ML models.

e. **Regulatory and ethical compliance:** The implementation of ML in healthcare must adhere to regulatory guidelines and ethical standards. Compliance

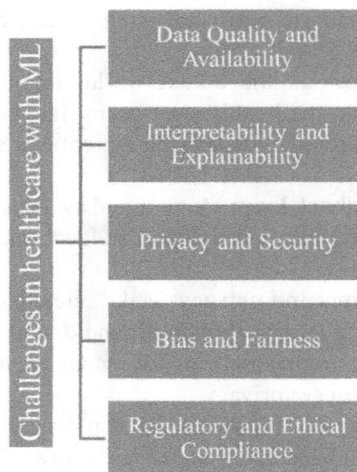

FIGURE 12.11   Issues in applying machine learning in healthcare.

with regulations, such as the general data protection regulation (GDPR) and ethical principles, including informed consent, transparency, and accountability, is crucial to ensure responsible use of ML technology.

SUCCESS STORIES AND CASE STUDIES

Several success stories and case studies highlight the transformative potential of ML in healthcare. For instance, ML algorithms have been successfully applied to improve the accuracy of breast cancer detection from mammograms, predict patient readmissions, enhance early diagnosis of Alzheimer's disease [27, 28], and optimize treatment plans for chronic conditions like diabetes [29]. These success stories demonstrate the positive impact. Numerous success stories and case studies that highlight ML's revolutionary impact have been published. It has demonstrated enormous promise for revolutionizing the healthcare industry. Here are a few noteworthy examples:

a. **Mammogram-based breast cancer detection**: ML algorithms have been used to analyse mammogram images in order to more accurately diagnose early-stage breast cancer. These algorithms can spot minor patterns and features in mammograms that human radiologists can find difficult to spot. Early diagnosis made possible by ML therefore has the potential to enhance patient outcomes and survival rates.

b. **Predicting patient readmissions:** To determine the likelihood of patient readmissions, ML algorithms have been employed to analyse patient data and EHRs. Healthcare professionals can take proactive steps to provide adequate post-discharge care and lower hospital readmission rates by identifying patients who are at high risk of readmission.

c. **Early diagnosis of Alzheimer's disease:** To identify the earliest indications of Alzheimer's disease [30], ML techniques have been used to analyse biomarker and brain imaging data (such as MRI and PET scans) [31]. These algorithms have the ability to spot tiny alterations in brain structure and function, enabling early diagnosis and treatment that may reduce the disease's course and enhance patient care.

d. **Improving care for chronic conditions**: In order to personalize treatment regimens for chronic illnesses like diabetes [32], ML algorithms have been used to analyse patient data, including medical history, genetics, and lifestyle factors. By adjusting interventions to the needs of each patient, these personalized treatment regimens can increase patient compliance and results.

e. **Clinical decision support systems:** ML models help clinicians make decisions by analysing patient data and suggesting the best course of therapy in light of the most recent scientific research.

f. **Optimizing healthcare resources:** To increase operational efficiency, ML algorithms are used to optimize hospital resource allocation, including bed management, personnel scheduling, and supply chain management.

g. **Enhancing the care of ICU patients:** To determine the possibility of sepsis or other critical situations, ML algorithms analyse real-time data from

ICU patients, such as vital signs and lab findings. This aids clinicians in making preemptive interventions and enhancing patient outcomes.

h. **Individualized prescription for drugs:** To personalize drug prescriptions, ML is applied to patient data, including genetics, medical history, and reaction to prior treatments. With this strategy, treatment effectiveness is increased while adverse effects are reduced.

i. **Chronic condition monitoring in real time:** To monitor patients with chronic illnesses like heart failure or COPD in real-time, ML and IoT devices are combined. In order to facilitate early intervention, the system informs healthcare providers to any major departures from baseline.

These case studies show the wide spectrum of uses for ML in healthcare, including disease diagnosis, therapy optimization, and enhancing patient care. For advancing medical research, enhancing patient outcomes, and revolutionizing healthcare delivery, ML in healthcare technologies has considerable promise.

FUTURE DIRECTIONS AND OPPORTUNITIES

ML in healthcare is an ever-evolving field with promising future directions and opportunities. Here are some areas that hold significant potential:

a. Developments in explainable AI techniques will enable healthcare practitioners to understand and trust the decisions made by ML models in healthcare. Transparent and interpretable models will promote accountability and facilitate the adoption of ML in clinical settings.

b. Federated learning and privacy-preserving techniques allow collaborative training of ML models while preserving the privacy of patient data. Advancements in privacy-preserving techniques will enable secure and efficient collaboration among healthcare institutions and researchers.

c. The integration of ML with wearable devices, IoT sensors, and mobile health applications will enable continuous and real-time monitoring of patients' health. This will enable timely interventions, personalized feedback, and remote healthcare delivery, leading to improved patient outcomes and reduced healthcare costs.

d. Integrating diverse data sources, such as medical imaging, genomics, EHRs, and patient-generated data, will provide a holistic view of patients' health. ML algorithms capable of handling and analysing such multi-modal data will unlock new insights and enable more accurate diagnoses, treatment recommendations, and disease management.

e. Advanced clinical decision support systems based on ML will assist healthcare professionals in real-time diagnosis, treatment planning, and patient management. These systems will incorporate patient-specific data, medical literature, guidelines, and real-world evidence to offer tailored recommendations and improve clinical outcomes.

Continuous ML research and development will enable WSNs and IoT to offer increasingly complex predictive and preventative healthcare solutions. Several future possibilities have the potential to have an even bigger influence as WSNs and IoT develop in healthcare. By combining ingestible devices with nanosensors, it may be possible to gain more precise and individualized insights into patient health at the cellular level. Moreover, to protect sensitive patient data from potential cyber threats, developments in data security and privacy safeguards will be essential. To facilitate seamless data interchange between various healthcare systems and devices and to promote a connected and effective healthcare ecosystem, interoperability and standardization challenges must be solved.

Future directions in the context of WSNs and the IoT in healthcare:

a. **Enhanced connectivity and interoperability**: In the future, WSN and IoT technologies in healthcare will focus on improving connectivity and interoperability. Efforts will be made to establish seamless communication and data exchange between different devices, systems, and platforms. This will enable efficient integration of data from various sources, such as wearable devices, medical sensors, EHRs, and healthcare databases, providing a comprehensive view of the patient's health status.

b. **Edge computing and real-time analytics**: With the increasing volume of data generated by WSN and IoT devices in healthcare, there will be a growing need for edge computing capabilities. Edge computing involves processing and analysing data closer to its source, reducing latency, and enabling real-time decision-making. Advanced analytics techniques, including ML and AI, will be applied at the edge to derive meaningful insights from the continuous streams of sensor data. This will facilitate faster and more accurate diagnosis, prognosis, and treatment decisions.

c. **Security and privacy measures**: As the adoption of WSN and IoT in healthcare expands, ensuring the security and privacy of sensitive health data will be of paramount importance [33]. Future directions will focus on developing robust security measures, including encryption, authentication, and access control mechanisms, to safeguard patient information from unauthorized access and cyber threats. Privacy-enhancing technologies, such as data anonymization and differential privacy, will also be employed to protect patient privacy while allowing for valuable data analysis and research.

d. **Predictive and prescriptive analytics**: Advancements in WSN and IoT technologies will enable the implementation of predictive and prescriptive analytics in healthcare. ML algorithms, combined with real-time sensor data, will enable the early identification of health risks and the prediction of adverse events [34]. These insights will support proactive interventions and personalized treatment plans, optimizing patient outcomes and reducing healthcare costs.

e. **Telehealth and remote patient monitoring**: WSN and IoT technologies will continue to drive the growth of telehealth and remote patient monitoring solutions. By leveraging wearable devices, home-based sensors, and connected healthcare platforms, healthcare providers can remotely monitor patients' vital signs, medication adherence, and overall health status. This approach enhances patient engagement, enables timely interventions, reduces hospital readmissions [35], and improves access to healthcare, especially for individuals in remote or underserved areas.

f. **Integration with EHR**: In the future, WSN and IoT technologies will be tightly integrated with EHR systems. Sensor data collected from WSN and IoT devices will be seamlessly integrated into patients' EHRs, providing a holistic view of their health history and current condition. This integration will enable healthcare professionals to make data-driven decisions based on comprehensive and up-to-date patient information, improving care coordination and continuity.

g. **Ethical and regulatory considerations**: As WSN and IoT technologies become more prevalent in healthcare, there will be a need for clear ethical guidelines and regulatory frameworks. Future directions will focus on addressing ethical considerations related to data ownership, patient consent, transparency, and algorithmic bias. Regulatory bodies will play a crucial role in establishing standards for data privacy, security, and the responsible use of WSN and IoT technologies in healthcare.

CONCLUSION

ML has the potential to transform healthcare through the utilization of data analysis and predictive modelling. Its impact can be seen in various areas such as disease diagnosis, analysis of medical imaging, personalized medicine, and public health monitoring, leading to improved patient care, better outcomes, and optimized healthcare delivery. Nevertheless, there are hurdles to overcome, including data quality, interpretability, privacy, and fairness, to ensure responsible and ethical implementation. By tackling these challenges and capitalizing on future prospects, the integration of ML into healthcare can create a more effective, tailored, and evidence-driven healthcare system, benefiting both patients and healthcare providers. Patient care and medical procedures have undergone a significant change as a result of the integration of WSNs and the IoT in the healthcare industry. WSNs and IoT have advanced healthcare to new levels through real-time data collecting, predictive analytics, and patient-centric applications, promising a future where medical treatments are individualized, effective, and available to all. The potential for improvements in healthcare is limitless as these technologies develop and tackle problems, creating a better society for future generations.

REFERENCES

1. Esteva, A., Kuprel, B., Novoa, R.A., Ko, J., Swetter, S.M., Blau, H.M., Thrun, S. (2017). Dermatologist-level classification of skin cancer with deep neural networks. *Nature*, 542(7639), 115–118. doi: 10.1038/nature21056.
2. Rajpurkar, P., Irvin, J., Zhu, K., Yang, B., Mehta, H., Duan, T., Ding, D., Bagul, A., Langlotz, C., Shpanskaya, K., Lungren, M.P., Ng, A.Y. (2017). CheXNet: Radiologist-level pneumonia detection on chest X-rays with deep learning. *arXiv*, 3–9.
3. Kaouk, J.H., Garisto, J., Eltemamy, M., Bertolo, R. (2019). Robot-assisted surgery for benign distal ureteral strictures: Step-by-step technique using the SP® surgical system. *BJU International*, 123(4), 733–739. doi: 10.1111/bju.14635.

4. Woldaregay, A.Z., Årsand, E., Botsis, T., Albers, D., Mamykina, L., Hartvigsen, G. (2019). Data-driven blood glucose pattern classification and anomalies detection: Machine-learning applications in type 1 diabetes. *Journal of Medical Internet Research*, 21(5), e11030. doi: 10.2196/11030.

5. Siddiqui, M.K., Morales-Menendez, R., Huang, X., Hussain, N. (2020). A review of epileptic seizure detection using machine learning classifiers. *Brain Informatics*, 7(1), 5. doi: 10.1186/s40708-020-00105-1.

6. Saini, P., Kaur, J., Lamba, S. (2021). A review on pattern recognition using machine learning. In Manik, G., Kalia, S., Sahoo, S.K., Sharma, T.K., Verma, O.P. (eds), *Advances in Mechanical Engineering. Lecture Notes in Mechanical Engineering*. Springer, Singapore. doi: 10.1007/978-981-16-0942-8_58

7. Kashyap, V., Saini, P., Rabina, Gautam, S. (2021). Mortality rate prediction for COVID-19 using machine learning technique. In *Advances in Mechanical Engineering: Select Proceedings of CAMSE 2020* (pp. 629–637). Springer, Singapore.

8. Alloghani, M., Al-Jumeily, D., Mustafina, J., Hussain, A., Aljaaf, A.J. (2020). A systematic review on supervised and unsupervised machine learning algorithms for data science. In Berry, M., Mohamed, A., Yap, B. (eds), *Supervised and Unsupervised Learning for Data Science* (pp. 3–21). Springer, US.

9. Habehh, H., Gohel, S. (2021). Machine learning in healthcare. *Current Genomics*, 22(4), 291–300. doi: 10.2174/1389202922666210705124359.

10. Sutton, R.S., Barto, A. (2018). *Reinforcement Learning: An Introduction*. MIT Press, Cambridge, MA.

11. McKinney, S.M., Sieniek, M., Godbole, V., et al. (2020). International evaluation of an AI system for breast cancer screening. *Nature*, 577(7788), 89–94. doi: 10.1038/s41586-019-1799-6.

12. Jabeen, T., Jabeen, I., Ashraf, H., Jhanjhi, N.Z., Yassine, A., Hossain, M.S. (2023). An intelligent healthcare system using IoT in wireless sensor network. *Sensors*, 23(11), 5055. doi: 10.3390/s23115055.

13. Jabeen, T., Ashraf, H., Ullah, A. (2021). A survey on healthcare data security in wireless body area network. *Journal of Ambient Intelligence and Humanized Computing*, 12, 9841–9854.

14. Ahmad, F.S., Ali, L. (2020). A hybrid machine learning framework to predict mortality in paralytic ileus patients using electronic health records (EHRs). *Journal of Ambient Intelligence and Humanized Computing*, 12(2), 3283–3293. doi: 10.1007/s12652-020-02509-7.

15. Shukla, P.K., Sandhu, J.K., Ahirwar, A., Ghai, D., Maheshwary, P. (2021). Multi-objective genetic algorithm and convolutional neural network based COVID-19 identification in chest X-ray images. *Mathematical Problems in Engineering*, 2021, 1–9.

16. Azeem, A.M., Ashraf, H., Alaboudi, A.A., Humayun, M., Jhanji, N.Z. (2021). Secure healthcare data aggregation and transmission in IoT—A survey. *IEEE Access*, 9, 16849–16865.

17. Tschandl, P., Rinner, C., Apalla, Z., et al. (2020). Human-computer collaboration for skin cancer recognition. *Nature Medicine*, 26, 1229–1234. doi: 10.1038/s41591-020-0942-0.

18. Brajer, N., Cozzi, B., Gao, M., et al. (2020). Prospective and external evaluation of a machine learning model to predict in-hospital mortality of adults at time of admission. *JAMA Network Open*, 3(2), e1920733. doi: 10.1001/jamanetworkopen.2019.20733.

19. Abdelrahman, L., Al Ghamdi, M., Collado-Mesa, F., Abdel-Mottaleb, M. (2021). Convolutional neural networks for breast cancer detection in mammography: A survey. *Computers in Biology and Medicine*, 131, 104248.

20. Vyas, D.A., James, A., Kormos, W., Essien, U.R. (2022). Revising the atherosclerotic cardiovascular disease calculator without race. *The Lancet Digital Health*, 4(1), e4–e5.

21. Panesar, A. (2019). *Machine Learning and AI for Healthcare* (pp. 1–73). Apress, Coventry, UK.

22. Adler-Milstein, J., Aggarwal, N., Ahmed, M., et al. (2022). Meeting the moment: Addressing barriers and facilitating clinical adoption of artificial intelligence in medical diagnosis. *NAM Perspective*, 2022. doi: 10.31478/202209c.

23. Carin, L., Pencina, M.J. (2018). On deep learning for medical image analysis. *JAMA*, 320(11), 1192–1193. doi: 10.1001/jama.2018.13316.

24. Alessa, A., Faezipour, M. (2019). Preliminary flu outbreak prediction using Twitter posts classification and linear regression with historical centers for disease control and prevention reports: Prediction framework study. *JMIR Public Health Surveillance*, 5(2), e12383. doi: 10.2196/12383.

25. Vamathevan, J., Clark, D., Czodrowski, P., et al. (2019). Applications of machine learning in drug discovery and development. *Nature Reviews Drug Discovery*, 18, 463–477. doi: 10.1038/s41573-019-0024-5.

26. Schwartz, B., Cohen, Z.D., Rubel, J.A., Zimmermann, D., Wittmann, W.W., Lutz, W. (2021). Personalized treatment selection in routine care: Integrating machine learning and statistical algorithms to recommend cognitive behavioral or psychodynamic therapy. *Psychotherapy Research*, 31(1), 33–51. doi: 10.1080/10503307.2020.1769219.

27. Sethi, M., Rani, S., Singh, A., Mazón, J. L. V. (2022). A CAD system for Alzheimer's disease classification using neuroimaging MRI 2D slices. *Computational and Mathematical Methods in Medicine*, 2022. 10.1155/2022/8680737

28 Li, R., Zhang, W., Suk, H. I., Wang, L., Li, J., Shen, D., Ji, S. (2014). Deep learning based imaging data completion for improved brain disease diagnosis. In *Medical Image Computing and Computer-Assisted Intervention—MICCAI 2014: 17th International Conference*, Boston, MA, September 14–18, 2014, Proceedings, Part III 17 (pp. 305–312). Springer International Publishing.

29. Chung, W. K., Erion, K., Florez, J. C., Hattersley, A. T., Hivert, M. F., Lee, C. G., ... Franks, P. W. (2020). Precision medicine in diabetes: A consensus report from the American Diabetes Association (ADA) and the European Association for the Study of Diabetes (EASD). *Diabetes Care*, 43(7), 1617–1635.

30. Faturrahman, M., Wasito, I., Hanifah, N., Mufidah, R. (2017). Structural MRI classification for Alzheimer's disease detection using deep belief network. In 2017 11th International Conference on Information Communication Technology and System (ICTS), pp. 37–42.

31. Abraham, A., Pedregosa, F., Eickenberg, M., Gervais, P., Mueller, A., Kossaifi, J., Gramfort, A., Thirion, B., Varoquaux, G. (2014). Machine learning for neuroimaging with scikit-learn. *Frontiers in Neuroinformatics*, 8, 14. doi: 10.3389/fninf.2014.00014.

32. Arcadu, F., Benmansour, F., Maunz, A., Willis, J., Haskova, Z., Prunotto, M. (2019). Deep learning algorithm predicts diabetic retinopathy progression in individual patients. *NPJ Digital Medicine*, 2(1), 92.

33. Avendi, M., Kheradvar, A., Jafarkhani, H. (2016). A combined deep-learning and deformable-model approach to fully automatic segmentation of the left ventricle in cardiac MRI. *Medical Image Analysis*, 30. doi: 10.1016/j.media.2016.01.005.

34. Ramesh, T. R., Lilhore, U. K., Poongodi, M., Simaiya, S., Kaur, A., Hamdi, M. (2022). Predictive analysis of heart diseases with machine learning approaches. *Malaysian Journal of Computer Science*, 2022 (Special Issue 1), 132–148.

35. Mohamed, I., Fouda, M.M., Hosny, K.M. (2022). Machine learning algorithms for COPD patients readmission prediction: A data analytics approach. *IEEE Access*, 10, 15279–15287.

Index

For Product Safety Concerns and Information please contact our EU
representative GPSR@taylorandfrancis.com
Taylor & Francis Verlag GmbH, Kaufingerstraße 24, 80331 München, Germany